Gene Mining and Germplasm Innovation for the Important Traits in Rice

Gene Mining and Germplasm Innovation for the Important Traits in Rice

Guest Editors

Yongmei Bao
Shimin Zuo

MDPI

Basel • Beijing • Wuhan • Barcelona • Belgrade • Novi Sad • Cluj • Manchester

Guest Editors
Yongmei Bao
College of Agriculture
Nanjing Agricultural
University
Nanjing
China

Shimin Zuo
College of Agriculture
Yangzhou University
Yangzhou
China

Editorial Office
MDPI AG
Grosspeteranlage 5
4052 Basel, Switzerland

This is a reprint of the Special Issue, published open access by the journal *International Journal of Molecular Sciences* (ISSN 1422-0067), freely accessible at: www.mdpi.com/journal/ijms/special_issues/VFZGS1II3C.

For citation purposes, cite each article independently as indicated on the article page online and using the guide below:

Lastname, A.A.; Lastname, B.B. Article Title. *Journal Name* **Year**, *Volume Number*, Page Range.

ISBN 978-3-7258-3572-0 (Hbk)
ISBN 978-3-7258-3571-3 (PDF)
https://doi.org/10.3390/books978-3-7258-3571-3

Contents

International Journal of *Molecular Sciences*

MDPI

Article

Physiological and Transcriptome Analysis Provide Insights into the Effects of Low and High Selenium on Methionine and Starch Metabolism in Rice Seedlings

Yang Yang [1,†], Jiarui Zhang [1,†], Lijuan Sun [2], Qin Qin [2], Shiyan Yang [2], Jun Wang [2], Yafei Sun [2,*] and Yong Xue [2,*]

[1] College of Fisheries and Life Science, Shanghai Ocean University, Shanghai 201306, China; rickyyoung2000@163.com (Y.Y.); 18363019947@163.com (J.Z.)

[2] ECO—Environment Protection Research Institute, Shanghai Academy of Agricultural Sciences, Shanghai 201403, China; sunliuliu2012@126.com (L.S.); qinqin19870987@126.com (Q.Q.); 20220703@saas.sh.cn (S.Y.); junwang2018@163.com (J.W.)

* Correspondence: sunyafei@saas.sh.cn (Y.S.); xueyong@saas.sh.cn (Y.X.)

† These authors contributed equally to this work.

Abstract: Selenium (Se) is an essential micronutrient for the human body and is closely linked to health. Rice (*Oryza sativa* L.), as a major staple food globally, is the primary source of Se intake for humans. To better achieve Se biofortification in rice, it is crucial to understand the molecular mechanisms behind rice's response to different Se concentrations. This study investigates the morphological and transcriptomic responses of rice seedlings to low (1 μM, LSe) and high (10 μM, HSe) Se concentrations compared to a control (CK). Morphological analyses revealed that LSe promoted growth, enhancing shoot and root length and biomass, whereas HSe treatment inhibited these parameters, indicating Se's dual role in rice growth. Notably, the most significant promotion of rice growth was observed at the Se concentration of 1 μM. The organic Se content and antioxidant enzyme activities (SOD, POD and CAT) in rice seedlings also reached their maximum values simultaneously. Total RNA was extracted for transcriptome sequencing, and differential gene expression analysis was conducted using DESeq2. Transcriptomic sequencing highlighted distinct responses under LSe and HSe conditions. Gene ontology (GO) enrichment analysis revealed significant involvement in processes related to oxidoreductase activity and cellular structures. KEGG pathway analysis emphasized that Se treatments notably enhanced the glutathione metabolism pathway, which is crucial for antioxidant defense. Additionally, significant changes were observed in starch and sucrose metabolism and cysteine (Cys) and methionine (Met) metabolism pathways, showing upregulation under LSe treatment and downregulation under HSe. Six key genes were validated using qRT-PCR, confirming their differential expression under varied Se treatments. The Cys, Met and starch content assays as well as qRT-PCR data demonstrated that LSe promoted the synthesis and accumulation of Cys, Met and starch, supporting enhanced growth and antioxidant capacity. Conversely, HSe inhibited the synthesis and accumulation of Cys, Met and starch in rice seedlings, and these data were also consistent with the physiological phenotype. These findings provide insights into the molecular mechanisms by which rice seedlings adapt to varying Se levels, with implications for Se biofortification and stress management strategies in crops.

Keywords: selenium; rice; transcriptome; physiological

Academic Editor: Karel Doležal

Received: 2 December 2024

Revised: 16 January 2025

Accepted: 4 February 2025

Published: 13 February 2025

Citation: Yang, Y.; Zhang, J.; Sun, L.; Qin, Q.; Yang, S.; Wang, J.; Sun, Y.; Xue, Y. Physiological and Transcriptome Analysis Provide Insights into the Effects of Low and High Selenium on Methionine and Starch Metabolism in Rice Seedlings. *Int. J. Mol. Sci.* **2025**, *26*, 1596. https://doi.org/10.3390/ijms26041596

1. Introduction

Selenium (Se), an essential trace element for human and animal health, plays critical structural and enzymatic roles in various physiological functions, encompassing antioxidant defense, anti-cancer activity, anti-inflammatory responses and immune regulation [1]. Se deficiency disrupts the normal functioning of selenoproteins and Se-dependent enzymes, resulting in a cascade of complications, including reproductive dysfunction, infertility, and mental retardation [2]. Dietary Se supplementation is recognized as a safe and efficient strategy to address Se deficiency [3]. Rice (*Oryza sativa* L.), a cornerstone of global agriculture, serves as a primary source of energy and micronutrients for over half of the world's population [4,5]. Since the late 1980s, several Asian countries have actively implemented Se fertilizer applications to cultivate Se-enriched rice, aiming to mitigate Se deficiency and enhance nutritional security [6]. Consequently, investigating the adaptive mechanisms and growth responses of rice under varying Se conditions holds profound implications for advancing food security and promoting human health.

Se is often described as a "double-edged sword" due to its dualistic biological behavior, exhibiting both essential and toxic properties depending on its concentration [7]. The threshold between Se deficiency and toxicity in living organisms is remarkably narrow, posing significant challenges for its management in biological and environmental systems [8]. At optimal concentrations, Se performs multiple pivotal roles in promoting plant growth, underscoring its essentiality in plant physiological process. Se enhances the activity and content of glutathione peroxidase (GSH-Px), a crucial enzyme in neutralizing reactive oxygen species (ROS) and lipid peroxides (LPO). This augmentation strengthens plants' antioxidant defenses, bolstering their resilience to environmental stressors and delaying senescence [9]. Se enhances plant resilience against diverse environmental stresses by scavenging excessive free radicals generated under extreme conditions, such as high radiation, extreme temperatures, drought, and high salinity, thereby fortifying plant tolerance to adverse environments [10,11]. At excessive concentrations, Se transitions from an antioxidant to a pro-oxidant, causing an accumulation of ROS and exacerbating oxidative stress, ultimately disrupting photosynthesis in plants [12]. Excess Se disrupts photosystem reaction centers, significantly reducing photosynthetic efficiency and impairing energy conversion processes in plants [13]. High selenate concentrations induce severe phytotoxicity, markedly decreasing the fresh biomass of crops such as broccoli and cabbage [14,15]. Additionally, high Se levels significantly diminish the accumulation of flavonoids and total phenols in mustard, thereby impairing the plant's metabolic functions [16]. Rao demonstrated that Se-binding protein 1 (SBP1) and sulfur deficiency-induced protein 2 (SDI2) are upregulated under high selenate concentrations, enhancing the tolerance of *Cardamine violifolia* to Se-induced toxicity [17]. Previous studies revealed that high Se concentrations significantly inhibit rice growth, particularly in roots, stems, and leaves, where dry biomass accumulation is notably reduced. This growth inhibition primarily arises from impaired photosynthesis and disrupted water metabolism, causing physiological drought-like symptoms in plants [18]. Excess Se significantly depletes chlorophyll, carotenoids, and protein content, visibly manifesting as chlorotic and necrotic leaf symptoms. Although antioxidant enzymes such as superoxide dismutase (SOD), perpxidase (POD) and catalase (CAT) are upregulated in response to Se toxicity, their activity is insufficient to fully mitigate the detrimental effects [19]. These findings lay a foundation for elucidating the physiological and biochemical responses of plants to high Se concentrations. However, the molecular mechanisms driving rice responses to varying Se concentrations remain inadequately understood. Consequently, further investigations into gene expression and metabolic shifts in rice are essential to unravel the molecular pathways governing rice adaptation to varying Se concentrations.

Se treatment regulates carbohydrate metabolism in rice kernels by influencing the expression of enzymes involved in sucrose and starch biosynthesis. This regulation leads to enhanced accumulation of starch and sucrose, ultimately promoting early maturity and increasing the yield of rice kernels [20]. Beyond serving as primary energy sources, starch and sucrose metabolism also play critical roles in orchestrating plant growth and developmental processes. For instance, during the early growth stages, sucrose translocation and catabolism provide essential substrates that support the development of roots and leaves, ensuring the establishment of a robust plant architecture. Moreover, these metabolic pathways interact synergistically with hormonal signals, such as abscisic acid and auxins, to modulate growth rates and shape the morphological structure of rice [21]. Under stress conditions, adjustments in starch and sucrose metabolism help sustain cellular metabolic demands and improve the resilience of rice, thereby enabling plants to better adapt to adverse environmental challenges [22].

Throughout the developmental stages of rice endosperm, amino acid metabolism —particularly Cys and Met pathways—not only facilitates seed growth but also synergistically interacts with starch and sucrose metabolism to form a nutrient-rich endosperm. This coordination is particularly prominent in polyploid rice seeds, which exhibit higher nutrient density compared to their diploid counterparts [23]. Se treatment has been shown to promote the biosynthesis of selenocysteine (SeCys) and selenomethionine (SeMet) in rice, thereby boosting the plant's antioxidant capacity and nutritional value [24]. Cys and Met serve as precursor amino acids for the synthesis of SeCys and SeMet, underpinning key metabolic pathways in rice. To mitigate the toxicity associated with Se amino acid substitutions in proteins, SeCys undergoes methylation to form methylselenocysteine (MeSeCys), which is further metabolized into the volatile compound dimethyl diselenide. Another fraction of SeCys is enzymatically converted into Se-homocysteine via cystathionine β-lyase and transported to the chloroplast, where Met synthase catalyzes its transformation into SeMet [25,26]. During this process, SeMet and MeSeCys partially substitute Cys and Met within the sulfur metabolism pathway, highlighting their functional integration into essential biochemical networks. Cys acts as a precursor for glutathione synthesis, a pivotal antioxidant that enables rice to mitigate oxidative stress and maintain cellular homeostasis [27]. S-adenosylmethionine (SAM), a metabolite derived from Met, exhibits potent antioxidant properties and serves as a key regulatory signaling molecule, bolstering rice resilience against abiotic stresses, including salinity and drought [28]. Met participates in methylation processes mediated by SAM, encompassing the methylation of DNA, RNA, and proteins. These epigenetic modifications are crucial for regulating gene expression, thereby influencing rice growth, development, differentiation, and environmental adaptation. Methylation additionally modulates chloroplast functionality, thereby optimizing photosynthetic efficiency and driving biomass accumulation [29]. Consequently, investigating the effects of varying Se treatments on starch and sucrose metabolism, as well as Cys and Met pathways, is vital for elucidating the molecular mechanisms governing rice growth and development under diverse Se conditions.

This study aims to conduct a comprehensive investigation of differential gene expression and pivotal metabolic pathways in rice seedlings exposed to low and high Se treatments. By centering on starch and sucrose metabolism, alongside Cys and Met metabolic pathways, this research aims to elucidate the molecular mechanisms underpinning rice growth and metabolic adaptations to varying Se concentrations. These findings aim to provide critical insights into Se's role in modulating plant health, enhancing stress resilience, and informing potential biofortification strategies to improve crop nutritional quality.

2. Results

2.1. The Effects of Low and High Se on the Phenotype of Rice Seedlings

The level of Se concentration is crucial for the growth and development of rice seedlings. It was found that low Se had a significant growth promotion effect on rice seedlings [30]. Our findings reveal that a low Se concentration of 1 μM is optimal for promoting rice seedling growth, as evidenced by significant increases in shoot and root length and biomass, which increased by an average of 20.9%, 16.4%, 30.4%, and 28.5%, respectively, compared with 0 μM. However, at HSe concentration (10 μM), all the phenotypes of rice seedlings were significantly lower than 0 μM, resulting in an average decrease of 50.8%, 54.9%, 40.2% and 29.4% (Figure 1). This suggests that low Se concentrations facilitate the growth and development of rice seedlings, while excessive Se may have toxic effects on plants and inhibit the growth and development of rice seedlings.

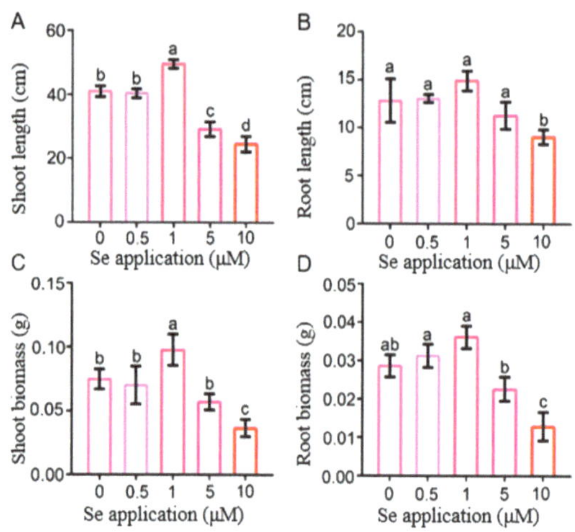

Figure 1. Effects of different Se concentration on the growth parameters of rice seedlings. (**A**) Shoot length, (**B**) root length, (**C**) shoot biomass, and (**D**) root biomass of rice seedlings under different Se treatments. Seedlings (14 d old) of the rice were grown hydroponically in nutrient rich solution and then under 0, 0.5, 1, 5 and 10 μM Se treatments for 14 d, respectively. Data are represented as the mean ± standard deviation (SD). Different letters indicate significant differences under different Se treatments ($p < 0.05$, Student's t-test).

2.2. The Effects of Low and High Se on the Concentration of Se in Rice Seedings

To investigate the effect of different Se concentrations on the Se content in rice seedlings, the total Se content in the seedlings was quantified (Figure 2A). The results revealed that Se application significantly enhanced the total Se content in the rice seedlings. The total Se content in the seedlings under the HSe treatment peaked at 1.51 times that of the LSe treatment and 17.83 times that of the CK treatment. Additionally, the inorganic and organic Se contents in the rice seedlings were also measured. The results indicated that both LSe and HSe treatments significantly elevated the inorganic Se content in the seedlings relative to the CK, with the most pronounced increase observed under the HSe treatment (Figure 2B). Moreover, Se application significantly increased the organic Se content in the rice seedlings, with the highest content observed under the LSe treatment, which was 1.16 times greater than that observed under the HSe treatment (Figure 2C).

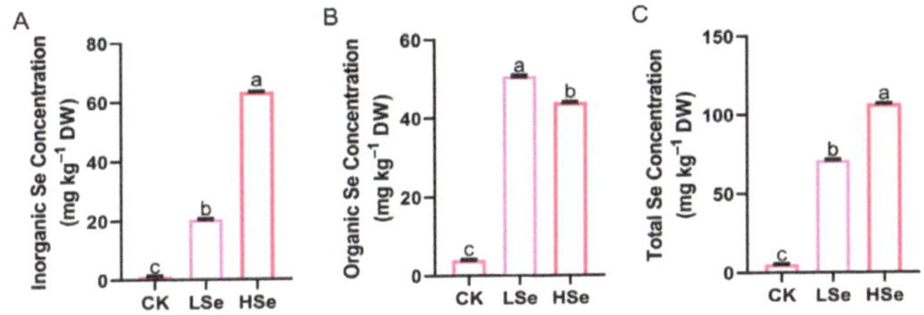

Figure 2. Se content of different species in rice seedlings under different Se treatments. (**A**) Total Se concentration, (**B**) inorganic Se concentration, and (**C**) organic Se concentration of rice seedlings under different Se treatments. Seedlings (14 d old) of the rice were grown hydroponically in nutrient rich solution and then under 0, 1, and 10 μM Se treatments for 14 d, respectively. Data are represented as the mean ± standard deviation (SD). Different letters indicate significant differences under different Se treatments ($p < 0.05$, Student's t-test).

2.3. The Effects of Low and High Se on the Antioxidant Enzyme Activities of Rice Seedings

Se exhibits a dual role in plants, acting as an antioxidant at low concentrations while inducing oxidative stress at high concentrations [7]. To better understand this dose-dependent effect of Se, we evaluated the activities of key antioxidant enzymes, SOD, POD, and CAT, in rice seedlings subjected to different Se treatments (Figure 3). These enzymes serve as reliable biochemical markers for assessing the oxidative balance and stress tolerance of plants. As the Se concentration increased, the activities of SOD and CAT in both the shoots and roots of rice seedlings showed a trend of increasing followed by decreasing (Figure 3A,C,D,F). Remarkably, under the HSe treatment, the SOD activity in the shoots and roots of rice seedlings was increased by 49.4% and 38.7%, respectively, compared to the CK, whereas the CAT activity in the shoots and roots under the same treatment decreased by 46.3% and 48.9%, respectively, compared to the CK (Figure 3B,E). As the Se application increased, the POD activity in the shoots and roots of rice seedlings showed a consistent upward trend, peaking under the HSe treatment, where the average increase was 18% and 8% higher than that observed in the CK and LSe, respectively.

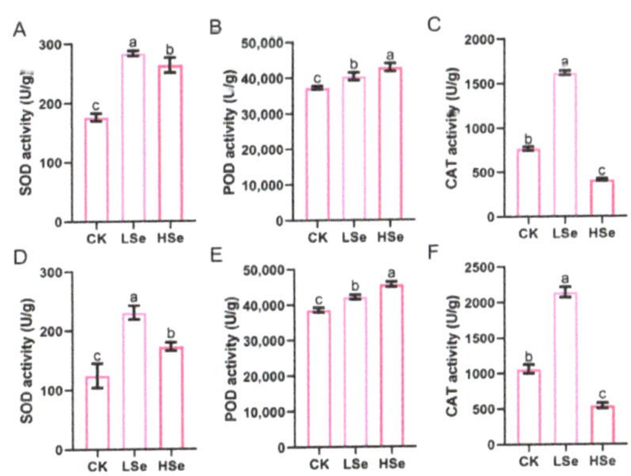

Figure 3. Effects of different Se concentration on the antioxidant enzyme activities of rice seedlings. (**A**,**D**) SOD activity, (**B**,**E**) POD activity, (**C**,**F**) CAT activity of rice seedlings under different Se treatments. (**A**–**C**) shows shoot data; (**D**–**F**) shows root data. Seedlings (14 d old) of the rice were grown hydroponically in nutrient rich solution and then under 0, 1, and 10 μM Se treatments for 14 d, respectively. Data are represented as the mean ± standard deviation (SD). Different letters indicate significant differences under different Se treatments ($p < 0.05$, Student's t-test).

2.4. The Different Impacts of Low and High Se on the Differential Gene Expression in Rice Seedlings

The mechanism of association between different Se concentrations and rice seedlings was analyzed by transcriptome sequencing. Table S2 provides a comprehensive summary of RNA-seq data and reads mapping across different sample conditions (CK, LSe, HSe) for both root and shoot tissues. The clean reads for all samples are consistently high, with Q30 values exceeding 95%, indicating high sequencing accuracy. GC content varies between 50.64% and 53.06%, reflecting consistency across the samples. The proportion of total mapped reads ranges from 92.85% to 98.25%, while uniquely mapped reads range from 88.82% to 93.54%, demonstrating robust alignment efficiency. These metrics collectively indicate high-quality RNA-seq data, suitable for reliable downstream analyses.

Principal component analysis (PCA) was conducted based on expression data (Figure S1), with principal component 1 (PC1) and principal component 2 (PC2) explaining 68.33% and 11.17% of the total variation, respectively. The six sample groups (CK-R, CK-S, HSe-R, HSe-S, LSe-R, LSe-S) were significantly separated in the two-dimensional principal component space, indicating notable differences in their characteristic variables. This distribution reveals significant differences among the treatment groups. The reproducibility and correlation of the three replicates under different treatment concentrations were high, ensuring the reliability of subsequent analyses.

The transcriptome sequencing results revealed that there were 2514 DEGs comprising 1161 upregulated genes and 1353 DEGs downregulated genes in the shoots and 1613 upregulated genes and 1022 downregulated genes in the roots under low Se treatment (Figure 4A). There were 2784 upregulated genes and 4111 downregulated genes in the shoots and 2271 upregulated genes and 2636 downregulated genes in the roots under high Se treatment. The above results indicated that high concentrations of Se account for an increase in the number of DEGs in rice seedlings. Notably, only the number of upregulated genes was more than downregulated genes in the roots under low Se treatment, while the number of downregulated genes was more than upregulated genes in all three other groups. The Venn analysis showed that the number of genes unique to each of the four treatment groups was as follows: 194 in the shoot of HSe vs. CK group, 3145 in the shoot of LSe vs. CK group, 642 in the root of LSe vs. CK group, and 1919 in the root of Hse vs. CK group (Figure 4B). There were 352 genes shared among the four groups which illustrated the cross-impact of different Se treatments on various metabolic pathways in rice.

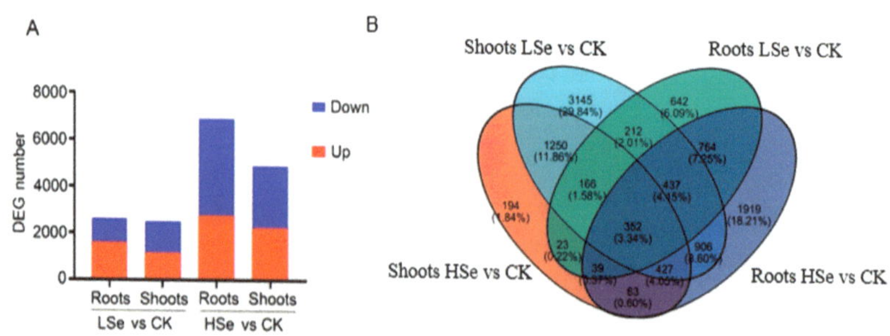

Figure 4. Differentially expressed genes (DEGs) in rice roots and shoots under Se treatments. (**A**) Number of upregulated and downregulated DEGs in roots and shoots of rice seedlings under LSe and HSe conditions compared to the CK. Red bars indicate upregulated genes, and blue bars indicate downregulated genes. (**B**) Venn diagram showing the overlap of DEGs among different comparisons (roots LSe vs. CK, shoots LSe vs. CK, roots HSe vs. CK, and shoots HSe vs. CK). Each section of the diagram represents the number and percentage of shared and unique DEGs in the respective conditions.

2.5. GO Enrichment Analysis of DEGs

Gene ontology (GO) enrichment analysis was carried out to further characterize the main biological functions of DEGs in rice seedlings under Se treatments. All the DEGs can be divided into three categories, including biological process, molecular function, and cellular component. The top 20 significantly enriched GO terms were shown in Figure 5. In the shoot of the LSe vs. CK groups, key terms included "response to stimulus" and "plant-type secondary cell wall biogenesis" under biological processes; "extracellular region", "plasma membrane", and "cellular anatomical entity" under cellular components, and "cellular processes", "oxidoreductase activity", and "protein serine/threonine kinase activity" under molecular function (Figure 5A). As shown in the shoot of the HSe vs. CK groups, the differential genes were involved in cellular component and molecular function, with molecular function accounting for the majority of them. "Cellular anatomical entity", "cellular component", and "plasma membrane" were the three most enriched subcategory in the cellular component. Oxidoreductase activity was the most enriched subcategory in the molecular function (Figure 5B).

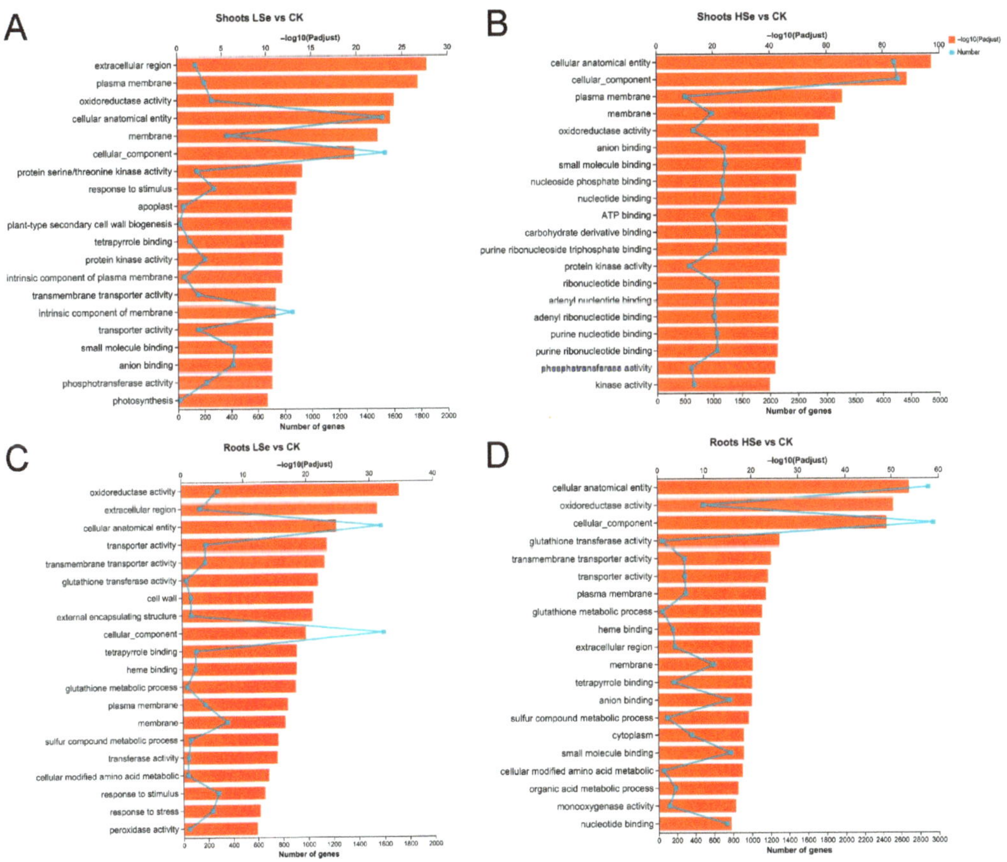

Figure 5. GO enrichment analysis of differentially expressed genes (DEGs) in rice shoots and roots under LSe and HSe treatments compared to CK. (**A**) Shoots LSe vs. CK, (**B**) shoots HSe vs. CK, (**C**) roots LSe vs. CK, and (**D**) roots HSe vs. CK. The x-axis shows the number of genes, and the y-axis lists enriched GO terms. The blue line represents the −log10 (Padjust) values, indicating the significance of enrichment.

Meanwhile, the DEGs are also divided into three categories in roots of rice. There were five, seven, and eight enriched subcategories belonging to the categories of biological process, molecular function and cellular component. The most enriched subcategory in biological process was glutathione metabolic process the LSe vs. CK groups, followed by "sulfur compound metabolic process" and cellular modified amino acid metabolic process. Extracellular region was the most enriched subcategory in cellular components. Among the molecular functions, the three most enriched subcategories were oxidoreductase activity, transporter activity and transmembrane transporter activity (Figure 5C). The three most enriched subcategories in the root of the HSe vs. CK groups were glutathione metabolic process, sulfur compound metabolic process and cellular modified amino acid metabolic process. Oxidoreductase activity was the most enriched subcategory in molecular function (Figure 5D).

2.6. *Kyoto Encyclopedia of Genes and Genomes (KEGG) Enrichment Analysis of DEGs*

In order to assign DEGs to cellular pathways, pathway enrichment analysis based on KEGG was performed. The DEGs significantly enriched pathways were calculated using hypergeometric distribution based on the whole genome. There are four KEGG pathway categories: organismal systems, metabolism, environmental information processing and genetic information processing. The top 20 significantly enriched KEGG terms were shown in Figure 6. Plant–pathogen interaction and plant hormone signal transduction were the only enriched processes in organismal systems and environmental information processing in the shoot of the LSe vs. CK groups (Figure 6A). In the category of metabolism, most of the pathways were highly overrepresented, including photosynthesis-antenna proteins, phenylpropanoid biosynthesis, glutathione metabolism, amino sugar and nucleotide sugar metabolism, diterpenoid biosynthesis and alpha-linolenic acid metabolism. No significant enrichment was found in genetic information processing. In addition, we found plant–pathogen interaction and plant hormone signal transduction were the only enriched processes in organismal systems and environmental information processing in the shoots of the HSe vs. CK groups which was the same in the shoot of the LSe vs. CK group. In the category of metabolism, glutathione metabolism was the most overrepresented, followed by amino sugar and nucleotide sugar metabolism, phenylpropanoid biosynthesis, diterpenoid biosynthesis and fatty acid biosynthesis (Figure 6B).

The most enriched category in the four groups was metabolism. "Plant–pathogen interaction" was the only enriched item in organismal systems in the root of the LSe vs. CK groups (Figure 6C). In regard to metabolism, Glutathione metabolism was the most highly overrepresented, followed by phenylpropanoid biosynthesis, Cys and Met metabolism, Diterpenoid biosynthesis, Sulfur metabolism, alpha-linolenic acid metabolism and Fatty acid biosynthesis. No enrichment was found in environmental information processing and genetic information processing. As shown in Figure 6D, the enriched items were involved in metabolism and genetic information processing, with metabolism accounting for the majority of them. "Glutathione metabolism", "Phenylpropanoid biosynthesis", and "Cys and Met metabolism" were the three most enriched item in the metabolism. Proteasome was the most enriched item in the genetic information processing. Interestingly, we found starch and sucrose metabolism and Cys and Met metabolism were upregulated in the root of the LSe vs. CK groups and downregulated in the root of the HSe vs. CK group. In addition, we further analyzed the DEGs in the roots to screen for genes that were upregulated under low Se treatment and downregulated under high Se treatment. The KEGG enrichment revealed that starch and sucrose metabolism and Cys and Met metabolism were the most enriched pathway (Figure S2).

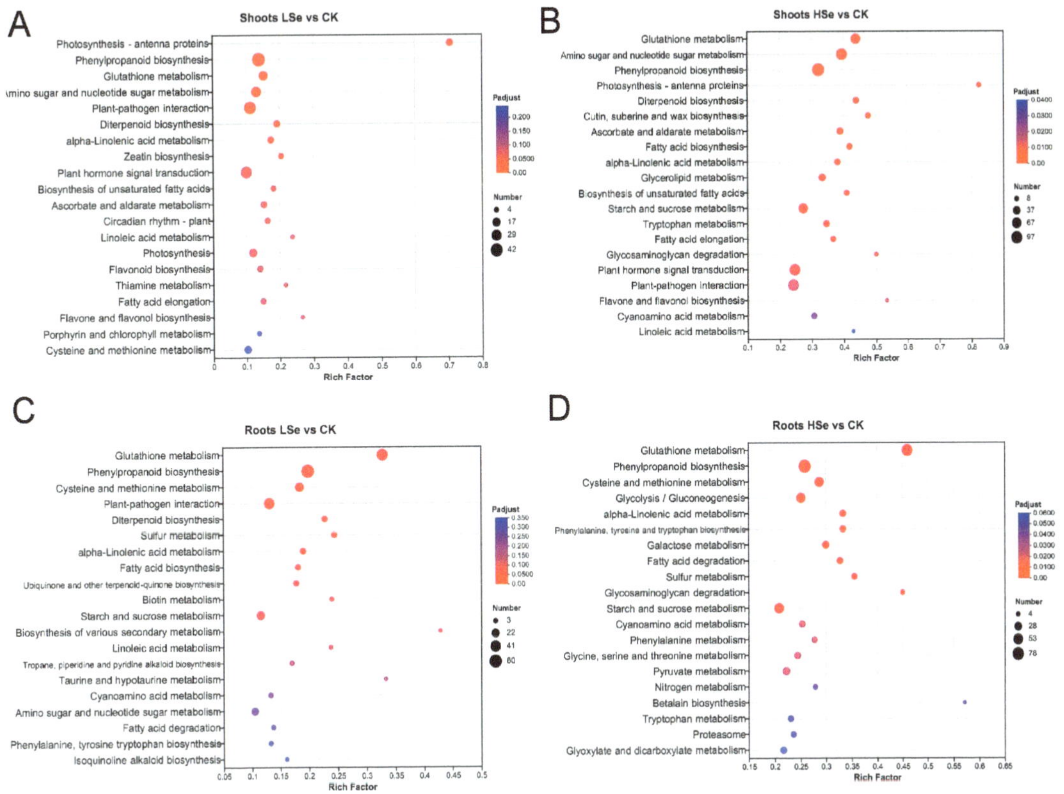

Figure 6. KEGG pathway enrichment analysis of differentially expressed genes (DEGs) in rice shoots and roots under different treatments. (**A**) Shoots LSe vs. CK, (**B**) shoots HSe vs. CK, (**C**) roots LSe vs. CK, and (**D**) roots HSe vs. CK. The x-axis represents the rich factor, indicating the ratio of DEGs in each pathway, while the y-axis lists the enriched KEGG pathways. Dot size reflects the number of DEGs, and the color gradient represents the significance level (Padjust), with red indicating higher significance.

In Cys and Met metabolism, two genes, LOC_Os09g25620 and LOC_Os02g19970 were all upregulated by low Se treatment and downregulated by high Se treatment (Figure 7A). In addition, LOC_Os03g11420, LOC_Os04g39880, LOC_Os04g43400 and LOC_Os02g51070 in starch and sucrose metabolism were upregulated by low Se treatment (Figure 7B), while they were all reduced by high Se treatment. These results imply these genes may play a significant role in the response of rice seedlings to different concentrations of Se treatments, and the pathways involving the expression of these genes are also crucial for the response of rice seedlings to Se stress.

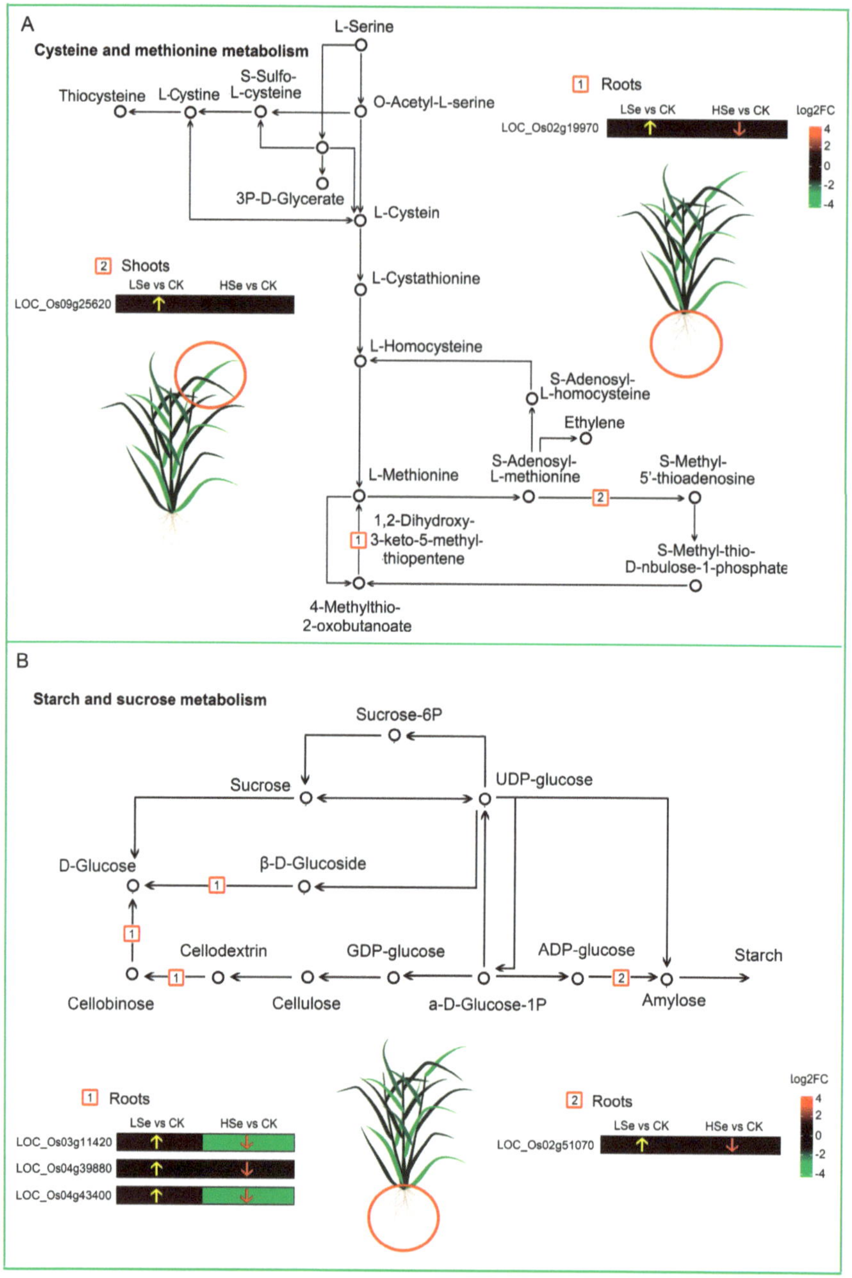

Figure 7. Schematic representation of Met, Cys, and starch metabolism pathways in rice under LSe and HSe treatment. The pathways show Cys and Met metabolism (**A**) and starch and sucrose metabolism (**B**), with corresponding differentially expressed genes highlighted by $\log_2$ fold change ($\log_2$FC) color-coded boxes. Seedlings (14-day-old) of rice were grown hydroponically in nutrient rich solution and then subjected to 7 d 0 (CK), 1 (LSe), and 10 (HSe) μM Se treatments. The yellow arrows highlight the upregulation of genes, while the red arrows emphasize the downregulation of genes. The red boxes and the different numbers within them represent various reaction pathways within that metabolism pathway.

2.7. The Effects of High and Low Se on the Accumulation of Starch, Cys and Met

The Met, Cys and starch contents in rice seedlings under different Se treatments were determined. The results indicate that Met concentration was highest under CK treatment, followed by LSe treatment, and was lowest under HSe treatment, showing a significant decreasing trend (Figure 8A). Cys concentration showed no significant difference between CK and LSe treatments, both of which were higher than the HSe treatment, which exhibited a significantly reduced concentration (Figure 8B). The distribution of starch concentration in both the shoots and roots of seedlings demonstrated that LSe treatment reached the highest levels, followed by CK treatment, while HSe treatment resulted in the lowest levels, showing a significant decrease (Figure 8C,D). These findings suggest that LSe significantly promoted the accumulation of starch, Cys, and Met in rice seedlings, whereas HSe inhibited the accumulation of these substances, highlighting the significant impact of Se concentration on the metabolism of rice seedlings.

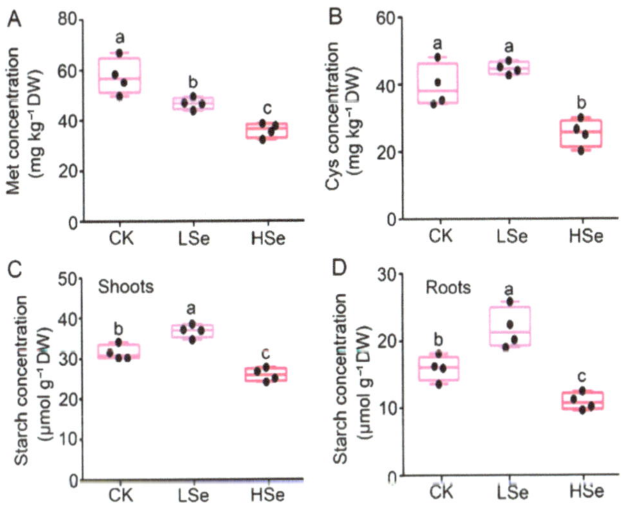

Figure 8. Concentrations of Met, Cys, and starch in rice seedlings under CK, LSe and HSe treatments. (**A**) Met concentration, (**B**) Cys concentration, (**C**) starch concentration in shoots, and (**D**) starch concentration in roots. Data are represented as the mean ± standard deviation (SD). Different letters indicate significant differences under different Se treatments ($p < 0.05$, Student's *t*-test).

2.8. Validation of RNA Sequencing Date by Quantative Real-Time PCR

A total of six genes were randomly selected and validated by quantitative real-time (qRT-PCR) to confirm the reliability of sequencing date (Figure 9). Among them, one gene encodes conserved peptide uORF-containing transcript (CPuORF8) (LOC_Os09g25620), one gene encodes aminotransferase, one gene encodes starch synthase (LOC_Os02g51070) and each of the three genes encodes *Oryza sativa* β-glucosidase6 (Os3bglu6), *Oryza sativa* β-glucosidase 12 (Os4bglu12) and *Oryza sativa* β-glucosidase 17 (Os4bglu17), members of the GH1 family (LOC_Os03g11420, LOC_Os04g39880, LOC_Os04g43400). The genes LOC_0s02g51070, LOC_0s03g11420, LOC_0s04g39880, LOC_0s04g43400, LOC_0s02g19970, and LOC_0s09g25620 exhibit significantly higher expression levels in the LSe treatment group compared to the CK and HSe groups. Overall, the trend indicated that these genes were upregulated under low Se conditions and significantly suppressed under HSe conditions. The results showed that the changes were basically consistent with the results of transcriptome sequencing. These selected and validated genes provide a more comprehensive understanding of the gene response patterns under Se stress and their contributions to the overall metabolic network.

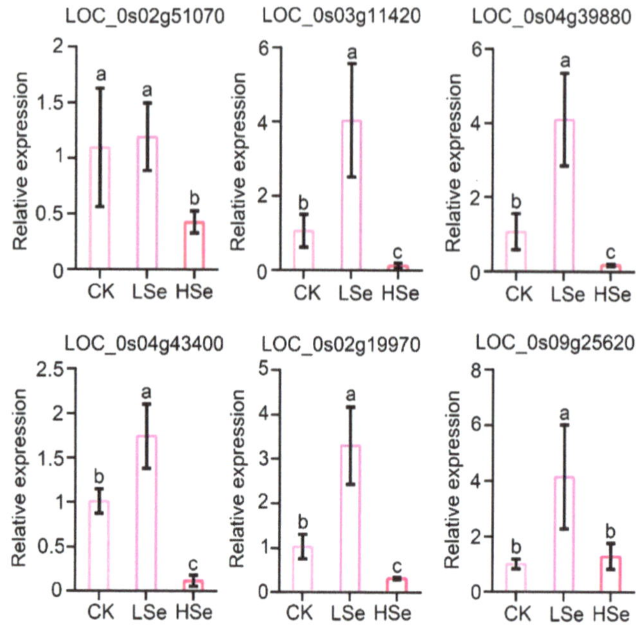

Figure 9. Relative expression levels of genes related to Cys/Met metabolism and starch/sucrose metabolism. Gene expression is shown as the mean ± standard deviation (SD). *OsACTIN* was used as an internal control. The relative expression of LOC_Os02g51070, LOC_Os03g11420, LOC_Os04g39880, LOC_Os04g43400, LOC_Os02g19970 and LOC_Os09g25620 under CK conditions was standardized to 1. Different letters indicate significant differences under different Se treatments ($p < 0.05$, Student's *t*-test).

3. Discussion

Se plays an important physiological role in plant growth. Although Se is not an essential element for plants, an appropriate amount of Se can enhance the plant's antioxidant capacity, improve stress resistance, promote growth, and enhance the nutritional quality of crops [31]. Se can enhance photosynthesis in rice, allowing the plant to accumulate more photosynthetic products, thus facilitating rapid shoot growth [32]. The results of this study indicate that low Se concentrations promote rice seedling growth, while high Se concentrations inhibit both biomass and root growth (Figure 1). This finding is consistent with previous studies, showing that 0.5–1 mg/kg Se promotes significant growth in rice seedlings [33]. Additionally, the oxidative damage induced by HSe concentrations can affect energy metabolism and reduce growth rates [34]. Our results further confirm the benefits of low Se on rice seedling growth, as well as the toxic effects of high Se, particularly 1 μM Se, which showed the most pronounced promotion of growth.

In this study, the effects of different Se concentrations on the Se content in rice seedlings were investigated. The results demonstrated that Se application significantly enhanced the total Se content in seedlings, with the highest accumulation observed under the HSe treatment (Figure 2A). This aligns with previous studies indicating that higher Se concentrations lead to greater Se uptake in rice [35]. Moreover, the low Se treatment promoted the synthesis of organic Se, as evidenced by a significantly higher organic Se content compared to high Se treatment (Figure 2C). This suggests that lower Se concentrations may favor the synthesis of organic Se species, such as SeMet and SeCys, which are more beneficial for human nutrition [36].

The antioxidant enzyme system plays a crucial role in alleviating oxidative stress in plants. Enzymes such as SOD, CAT, and POD can scavenge ROS and protect cells from damage caused by oxidative stress [37]. It was found that the activities of SOD and CAT in both shoots and roots displayed a trend of increasing followed by decreasing as Se concentration increased (Figure 3A,C,D,F). Under HSe treatment, SOD activity in shoots and roots increased by 49.4% and 38.7%, respectively, compared to CK, indicating an enhanced scavenging of superoxide radicals. However, CAT activity decreased by 46.3% and 48.9%, respectively, under the same treatment, suggesting that excessive Se may overwhelm the enzyme system, leading to its inactivation or degradation. POD activity, in contrast, exhibited a consistent upward trend with increasing Se levels, peaking under HSe treatment (Figure 3B,E). This aligns with studies showing that POD activity often increases under stress conditions to compensate for the loss of other antioxidant enzymes [38].

Transcriptome analysis revealed significant differences in DEGs in different tissues of rice under high and low Se treatments (Figure 4). Low Se treatment promoted rice seedling growth, leading to more upregulated genes than downregulated ones, thus altering the biological processes. GO enrichment analysis is instrumental in highlighting key biological processes involved in response to stress conditions. For instance, in *Pueraria lobata*, transcriptome analysis showed that Se stimulated the upregulation of genes involved in signal transduction and stress response, such as phosphate transporter and sulfate transporter genes, which regulate Se metabolism and accumulation [39]. In celery, Se treatment enriched secondary metabolic pathways, especially genes related to phenylpropanoid biosynthesis, indicating that Se may enhance plant resistance through secondary metabolic pathways [40]. Our research found that GO enrichment analysis showed significant enrichment of terms like "cellular anatomical entity" and "oxidoreductase activity" in both roots and shoots under low Se treatment, while high Se treatment enriched pathways related to "glutathione metabolism" in roots (Figure 5). Notably, the enrichment of glutathione metabolism in roots suggests that roots are the primary organs for Se uptake and conversion in plants and that they may require additional antioxidant mechanisms to mitigate Se stress and maintain cellular redox homeostasis when exposed to exogenous Se. Previous studies have also shown similar trends, confirming that appropriate Se may generate ROS during root metabolism, which can activate the glutathione metabolic pathway in roots [41].

KEGG pathway analysis further revealed that Se treatment significantly upregulated pathways related to "glutathione metabolism", "phenylpropanoid biosynthesis", and "alpha-linolenic acid metabolism" in both shoots and roots (Figure 6). These pathways play key roles in the antioxidant response and stress tolerance under Se treatment [3,42,43]. Additionally, genes related to starch and sucrose metabolism were significantly upregulated under low Se treatment (Figure S2). Starch and sucrose metabolism play a crucial role in plant growth. Under drought and salt stress conditions, plants regulate sucrose metabolism to increase the accumulation of storage substances, thereby enhancing stress resistance [44]. Previous study also showed that Se treatment promotes the accumulation of starch and soluble sugars in potatoes, indicating a positive effect of Se on carbon metabolism and yield [45]. Se treatment aids potato growth under heavy metal stress by regulating carbohydrate metabolism, including starch and sucrose metabolism [46]. This indicates that the upregulation of genes related to starch and sucrose metabolism under low Se treatment may lead to increased carbohydrate synthesis and accumulation. Plants may support growth and development by enhancing the storage of carbohydrates. This view was corroborated by the determination of starch content in rice (Figure 8C,D). While high Se levels may result in oxidative stress or metabolic disorders, inhibiting starch and sucrose synthesis. This could be a protective mechanism by which plants reduce energy consumption or limit the accumulation of harmful metabolic by products.

Meanwhile, Cys and Met metabolism showed a trend similar to starch and sucrose metabolism. Cys contributes to the synthesis of glutathione (GSH), helping plants cope with oxidative stress and protecting cells from damage by free radicals. Additionally, Cys participates in processes such as photosynthesis and sulfur metabolism, thereby regulating plant growth, development, and stress resistance [47]. Met, beyond being a fundamental component of proteins, is involved in various critical methylation reactions through S-adenosylmethionine (AdoMet), regulating DNA methylation and plant hormones like ethylene [48]. In this study, we found that genes related to Cys and Met metabolism were upregulated in the root of the LSe vs. CK groups and downregulated in the root of the HSe vs. CK groups (Figure 7A), supported by measurements of Cys and Met content (Figure 8A,B). This indicates that low Se enhances tolerance to ROS and supports stress adaptation, consistent with changes in antioxidant enzyme activities (Figure 3). Conversely, the oxidative stress caused by high Se may suppress Cys and Met metabolic pathways, likely as a protective mechanism to prevent overaccumulation of toxic Se derivatives. While the reduced Met content may stem from Se-induced stress and metabolic disorders. These findings provide new insights into the adaptive metabolic strategies under Se stress and highlight the Cys and Met metabolism pathway as a potential target for enhancing Se tolerance in crops.

Starch synthase is primarily responsible for synthesizing different types of starch in rice to ensure energy storage during grain development. Studies have shown that in vegetative organs such as rice leaves, enzymes like soluble starch synthase IIIb (SSIIIb) are involved in the synthesis of temporary starch, which can be converted into energy under stress or high growth demand [49]. Under drought stress, the expression and activity of AGPase and sucrose synthase genes in rice grains increase to promote the accumulation of storage starch, ensuring grain development under water-deficient conditions [50]. Our study found that genes encoding starch synthases were upregulated under low Se treatment and downregulated under high Se treatment. This indicates that changes in the expression of starch synthase-encoding genes help modulate starch synthesis under different Se treatments, optimizing energy storage and utilization in rice.

GH1 β-glucosidase is involved in hormone regulation, stress response, and ell wall remodeling in rice [51,52]. GH1 enzymes promote tissue development and help maintain plant structure. In Arabidopsis, GH1 enzymes are upregulated under high salinity, aiding in salt stress tolerance [53]. Our study found that Os3BGlu6, Os4BGlu12 and Os4BGlu17, which are members of the GH1 family, were predominantly upregulated in rice shoots under low Se treatment and downregulated under high Se treatment. Ref. [54] showed that Os3BGlu6 enhances drought resistance and photosynthetic efficiency by hydrolyzing ABA-glucose ester. Impaired expression of Os3BGlu6 leads to decreased ABA levels, reducing drought resistance and photosynthetic efficiency. Additionally, Os4BGlu12 is highly expressed during germination and maturation, playing a key role in these stages [55]. The significant differential expression of GH1 genes in rice roots under varying Se treatments may underscore their critical role in plant defense and adaptation to stress.

Aminotransferases play a crucial role in amino acid metabolism, osmotic regulation, and antioxidant functions, contributing to enhanced stress tolerance and growth stability in plants [56]. Se treatment promotes metabolic product synthesis, stress resistance, and protein content in grains [57]. The study of naturally Se-enriched rice and non-Se-enriched rice demonstrated that Se-enriched rice exhibited higher antioxidant activity and amino acid metabolism levels, highlighting the role of aminotransferases in amino acid metabolism and stress resilience [58]. Our research found that aminotransferase genes were upregulated under low Se treatment and downregulated under high Se treatment. This suggests that lower Se concentrations enhance rice growth and metabolism by stimulating aminotransferase activity. Conversely, higher Se concentrations may impair protein synthesis pathways, resulting in the downregulation of aminotransferases and associated metabolic pathways, thus slowing down growth and adaptability. This observation is consistent with previous studies indicating that Se treatment enhances the synthesis and metabolism of essential amino acids in rice, likely by stimulating aminotransferase activity. This metabolic modulation supports cellular growth and improves stress tolerance [59].

Conserved Peptide Upstream Open Reading Frames (CPuORFs) are a special class of upstream open reading frames (uORFs) whose encoded peptides are conserved across different species and are primarily involved in regulating gene translation [60]. Studies have shown that CPuORFs can play a regulatory role in gene expression through specific conserved peptide sequences, often in response to small molecule signals and environmental stress [61]. Our research identified that CPuORF8, a member of the CPuORF gene family, may play an important role in rice in response to Se treatment as an environmental signal. Prior to this study, no research had reported the regulation of these genes in rice under Se treatment. Our findings provide a reference for future investigations.

4. Materials and Methods

4.1. Rice Cultivation and Treatment Methods

The experiment was conducted in April–June 2024 in the greenhouse of the ECO—Environment Protection Research Institute, Shanghai Academy of Agricultural Sciences, Shanghai, China. The rice seeds (*Oryza sativa* L., Nipponbare) were soaked at 30 °C for 48 h, washed with aseptic water for three times, germinated at 35 °C in the dark for 24 h, and transferred into a cultivation box to raise seedlings. The Seedlings were then transplanted and cultured in nutrient solution containing different concentrations of Se. The nutrient solution was composed of ($\mu mol \cdot L^{-1}$): $Ca(NO_3)_2 \cdot 4H_2O$ 183, KNO_3 91, KH_2PO_4 100, $(NH_4)_2SO_4$ 183, $MgCl_2 \cdot 6H_2O$ 274, $MnSO_4 \cdot H_2O$ 1, $(NH_4)_6Mo_7O_{24} \cdot 4H_2O$ 1, $ZnSO_4 \cdot 7H_2O$ 1, $CuSO_4 \cdot 5H_2O$ 0.2, H_3BO_3 3, Fe(III)-EDTA 60. The Se treatments included a control (0 μM Se) and increasing concentrations (1 and 10 μM Se) as sodium selenite. Plants were grown in a greenhouse with a 16/8-h light/dark cycle at 28 °C/22 °C and 60–70% relative humidity. Change the nutrient solution treated by hydroponics every 3 days, and pH is adjusted to 5.5 (±0.1). After 14 days of treatment, the samples were taken and the indexes were determined and analyzed. Shoot and root were harvested and frozen with liquid nitrogen, respectively, and then stored in the refrigerator at −80 °C for gene expression analysis.

4.2. Detection of Total Se and Different Forms of Se in Rice Seedlings

Total Se content: the samples were placed in a 105 °C oven for 20 min to deactivate enzymes and then dried at 90 °C to a constant weight. Afterward, the samples were ground and passed through a 0.15 mm sieve. The total Se content was determined using the hydride generation-atomic fluorescence spectrometry (HG-AFS) method.

Organic Se content: Following the method described by [62], 3 g of the sample was weighed and mixed with 3.5 mL of 4 mol/L hydrochloric acid solution. The mixture was heated in a boiling water bath at 100 °C for 10 min, followed by centrifugation at 4 °C at a speed of 2500 r/min for 10 min. The supernatant was diluted to 10 mL with ultrapure water, and the inorganic selenium content was determined using HG-AFS. The organic Se content was calculated as the difference between the total Se content and the inorganic Se content.

4.3. Determination of Antioxidant Enzyme Activities

Rice seedling samples were collected under 2 weeks different Se concentration treatments for transcriptome sequencing analysis. The rice seeding samples were divided into shoot and root, frozen quickly with liquid nitrogen and stored at −80 °C. The activities of enzymes including SOD, POD and CAT were determined through biochemical analysis kits (Sangon Biotech, Shanghai, China) according to the manufacturer's instructions.

4.4. Transcriptome Sequencing Analysis

4.4.1. Sampling and RNA Extraction

Rice seedling samples were collected under 2 weeks different Se concentration treatments for transcriptome sequencing analysis. The rice seeding samples were divided into shoot and root, frozen quickly with liquid nitrogen and stored at −80 °C for RNA extraction.

Total RNA was extracted from the tissue using TRIzol® Reagent according the manufacturer's instructions. Then, RNA quality was determined by 5300 Bioanalyser (Agilent Technologies, Santa Clara, CA, USA) and quantified using the ND-2000 (NanoDrop Technologies, Wilmington, DE, USA). All the RNAs were processed using the following parameters: concentration > 10 ng, OD260/280 = 1.8~2.2, 28S:18S $\geq$ 1.0 and RIN > 6.5 or concentration > 1 μg, OD260/280 = 1.8~2.2, OD260/230 $\geq$ 2.0, RQN $\geq$ 6.5, 28S:18S $\geq$ 1.0.

4.4.2. cDNA Library Construction and RNA Sequencing

RNA purification, reverse transcription, library construction and sequencing were performed at Shanghai Majorbio Bio-pharm Biotechnology Co., Ltd. (Shanghai, China) according to the manufacturer's instructions. The rice RNA-seq transcriptome library was prepared following Illumina® Stranded mRNA Prep, Ligation (San Diego, CA, USA) using 1 μg of total RNA. Briefly, messenger RNA was isolated according to polyA selection method by oligo(dT) beads and then fragmented by fragmentation buffer firstly. Secondly, double-stranded cDNA was synthesized with random hexamer primers. Then, the synthesized cDNA was subjected to end-repair, phosphorylation and adapter addition according to library construction protocol. Libraries were size selected for cDNA target fragments of 300–400 bp use magnetic beads followed by PCR amplified for 10–15 PCR cycles. After quantified by Qubit 4.0, the sequencing library was performed on NovaSeq X Plus platform (PE150) using NovaSeq Reagent Kit (Illumina, San Diego, CA, USA).

4.4.3. Sequence Assembly and Gene Annotation and Quantification

The raw paired end reads were trimmed and quality controlled by fastp [63] with default parameters. Then, clean reads were separately aligned to reference genome with orientation mode using HISAT2 (version 2.2.1) [64] software. The mapped reads of each sample were assembled by StringTie (version 2.2.1) [65] in a reference-based approach.

4.4.4. Differentially Expressed Genes (DEGs) and Their GO and KEGG Pathway Enrichment Analyses

To identify DEGs (differential expression genes) between two different samples, the expression level of each transcript was calculated according to the transcripts per million reads (TPM) method. RSEM [66] was used to quantify gene abundances. Essentially, differential expression analysis was performed using the DESeq2 (version 1.26.0) [67]. DEGs with $|\log_2 FC| \geq 1$ and FDR < 0.05 (DESeq2) were considered to be significantly different expressed genes. In addition, functional-enrichment analysis including GO and KEGG were performed to identify which DEGs were significantly enriched in GO terms and metabolic pathways at Bonferroni-corrected p-value < 0.05 compared with the whole-transcriptome background. GO functional enrichment and KEGG pathway analysis were carried out by Goatools and Python scipy.

4.5. qRT-PCR Analysis

Six DEGs involved in starch and sucrose metabolism and Cys and Met metabolism were selected for qRT-PCR and used in validating the RNA-seq results. Primers were designed with Primer Premier 5 software (Premier Biosoft International, Palo Alto, CA, USA). The primers are listed in Table S1. Detailed information for qRT-PCR was the same as that of our previous study [68]. 18s-rRNA gene was selected as the internal reference gene, and the $2^{-\Delta\Delta Ct}$ method [69] was used in determining the relative expression levels of target genes.

4.6. The Analysis of Amino Acid and Starch of Rice Seedlings

The samples were prepared as follows: 2 g of frozen rice sample was taken, 10 mL solution (70% (v/v) chloroform and 30% (v/v) methanol) was added, it was grinded into homogenate under liquid N condition, free amino acid was extracted with 8 mL of water, the water layer was extracted after 3 times, it was let to stand on ice for 30 min, then 500 µL supernatant was added and absorbed. The supernatant was filtered through a 0.45-µm micropore filter and loaded into a sample vial. Then, the automatic pre-column derivatization was used and single free amino acid concentrations were measured using Agilent 1260 High Performance Liquid Chromatography (Palo Alto, CA, USA).

The rice seedlings were freezedried and weighted. The dry seeds were extensively ground into flour. The soluble sugar was extracted from flour in 80% (v/v) ethanol for 30 min at 80 °C and then centrifuged at 5000× g for 10 min. The resulting pellet was further washed two times with 80% ethanol. The ethanol-insoluble residue was used for total starch content determination by using Megazyme Total Starch Assay Kit (K-TSTA) (Megazyme, Bray, County Wicklow, Ireland).

4.7. Statistical Analysis

The experimental data are the mean of 3 replicates and standard error (mean ± SE). OAPnuke 1.5.2 software, HISAT 2.1.0 software, Bowtie 2 2.2.5, RSEM 2.0 software and KOBAS 2.0 software were used in data analysis. Origin 2020 and Graphpad Prism 8 were used for drawing. Pearson correlation analysis were performed on Se and associated metals in rice, and the significance level was set at a p-value of <0.05.

5. Conclusions

This study provides a comprehensive investigation into the physiological and transcriptomic effects of varying Se concentrations on rice seedlings, with a particular focus on starch and sucrose, Cys and Met metabolic pathways. The findings confirm the dual role of Se in plant growth, where a low Se concentration (1 µM) enhances shoot and root growth, promotes antioxidant activity, and upregulates genes involved in critical metabolic pathways, contributing to improved carbohydrate and amino acid synthesis. In contrast, a high Se concentration (10 µM) exerts toxic effects, inhibiting growth, suppressing metabolic activities, and downregulating key genes, highlighting the delicate balance required for Se supplementation in rice cultivation. Key metabolic pathways, including glutathione metabolism, starch and sucrose metabolism, and Cys and Met metabolism, showed significant differential gene expression under varying Se treatments. Several key genes were also identified in these metabolic processes. This study underscores the potential of low Se treatment for biofortification strategies to improve nutritional quality while minimizing toxic effects. These findings advance the understanding of Se-induced molecular mechanisms in rice and offer valuable insights for enhancing crop stress resilience and nutritional value through optimized Se management.

Supplementary Materials: The following supporting information can be downloaded at: https://www.mdpi.com/article/10.3390/ijms26041596/s1.

Author Contributions: Y.Y. and J.Z. conducted the research work and prepared this manuscript. Y.S. helped to revise this manuscript. L.S., Q.Q. and S.Y. helped in transcriptome data analysis. J.W. was involved in amino acid and starch content analysis. Y.S. and Y.X. are the supervisors of the research work and also the corresponding authors. All authors have read and agreed to the published version of the manuscript.

Funding: This work was supported by Shanghai Agriculture Applied Technology Development Program, China (Grant No. 2022-02-08-00-12-F01202), the Natural Science Foundation of Shanghai (23ZR1469200), the Chinese National Natural Science Foundation (31902102), the Shanghai Rising-Star Program (23QB1405900), the Flagship Project of Eco-Environmental Protection Research Institute, SAAS (SK-JC 2023-1), and the National Agricultural Experimental Station for Agricultural Environment, Fengxian (grant number: NAES035AE03).

Institutional Review Board Statement: Not applicable.

Informed Consent Statement: Not applicable.

Data Availability Statement: The data of this study are available from the corresponding author, [Y.X.], upon request.

Conflicts of Interest: The authors declare that they have no known competing financial interests or personal relationships that could have appeared to influence the work reported in this paper.

References

1. Roman, M.; Jitaru, P.; Barbante, C. Selenium biochemistry and its role for human health. *Metallomics* **2013**, *6*, 25–54. [CrossRef] [PubMed]
2. Zhang, L.; Song, H.; Guo, Y.; Fan, B.; Huang, Y.; Mao, X.; Liang, K.; Hu, Z.; Sun, X.; Fang, Y.; et al. Benefit-risk assessment of dietary selenium and its associated metals intake in China (2017–2019): Is current selenium-richagro-food safe enough? *J. Hazard. Mater.* **2020**, *398*, 123224. [CrossRef] [PubMed]
3. Wu, L.; Liu, T.; Xu, Y.; Chen, W.; Liu, B.; Zhang, L.; Liu, D.; Zhang, H. Comparative transcriptome analysis of two selenium-accumulating genoty pes of Aegilops tauschii Coss. in response to selenium. *BMC Genet.* **2019**, *20*, 9. [CrossRef]
4. Lucca, P.; Poletti, S.; Sautter, C. Genetic engineering approaches to enrich rice with iron and vitamin A. *Physiol. Plant.* **2006**, *126*, 291–303. [CrossRef]
5. Muthayya, S.; Sugimoto, J.D.; Montgomery, S.; Maberly, G.F. An overview of global rice production, supply, trade, and consumption. *Ann. N. Y. Acad. Sci.* **2014**, *1324*, 7–14. [CrossRef] [PubMed]
6. Yoshida, M.; Yasumoto, K. Selenium contents of rice grown at various sites in Japan. *J. Food Compos. Anal.* **1987**, *1*, 71–75. [CrossRef]
7. Hawrylak-Nowak, B.; Matraszek, R.; Pogorzelec, M. The dual effects of two inorganic selenium forms on the growth, selected physiological parameters and macronutrients accumulation in cucumber plants. *Acta Physiol. Plant.* **2015**, *37*, 41. [CrossRef]
8. Supriatin, S.; Weng, L.; Comans, R.N. Selenium speciation and extractability in Dutch agricultural soils. *Sci. Total. Environ.* **2015**, *532*, 368–382. [CrossRef]
9. Khan, Z.; Thounaojam, T.C.; Chowdhury, D.; Upadhyaya, H. The role of selenium and nano selenium on physiological responses in p lant: A review. *Plant Growth Regul.* **2023**, *100*, 409–433. [CrossRef]
10. Malerba, M.; Cerana, R. Effect of Selenium on the Responses Induced by Heat Stress in Plant Cell Cultures. *Plants* **2018**, *7*, 64. [CrossRef]
11. Han, D.; Li, X.; Xiong, S.; Tu, S.; Chen, Z.; Li, J.; Xie, Z. Selenium uptake, speciation and stressed response of *Nicotiana tabacum* L. *Environ. Exp. Bot.* **2013**, *95*, 6–14. [CrossRef]
12. Gupta, M.; Gupta, S. An Overview of Selenium Uptake, Metabolism, and Toxicity in Plants. *Front. Plant Sci.* **2017**, *7*, 2074. [CrossRef]
13. Yao, X.; Chu, J.; Wang, G. Effects of selenium on wheat seedlings under drought stress. *Biol. Trace Element Res.* **2009**, *130*, 283–290. [CrossRef] [PubMed]
14. Rao, S.; Gou, Y.; Yu, T.; Cong, X.; Gui, J.; Zhu, Z.; Zhang, W., Liao, Y.; Ye, J.; Cheng, S.; et al. Effects of selenate on Se, flavonoid, and glucosinolate in broccoli florets by combined transcriptome and metabolome analyses. *Food Res. Int.* **2021**, *146*, 110463. [CrossRef]

15. Yang, X.; Liao, X.; Yu, L.; Rao, S.; Chen, Q.; Zhu, Z.; Cong, X.; Zhang, W.; Ye, J.; Cheng, S.; et al. Combined metabolome and transcriptome analysis reveal the mechanism of selenate influence on the growth and quality of cabbage (*Brassica oleracea* var. capitata L.). *Food Res. Int.* **2022**, *156*, 111135. [CrossRef]
16. Li, L.; Wu, S.; Wang, S.; Shi, X.; Cheng, S.; Cheng, H. Molecular Mechanism of Exogenous Selenium Affecting the Nutritional Quality, Species and Content of Organic Selenium in Mustard. *Agronomy* **2023**, *13*, 1425. [CrossRef]
17. Rao, S.; Yu, T.; Cong, X.; Lai, X.; Xiang, J.; Cao, J.; Liao, X.; Gou, Y.; Chao, W.; Xue, H.; et al. Transcriptome, proteome, and metabolome reveal the mechanism of tolerance to selenate toxicity in Cardamine violifolia. *J. Hazard. Mater.* **2021**, *406*, 124283. [CrossRef]
18. Mostofa, M.G.; Hossain, M.A.; Siddiqui, N.; Fujita, M.; Tran, L.-S.P. Phenotypical, physiological and biochemical analyses provide insight into selenium-induced phytotoxicity in rice plants. *Chemosphere* **2017**, *178*, 212–223. [CrossRef]
19. Gouveia, G.C.C.; Galindo, F.S.; Lanza, M.G.D.B.; Silva, A.C.d.R.; Mateus, M.P.d.B.; da Silva, M.S.; Tavanti, R.F.R.; Tavanti, T.R.; Lavres, J.; dos Reis, A.R. Selenium toxicity stress-induced phenotypical, biochemical and physiological responses in rice plants: Characterization of symptoms and plant metabolic adjustment. *Ecotoxicol. Environ. Saf.* **2020**, *202*, 110916. [CrossRef]
20. Thitisaksakul, M.; Tananuwong, K.; Shoemaker, C.F.; Chun, A.; Tanadul, O.; Labavitch, J.M.; Beckles, D.M. Effects of timing and severity of salinity stress on rice (*Oryza sativa* L.) yield, grain composition, and starch functionality. *J. Agric. Food Chem.* **2015**, *63*, 2296–2304. [CrossRef]
21. Kumari, M.; Asthir, B. Transformation of Sucrose to Starch and Protein in Rice Leaves and Grains under Two Establishment Methods. *Rice Sci.* **2016**, *23*, 255–265. [CrossRef]
22. Li, G.; Hu, Q.; Shi, Y.; Cui, K.; Nie, L.; Huang, J.; Peng, S. Low Nitrogen Application Enhances Starch-Metabolizing Enzyme Activity and Improves Accumulation and Translocation of Non-structural Carbohydrates in Rice Stems. *Front. Plant Sci.* **2018**, *9*, 1128. [CrossRef] [PubMed]
23. Xian, L.; Tian, J.; Long, Y.; Ma, H.; Tian, M.; Liu, X.; Yin, G.; Wang, L. Metabolomics and transcriptomics analyses provide new insights into the nutritional quality during the endosperm development of different ploidy rice. *Front. Plant Sci.* **2023**, *14*, 1210134. [CrossRef] [PubMed]
24. Nguyen, H.C.; Hoefgen, R.; Hesse, H. Improving the nutritive value of rice seeds: Elevation of cysteine and methionine contents in rice plants by ectopic expression of a bacterial serine acetyltransferase. *J. Exp. Bot.* **2012**, *63*, 5991–6001. [CrossRef] [PubMed]
25. Wang, F.; Zhang, J.; Xu, L.; Ma, A.; Zhuang, G.; Huo, S.; Zou, B.; Qian, J.; Cui, Y. Selenium volatilization in plants, microalgae, and microorganisms. *Heliyon* **2024**, *10*, e26023. [CrossRef] [PubMed]
26. Cheng, R.; Zhang, Z.; Zhan, C.; Qin, T.; Wang, L.; Zhang, X. Environmentally relevant concentrations of selenite trigger reproductive toxicity by affecting oocyte development and promoting larval apoptosis. *Environ. Pollut.* **2022**, *316 Pt 1*, 120648. [CrossRef]
27. Chan, K.X.; Phua, S.Y.; Van Breusegem, F. Secondary sulfur metabolism in cellular signalling and oxidative stress responses. *J. Exp. Bot.* **2019**, *70*, 4237–4250. [CrossRef]
28. Mishra, P.; Mishra, V.; Takabe, T.; Rai, V.; Singh, N.K. Elucidation of salt-tolerance metabolic pathways in contrasting rice g enotypes and their segregating progenies. *Plant Cell Rep.* **2016**, *35*, 1273–1286. [CrossRef]
29. Whitcomb, S.J.; Rakpenthai, A.; Brückner, F.; Fischer, A.; Parmar, S.; Erban, A.; Kopka, J.; Hawkesford, M.J.; Hoefgen, R. Cysteine and Methionine Biosynthetic Enzymes Have Distinct Effects on Seed Nutritional Quality and on Molecular Phenotypes Associated With Accumulation of a Methionine-Rich Seed Storage Protein in Rice. *Front. Plant Sci.* **2020**, *11*, 1118. [CrossRef]
30. Malheiros, R.S.P.; Costa, L.C.; Ávila, R.T.; Pimenta, T.M.; Teixeira, L.S.; Brito, F.A.L.; Zsögön, A.; Araújo, W.L.; Ribeiro, D.M. Selenium downregulates auxin and ethylene biosynthesis in rice seedlings to modify primary metabolism and root architecture. *Planta* **2019**, *250*, 333–345. [CrossRef]
31. Chao, W.; Rao, S.; Chen, Q.; Zhang, W.; Liao, Y.; Ye, J.; Cheng, S.; Yang, X.; Xu, F. Advances in Research on the Involvement of Selenium in Regulating Plant Ecosystems. *Plants* **2022**, *11*, 2712. [CrossRef] [PubMed]
32. Lu, Y.; Wu, J.; Li, J. The alleviating effects and underlying mechanisms of exogenous selenium on both Sb(III) and Sb(V) toxicity in rice seedlings (*Oryza sativa* L.). *Environ. Sci. Pollut. Res.* **2023**, *30*, 89927–89941. [CrossRef]
33. Yan, J.; Chen, X.; Zhu, T.; Zhang, Z.; Fan, J. Effects of Selenium Fertilizer Application on Yield and Selenium Accumulation Characteristics of Different Japonica Rice Varieties. *Sustainability* **2021**, *13*, 10284. [CrossRef]
34. Liang, Y.; Su, Y.; Li, L.; Huang, X.; Panhwar, F.H.; Zheng, T.; Tang, Z.; Ei, H.H.; Farooq, M.U.; Zeng, R.; et al. Quick selenium accumulation in the selenium-rich rice and its physiological responses in changing selenium environments. *BMC Plant Biol.* **2019**, *19*, 559. [CrossRef] [PubMed]
35. Su, Y.; Huang, X.; Li, L.; Muhammad, Z.A.; Li, M.; Zheng, T.; Guo, Z.; Zhang, Y.; Luo, D.; Ye, X.; et al. Comparative Responses of Silicon to Reduce Cadmium and Enrich Selenium in Rice Varieties. *Foods* **2023**, *12*, 1656. [CrossRef]
36. Kiełczykowska, M.; Kocot, J.; Paździor, M.; Musik, I. Selenium—A fascinating antioxidant of protective properties. *Adv. Clin. Exp. Med. Off. Organ Wroc. Med. Univ.* **2018**, *27*, 245–255. [CrossRef]

37. Gill, S.S.; Tuteja, N. Reactive oxygen species and antioxidant machinery in abiotic stress to lerance in crop plants. *Plant Physiol. Biochem. PPB* **2010**, *48*, 909–930. [CrossRef]
38. Luo, H.; Xing, P.; Liu, J.; Pan, S.; Tang, X.; Duan, M. Selenium improved antioxidant response and photosynthesis in fragrant rice (*Oryza sativa* L.) seedlings during drought stress. *Physiol. Mol. Biol. Plants* **2021**, *27*, 2849–2858. [CrossRef]
39. Guo, K.; Yao, Y.; Yang, M.; Li, Y.; Wu, B.; Lin, X. Transcriptome sequencing and analysis reveals the molecular response to selenium stimuli in *Pueraria lobata* (willd.) Ohwi. *PeerJ* **2020**, *8*, e8768. [CrossRef]
40. Zhang, C.; Xu, B.; Zhao, C.-R.; Sun, J.; Lai, Q.; Yu, C. Comparative de novo transcriptomics and untargeted metabolomic analyses elucidate complicated mechanisms regulating celery (*Apium graveolens* L.) responses to selenium stimuli. *PLoS ONE* **2019**, *14*, e0226752. [CrossRef]
41. Ren, H.; Li, X.; Guo, L.; Wang, L.; Hao, X.; Zeng, J. Integrative Transcriptome and Proteome Analysis Reveals the Absorption and Metabolism of Selenium in Tea Plants [*Camellia sinensis* (L.) O. Kuntze]. *Front. Plant Sci.* **2022**, *13*, 848349. [CrossRef] [PubMed]
42. Jia, J.; Liu, Q.; Liu, H.; Yang, C.; Zhao, Q.; Xu, Y.; Wu, W. Structure characterization and antioxidant activity of abalone viscera l peptides-selenium in vitro. *Food Chem.* **2024**, *433*, 137398. [CrossRef]
43. Wang, F.; Yang, J.; Hua, Y.; Wang, K.; Guo, Y.; Lu, Y.; Zhu, S.; Zhang, P.; Hu, G. Transcriptome and Metabolome Analysis of Selenium Treated Alfalfa Reve als Influence on Phenylpropanoid Biosynthesis to Enhance Growth. *Plants* **2023**, *12*, 2038. [CrossRef] [PubMed]
44. Cui, G.; Zhang, Y.; Zhang, W.; Lang, D.; Zhang, X.; Li, Z.; Zhang, X. Response of Carbon and Nitrogen Metabolism and Secondary Metabolites to Drought Stress and Salt Stress in Plants. *J. Plant Biol.* **2019**, *62*, 387–399. [CrossRef]
45. Turakainen, M.; Hartikainen, H.; Seppänen, M.M. Effects of selenium treatments on potato (*Solanum tuberosum* L.) growth and concentrations of soluble sugars and starch. *J. Agric. Food Chem.* **2024**, *52*, 5378–5382. [CrossRef]
46. Shahid, M.A.; Balal, R.M.; Khan, N.; Zotarelli, L.; Liu, G.D.; Sarkhosh, A.; Fernández-Zapata, J.C.; Nicolás, J.J.M.; Garcia-Sanchez, F. Selenium impedes cadmium and arsenic toxicity in potato by modulating carbohydrate and nitrogen metabolism. *Ecotoxicol. Environ. Saf.* **2019**, *180*, 588–599. [CrossRef]
47. Ingrisano, R.; Tosato, E.; Trost, P.; Gurrieri, L.; Sparla, F. Proline, Cysteine and Branched-Chain Amino Acids in Abiotic Stress Res ponse of Land Plants and Microalgae. *Plants* **2023**, *12*, 3410. [CrossRef]
48. Sauter, M.; Moffatt, B.; Saechao, M.C.; Hell, R.; Wirtz, M. Methionine salvage and S-adenosylmethionine: Essential links between sulfur, ethylene and polyamine biosynthesis. *Biochem. J.* **2013**, *451*, 145–154. [CrossRef]
49. Morita, R.; Crofts, N.; Miura, S.; Ikeda, K.-I.; Aoki, N.; Fukayama, H.; Fujita, N. Characterization of the Functions of Starch Synthase IIIb Expressed in the Vegetative Organs of Rice (*Oryza sativa* L.). *Plant Cell Physiol.* **2022**, *64*, 94–106. [CrossRef]
50. Prathap, V.; Tyagi, A. Correlation between expression and activity of ADP glucose pyrophosphorylase and starch synthase and their role in starch accumulation during grain filling under drought stress in rice. *Plant Physiol. Biochem.* **2020**, *157*, 239–243.
51. Opassiri, R.; Pomthong, B.; Onkoksoong, T.; Akiyama, T.; Esen, A.; Cairns, J.R.K. Analysis of rice glycosyl hydrolase family 1 and expression of Os4bglu 12 beta-glucosidase. *BMC Plant Biol.* **2006**, *6*, 33. [CrossRef] [PubMed]
52. Seshadri, S.; Akiyama, T.; Opassiri, R.; Kuaprasert, B.; Cairns, J.K. Structural and enzymatic characterization of Os3BGlu6, a rice beta-glucosidase hydrolyzing hydrophobic glycosides and (1->3)- and (1->2)-linked disaccharides. *Plant Physiol.* **2009**, *151*, 47–58. [CrossRef] [PubMed]
53. Cao, Y.-Y.; Yang, J.-F.; Liu, T.-Y.; Su, Z.-F.; Zhu, F.-Y.; Chen, M.-X.; Fan, T.; Ye, N.-H.; Feng, Z.; Wang, L.-J.; et al. A Phylogenetically Informed Comparison of GH1 Hydrolases between Arabidopsis and Rice Response to Stressors. *Front. Plant Sci.* **2017**, *8*, 350. [CrossRef] [PubMed]
54. Wang, C.; Chen, S.; Dong, Y.; Ren, R.; Chen, D.; Chen, X. Chloroplastic Os3BGlu6 contributes significantly to cellular ABA pools and impacts drought tolerance and photosynthesis in rice. *New Phytol.* **2020**, *226*, 1042–1054. [CrossRef]
55. Opassiri, R.; Maneesan, J.; Akiyama, T.; Pomthong, B.; Jin, S.; Kimura, A.; Cairns, J.R.K. Rice Os4BGlu12 is a wound-induced β-glucosidase that hydrolyzes cell wall-β-glucan-derived oligosaccharides and glycosides. *Plant Sci.* **2010**, *179*, 273–280. [CrossRef]
56. Ge, G.; Liu, Z.; Yu, T.; Zhou, L.; Sun, X.; Li, Z.; Zheng, Y. Aminotransferase Class I and II Gene Family in the Jinjiang Oyster (*Crassostrea ariakensis*): Genomewide Identification, Phylogenetic Analysis and Expression Profiles after Salinity Stress. *Fishes* **2023**, *8*, 459. [CrossRef]
57. Luo, H.; He, L.; Du, B.; Pan, S.; Mo, Z.; Duan, M.; Tian, H.; Tang, X. Biofortification with chelating selenium in fragrant rice: Effects on photosynthetic rates, aroma, grain quality and yield formation. *Field Crop. Res.* **2020**, *255*, 107909. [CrossRef]
58. Zeng, R.; Farooq, M.U.; Wang, L.; Su, Y.; Zheng, T.; Ye, X.; Jia, X.; Zhu, J. Study on Differential Protein Expression in Natural Selenium-Enriched and Non-Selenium-Enriched Rice Based on iTRAQ Quantitative Proteomics. *Biomolecules* **2019**, *9*, 130. [CrossRef]
59. Khaliq, A.; Aslam, F.; Matloob, A.; Hussain, S.; Geng, M.; Wahid, A.; Rehman, H.U. Seed Priming with Selenium: Consequences for Emergence, Seedling Growth, and Biochemical Attributes of Rice. *Biol. Trace Element Res.* **2015**, *166*, 236–244. [CrossRef]
60. A Hayden, C.; Bosco, G. Comparative genomic analysis of novel conserved peptide upstream open reading frames in Drosophila melanogaster and other dipteran species. *BMC Genom.* **2008**, *9*, 61. [CrossRef]

61. Causier, B.; Hopes, T.; McKay, M.; Paling, Z.; Davies, B. Plants utilise ancient conserved peptide upstream open reading frames in stress-responsive translational regulation. *Plant Cell Environ.* **2022**, *45*, 1229–1241. [CrossRef] [PubMed]
62. Wang, Y.-D.; Wang, X.; Wong, Y.-S. Generation of selenium-enriched rice with enhanced grain yield, seleni um content and bioavailability through fertilisation with selenite. *Food Chem.* **2013**, *141*, 2385–2393. [CrossRef] [PubMed]
63. Chen, S.; Zhou, Y.; Chen, Y.; Gu, J. fastp: An ultra-fast all-in-one FASTQ preprocessor. *Bioinformatics* **2018**, *34*, i884–i890. [CrossRef]
64. Kim, D.; Langmead, B.; Salzberg, S.L. HISAT: A fast spliced aligner with low memory requirements. *Nat. Methods* **2015**, *12*, 357–360. [CrossRef]
65. Pertea, M.; Pertea, G.M.; Antonescu, C.M.; Chang, T.-C.; Mendell, J.T.; Salzberg, S.L. StringTie enables improved reconstruction of a transcriptome from RNA-seq reads. *Nat. Biotechnol.* **2015**, *33*, 290–295. [CrossRef]
66. Li, B.; Dewey, C.N. RSEM: Accurate transcript quantification from RNA-Seq data with or wit hout a reference genome. *BMC Bioinform.* **2011**, *12*, 323. [CrossRef]
67. Love, M.I.; Huber, W.; Anders, S. Moderated estimation of fold change and dispersion for RNA-seq data with DESeq2. *Genome Biol.* **2014**, *15*, 550. [CrossRef]
68. Sun, Y.; Qin, Q.; Song, K.; Sun, L.; Jiang, T.; Yang, S.; Li, Z.; Xu, G.; Sun, S.; Xue, Y. Does Sulfoquinovosyl Diacylglycerol Synthase OsSQD1 Affect the Composition of Lipids in Rice Phosphate-Deprived Root? *Int. J. Mol. Sci.* **2022**, *24*, 114. [CrossRef]
69. Livak, K.J.; Schmittgen, T.D. Analysis of relative gene expression data using real-time quantitative PCR and the 2(-Delta Delta C(T)) Method. *Methods* **2001**, *25*, 402–408. [CrossRef]

International Journal of *Molecular Sciences*

MDPI

Article

Genome-Wide Analysis of Heat Shock Protein Family and Identification of Their Functions in Rice Quality and Yield

Hong Wang †, Sidra Charagh †, Nannan Dong, Feifei Lu, Yixin Wang, Ruijie Cao, Liuyang Ma, Shiwen Wang, Guiai Jiao, Lihong Xie, Gaoneng Shao, Zhonghua Sheng, Shikai Hu, Fengli Zhao, Shaoqing Tang, Long Chen, Peisong Hu * and Xiangjin Wei *

State Key Laboratory of Rice Biology, China National Center for Rice Improvement, China National Rice Research Institute, Hangzhou 311400, China; benxiwanghong@163.com (H.W.); sidracharagh005@gmail.com (S.C.); 13126531292@163.com (N.D.); lufeifei0506@163.com (F.L.); 13835410849@163.com (Y.W.); caoruijie@caas.cn (R.C.); maliuyang2018@163.com (L.M.); 15309894079@163.com (S.W.); jiaoguiai@caas.cn (G.J.); xielihong@caas.cn (L.X.); shaogaoneng@caas.cn (G.S.); shengzhonghua@caas.cn (Z.S.); hushikai@caas.cn (S.H.); zhaofengli@caas.cn (F.Z.); tangshaoqing@caas.cn (S.T.); chenlong@caas.cn (L.C.)

* Correspondence: hupeisong@caas.cn (P.H.); weixiangjin@caas.cn (X.W.)

† These authors contributed equally to this work.

Abstract: Heat shock proteins (Hsps), acting as molecular chaperones, play a pivotal role in plant responses to environmental stress. In this study, we found a total of 192 genes encoding Hsps, which are distributed across all 12 chromosomes, with higher concentrations on chromosomes 1, 2, 3, and 5. These Hsps can be divided into six subfamilies (sHsp, Hsp40, Hsp60, Hsp70, Hsp90, and Hsp100) based on molecular weight and homology. Expression pattern data indicated that these *Hsp* genes can be categorized into three groups: generally high expression in almost all tissues, high tissue-specific expression, and low expression in all tissues. Further analysis of 15 representative genes found that the expression of 14 *Hsp* genes was upregulated by high temperatures. Subcellular localization analysis revealed seven proteins localized to the endoplasmic reticulum, while others localized to the mitochondria, chloroplasts, and nucleus. We successfully obtained the knockout mutants of above 15 *Hsps* by the CRISPR/Cas9 gene editing system. Under natural high-temperature conditions, the mutants of eight *Hsps* showed reduced yield mainly due to the seed setting rate or grain weight. Moreover, the rice quality of most of these mutants also changed, including increased grain chalkiness, decreased amylose content, and elevated total protein content, and the expressions of starch metabolism-related genes in the endosperm of these mutants were disturbed compared to the wild type under natural high-temperature conditions. In conclusion, our study provided new insights into the HSP gene family and found that it plays an important role in the formation of rice quality and yield.

Keywords: heat shock protein; gene family; heat resistance; rice quality and yield

Citation: Wang, H.; Charagh, S.; Dong, N.; Lu, F.; Wang, Y.; Cao, R.; Ma, L.; Wang, S.; Jiao, G.; Xie, L.; et al. Genome-Wide Analysis of Heat Shock Protein Family and Identification of Their Functions in Rice Quality and Yield. *Int. J. Mol. Sci.* **2024**, *25*, 11931. https://doi.org/10.3390/ijms252211931

Academic Editor: Abir U. Igamberdiev

Received: 26 September 2024
Revised: 25 October 2024
Accepted: 1 November 2024
Published: 6 November 2024

1. Introduction

Heat shock proteins (Hsps) are synthesized in organisms in response to elevated temperatures, and their production can also be triggered by other stressors such as cold, salinity, drought, water stress, and oxidative stress [1]. Hsps are highly conserved across both prokaryotes and eukaryotes, with homology between these domains reaching up to 65% [1,2]. Hsps localized in different organelles exhibit diverse biological functions [3–5]. As molecular chaperones, Hsps play crucial roles in the transport, folding, assembly, and localization of newly synthesized peptides, as well as in refolding and degrading denatured proteins [6]. In addition to their chaperone functions, Hsps are involved in maintaining signal transduction and other biological processes essential for tissue and organ differentiation and development [7–10].

Hsps can be classified into various groups based on molecular weight, amino acid sequence, cellular localization, and function [6]. They are primarily grouped into small Hsp (sHsp), Hsp40, Hsp60, Hsp70, Hsp90, and Hsp100 families based on molecular weight [5,11]. The functions of these groups differ due to their distinct conserved domains. The sHsps, with molecular weights from 12 to 42 kDa, contain an α-crystallin domain of approximately 80–100 amino acids and act as ATP-independent chaperones, preventing protein aggregation under stress [12,13]. Hsp40 proteins, with molecular weights of 40–50 kDa, contain a "J-domain" and are homologous to the *E. coli* DnaJ protein, which is why they are also called the DnaJ subfamily [10]. Based on the presence or absence of a Zinc-finger domain and key HPD (His-Pro-Asp) tripeptides, the Hsp40 subfamily is further categorized into three subgroups: DnaJA, DnaJB, and DnaJC [10,14]. Hsp40 proteins serve as co-chaperones for the HSP70 family, enhancing their ATPase activity [15]. Hsp60, with a molecular weight of 60 kDa, includes the GroEL domain and exists in two forms: one found in bacteria, mitochondria, and chloroplasts (GroE/Cpn60), and the other in the cytoplasm of archaea and eukaryotes (TCP-1) [16]. Hsp60 proteins rely on energy for peptide synthesis and folding under stress [17–19]. The Hsp70 family (DnaK in bacteria) is one of the most thoroughly studied chaperone groups, featuring a conserved N-terminal ATPase domain (~44 kDa), a substrate-binding domain (~18 kDa), and a C-terminal "lid" (~10 kDa) [20]. Some members of the Hsp70 family, known as Hsp110, exhibit molecular weights of 100–170 kDa due to an acidic insertion or extended C-terminus [21,22]. Hsp90 proteins have three main structural domains: a highly conserved N-terminal ATPase domain, an intermediate domain, and a C-terminal domain featuring a dimeric interface [23]. These proteins play crucial roles in stress signal transduction, maintaining protein activity under stress conditions. Finally, Hsp100 proteins (also known as ClpB) have molecular weights of ~100 kDa, consisting of three distinct regions, the N-terminal domain, the ATPase domain, and the C-terminal domain, and are involved in protein folding and depolymerization [24–26]. Different Hsp proteins work together to form a network responsible for protein folding, translocation, degradation, and stress response, thus supporting normal growth and development under stress [27].

The biological functions of many Hsp genes have been identified across various species, with a significant impact on plant growth, especially under environmental stress. For instance, AtHSP17.4 is essential for seed maturation in *Arabidopsis* [28], while GhHSP24.7 in cotton regulates seed germination in a temperature-dependent manner [29]. The AtDjC21 and AtDjC29 proteins are vital for *Arabidopsis* growth, with mutations leading to embryo lethality [30]. *Arabidopsis* Hsp40 family member TMS1 is critical for pollen tube thermotolerance [31]. Enhanced expression of multiple Hsp70 proteins promotes bud burst in dormant poplar trees [32], and HSP70 (b-70) aids in gliadin assembly in maize endosperm [33]. Tobacco nbHSP90 regulates apoptosis signals to mitochondria [34]. In wheat, interference with HSP90.2 reduces leaf photosynthesis, grain weight, and yield, while its overexpression increases grain weight [35]. HSP90.3 in *Arabidopsis* mediates temperature-dependent defense responses [36], and HSP101 regulates chloroplast development [37]. These findings suggest that Hsps perform diverse biological roles in plant growth, development, and stress responses. In rice, four Hsp subfamilies have been identified, including 23 *sHsp* genes, 104 *Hsp40* genes, 32 *Hsp70* genes, and 3 *Hsp100* genes [9,10,20,26]. However, the function of only a few rice *Hsp* genes has been reported. For example, *DjA7* and *DjA8* are involved in chloroplast development [38], and *OSHSP40* regulates pollen development and seed setting [39]. Overexpression of *OsHsp17.0* and *OsHsp23.7* enhances drought and salt tolerance [40,41]. The Hsp60 family member *TCD9* influences chloroplast development under high temperatures [42], while *HSP60-3B* regulates male fertility by affecting starch particle formation [43]. Rice *cpHsp70-2* plays a key role in amyloplast development and chalkiness regulation under high-temperature stress [44,45], and *mtHsp70-1* inhibits heat-induced programmed cell death [46]. Finally, *Hsp101* is involved in starch biosynthesis and endosperm development under high temperatures [47]. However, the panoramic study of rice Hsp family genes is still lacking.

In order to fully understand the functions of rice Hsp family genes, especially their functions in regulating rice yield and quality, we carried out a comprehensive study on the molecular biochemical characteristics and biological functions of rice HSP family genes. We identified 22 more Hsp60 subfamily members and 8 more Hsp90 subfamily members, along with the reported members from 4 subfamilies; there are a total of 192 *Hsp* genes in rice across 6 subfamilies. The bioinformatics of these rice Hsps were performed. Subsequently, we conducted an in-depth analysis of 15 representative genes from five subfamilies. Using CRISPR/Cas9, we obtained the mutants for these 15 genes and evaluated yield and quality traits. We found that loss-of-function of most of these *Hsp* genes resulted in reduced yield and quality; they can regulate starch synthesis by influencing the expression of starch synthesis-related genes, highlighting their essential role in rice seed development and stress response. Our findings provide new insights into the mechanism of *Hsps* regulation of rice yield and quality, as well as a new theoretical basis and gene resources for the genetic improvement of rice yield and quality.

2. Results

2.1. Classification, Distribution, and Gene Structural Analysis of Hsp Family in Rice

We found that the rice genome contains a total of 192 Hsps, categorized into six subfamilies based on molecular weight and amino acid sequence: sHsp, Hsp40, Hsp60, Hsp70, Hsp90, and Hsp100, each with 23, 104, 22, 32, 8, and 3 members, respectively (Figure 1). Each subfamily has distinct conserved domains. The sHsp protein has an α-crystalline protein domain (Figure 1A). Hsp40 members contain a Dnaj domain structure and are further classified into three types: A (DjA1–DjA12), B (DjB1–DjB9), and C (DjC1–DjC83) (Figure 1B). Hsp60 proteins have a GroEL domain (Figure 1C). The Hsp70 subfamily consists of thirty-two members, including eighteen Hsp70 members, six Bip (Binding immunoglobulin Protein) members, and eight Hsp110 members, each of which contains domains of NBD, SBD, and the C-terminal "lid" (Figure 1D). Hsp90 subfamily members harbor N-terminal ATPase, an intermediate, and a C-terminal domain (Figure 1E). Hsp100 proteins contain the ClpB domain (Figure 1F). Conserved motifs were identified within subfamilies (Figure S1).

Phylogenetic analysis using aligned amino acid sequences showed that most subfamily members cluster together (Figure 2A), with sHsps concentrated in two regions and Hsp70 members in a single group. Hsp40, the largest subfamily, showed interspersion with other families. This analysis highlighted the close evolutionary relationships among different Hsp subfamilies. The 192 *Hsp* genes are distributed across all 12 chromosomes, with higher concentrations on chromosomes 1, 2, 3, and 5 (Figure 2B). Gene structure analysis revealed diversity in intron/exon numbers among subfamilies (Figure 3). Most *sHsp* genes (78%) lacked introns, while *Hsp60* and *Hsp100* exhibited multiple introns. Variations were seen in *Hsp40, Hsp70,* and *Hsp90* members, with some genes containing up to a dozen introns.

2.2. Differential Expression and Subcellular Localization of Hsp Genes

Based on the predicted expression patterns from the website https://www.ricedata.cn/gene/index.htm (accessed on 19 April 2024), the 192 *Hsp* genes can be categorized into three groups (Figure S2): generally high expression in almost all tissues, high tissue-specific expression, and generally low expression in almost all tissues. We selected 15 representative genes from five subfamilies for further functional characterization analysis. RT-qPCR confirmed that the expression levels of all 15 genes were high in 10d grains after flowering; 8 genes (*Hsp16.9A, DjB6, DjB7, mtHsp70-3, cHsp70-6, cHsp70-7, Hsp110-2,* and *Hsp110-8*) exhibited high expression levels in grains of multiple periods (Figure 4A). It is evident that these *Hsp* genes exhibit a trend of preferential expression in grains.

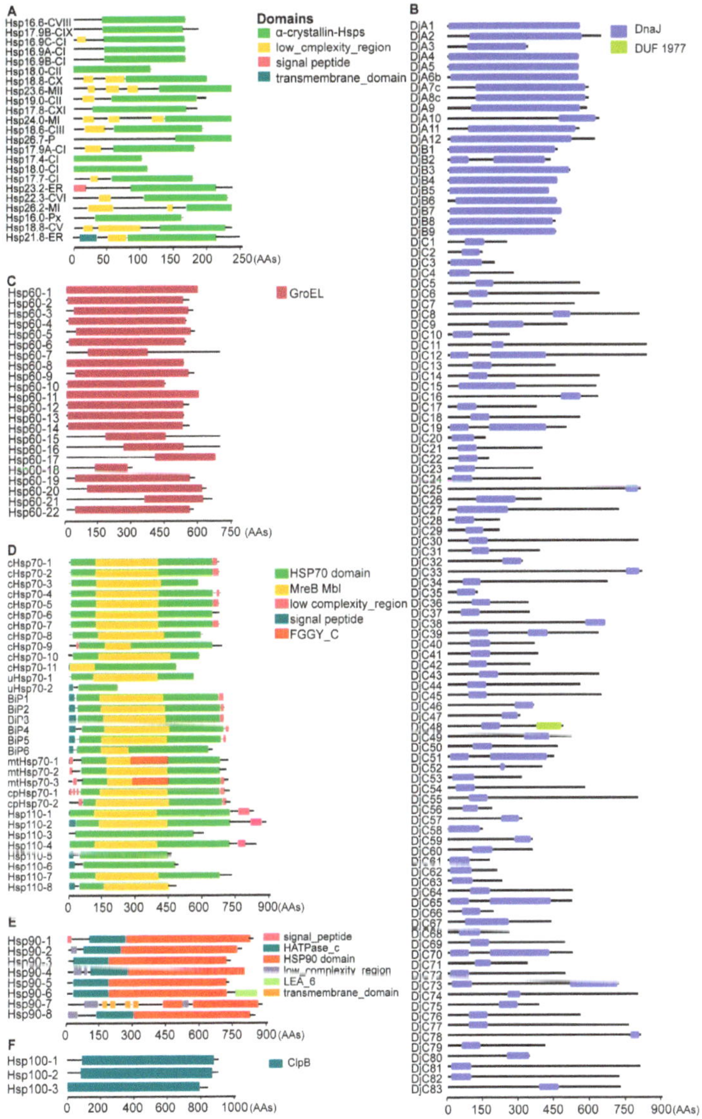

Figure 1. Conservative domains of 6 Hsp subfamilies in rice: (**A**) 23 small Hsp (sHsp) members: protein name suffix is subcellular localization, and each protein has the α-crystallin domain, which is the characteristic domain of sHsp; (**B**) 104 Hsp40 members, including three types, A (DjA1-DjA12), B (DjB1-DjB9), and C (DjC1-DjC83); each member has the DnaJ domain; (**C**) 22 Hsp60 members: each protein has a GroEL domain, and these proteins were named according to their positions on the chromosome; (**D**) 32 Hsp70 members, including 18 proteins whose name contain Hsp70, 6 Bips (Binding immunoglobulin proteins), and 8 Hsp110 members, each with an Hsp70 domain; (**E**) 8 Hsp90 members, each containing the Hsp90 domain; (**F**) 3 Hsp100 members, each harboring the ClpB domain specific to Hsp100. The scale at the bottom represents the number of amino acids. The conserved domain data of sHsp, Hsp70, and Hsp90 were downloaded from the Pfam database and visualized using the TBtools-II software. For Hsp40, Hsp60, and Hsp100, the conserved domain data were obtained from NCBI and visualized using a website.

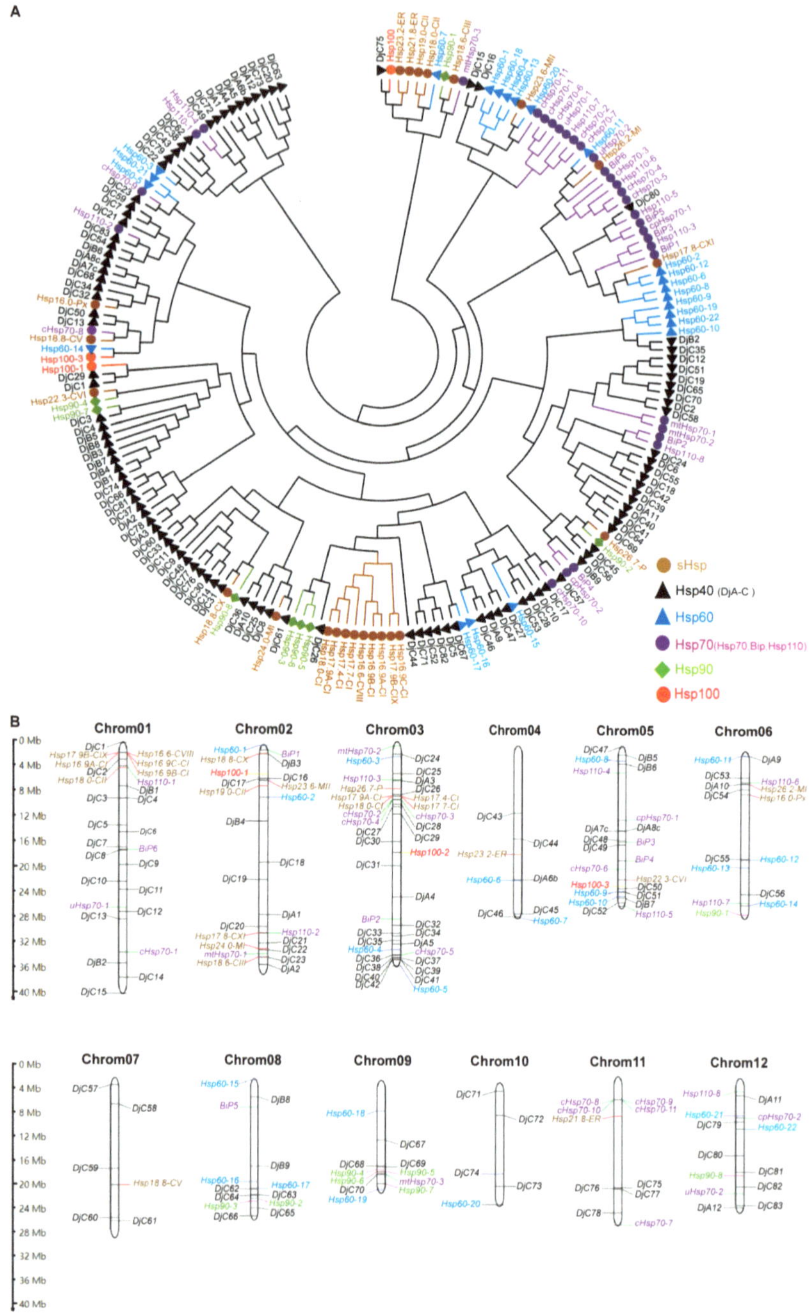

Figure 2. Phylogenetic tree and chromosomal localization analysis of 6 Hsp subfamilies. (**A**) Phylogenetic tree illustrating relationships among 192 Hsps. (**B**) Chromosomal distribution of 192 *Hsp* genes. The scale on the left is chromosome physical distance.

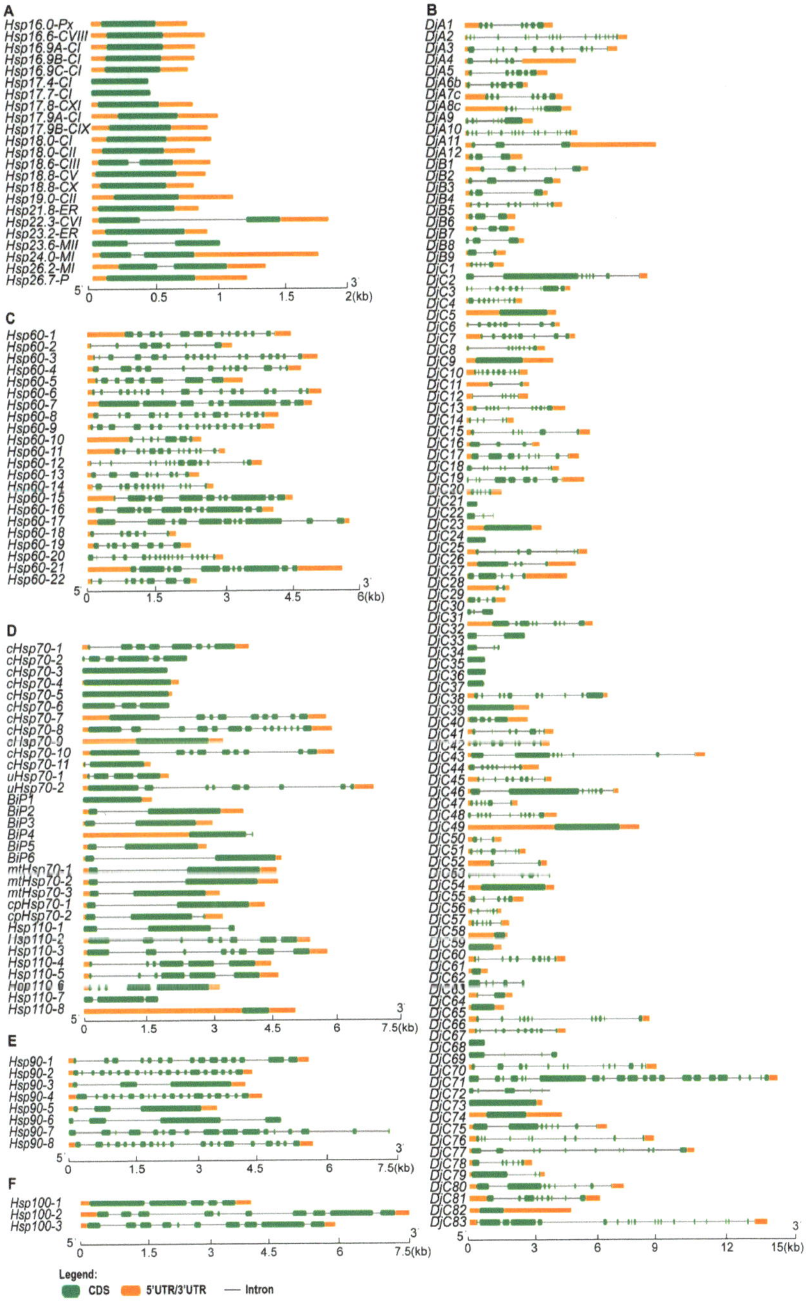

Figure 3. Gene structure of 192 *Hsp* genes in rice. (**A–F**) Gene structure depicting *sHsp*, *Hsp40*, *Hsp60*, *Hsp70*, *Hsp90*, and *Hsp100*. CDS indicates coding sequence for protein; UTR denotes untranslated region. The scale is the number of gene bases.

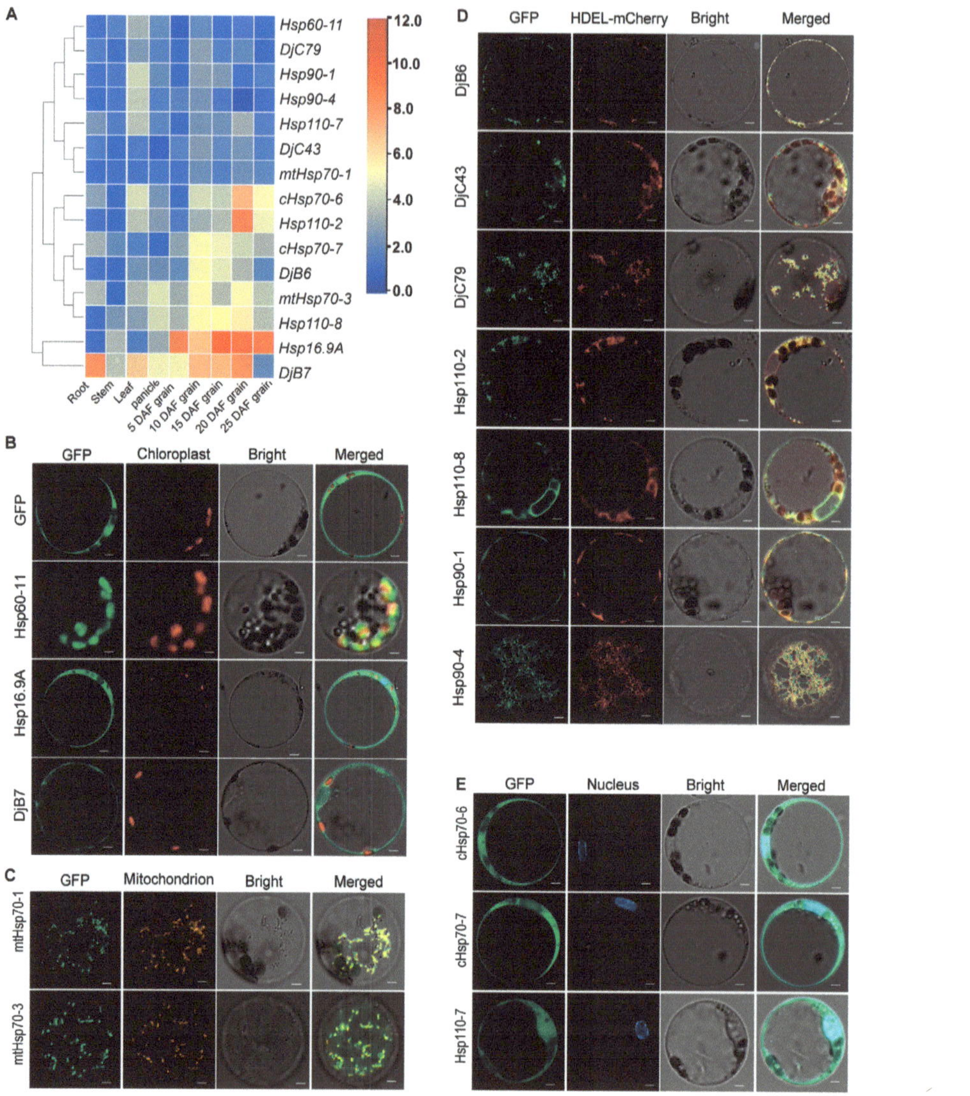

Figure 4. Expression pattern of 15 *Hsp* genes and subcellular localization analysis of 15 Hsp proteins. (**A**) The relative expression levels of 15 *Hsps* in roots, stems, leaves, panicles, and seeds at 5, 10, 15, 20, and 25 days after flowering. The heat map shows the expression level determined by TBtools-II (https://github.com/CJ-Chen/TBtools-II, (accessed on 31 March 2024)); the range of blue to red indicates the expression levels from low to high. Clustering is according to the expression level in each tissue. (**B–E**) Subcellular localization of Hsp proteins. Free green fluorescent protein (GFP) and full-length Hsp fusion proteins (Hsp-GFP) were transiently expressed in rice protoplasts. Hsp60-11-GFP co-localized with the chloroplast autofluorescence, Hsp16.9A and DjB7 localized in the cytoplasm (**B**). mtHsp70-1 and mtHsp70-3 co-localized with mitochondria, and the yellow signal represents the mitochondria dyed by Mito-Tracker Red (**C**). DjB6, DjC43, DjC79, Hsp110-2, Hsp110-8, Hsp90-1, and Hsp90-4 co-localized with HDEL-mCherry signals of the endoplasmic reticulum (**D**). cHsp70-6, cHsp70-7 and Hsp110-7 co-localized with the cytoplasm and DAPI signals of the nucleus (**E**). Scale bars, 5 μm.

The subcellular localization of Hsp proteins was examined using HSP-GFP fusion constructs in rice protoplasts (Figure 4B–E). Among the 15 Hsp proteins, 7 proteins (DjB6, DjC43, DjC79, Hsp110-2, Hsp110-8, Hsp90-1, and Hsp90-4) from three subfamilies were co-localized with the HDEL-mCherry signal of the endoplasmic reticulum (ER) (Figure 4D). Hsp60-11 co-localized with chloroplasts (Figure 4B). Hsp16.9A and DjB7 localized in the cytoplasm (Figure 4B). mtHsp70-1 and mtHsp70-3 co-localized with mitochondria (Figure 4C). Additionally, three proteins (cHsp70-6, cHsp70-7, and Hsp110-7) co-localized with DAPI signals of the nucleus and cytoplasm (Figure 4E).

2.3. *The Expression Levels of Hsp Genes Are More Easily and Strongly Induced by Heat Stress*

To investigate the response of *Hsp* genes to adversity stresses, the induced expression levels of 15 genes under high-temperature, low-temperature, high-salt, and drought conditions were assessed by the RT-qPCR technique. The results showed that 14 genes exhibited increased expression under high temperatures, the expression levels of 7 genes were upregulated by low temperatures, 2 gene (*DjC43* and *Hsp110-7*) were upregulated in high-salt conditions, and 4 genes (*DjB7*, *Hsp110-2*, *Hsp110-7*, and *Hsp90-4*) showed increased expression under drought stress (Figure 5). This indicates that some *Hsp* genes were also induced by other stresses except for heat stress, though the induced expression of the same gene under high-temperature conditions was significantly higher than that under other stresses. Specifically, high temperature resulted in a hundreds-of-fold increase in the synthesis of *Hsp16.9A* and *cHsp70-6* within just 10 min. The expression of *DjB7*, *DjC43*, *Hsp60-11*, *mtHsp70-1*, *Hsp110-7*, and *hsp90-1* increased dozens of times under heat stress. While the expression of only one gene (*Hsp90-4*) exhibited a 100-fold increase under other stress conditions (drought), the expression of other genes increased by twenty or several times at most.

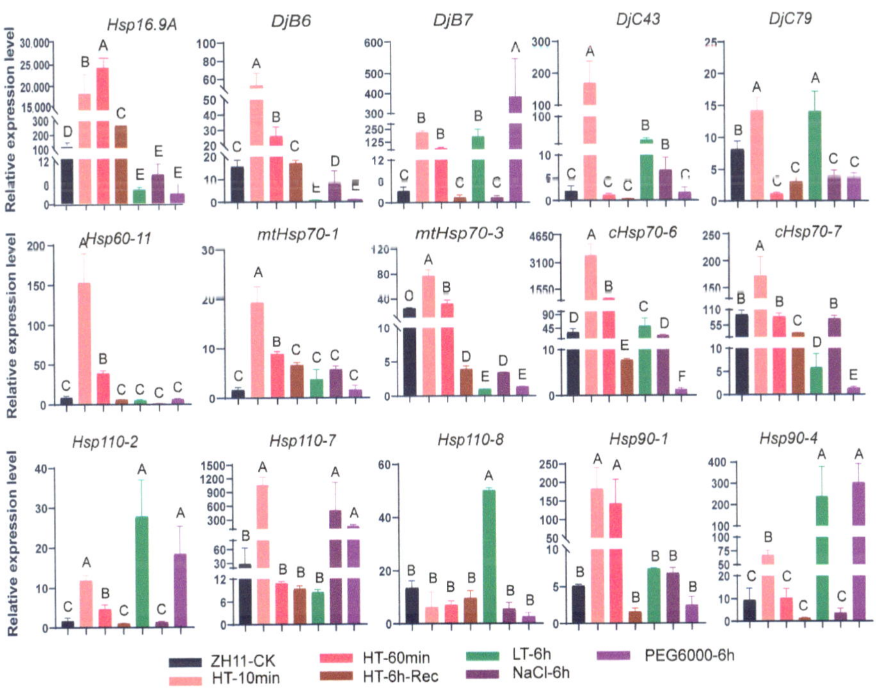

Figure 5. The induced expression levels of 15 *Hsp* genes in the seedlings of ZH11 under different stresses. ZH11-CK represents the normal seedlings as control with no treatment. HT-10min and

HT-60min represent the seedlings that were moved to a high temperature of 42 °C from normal temperature for 10 and 60 min. HT-6h-Rec represents the seedlings that recovered at a normal temperature for 6 h after 60 min of heat shock. LT-6h represents the seedlings that were treated at 5 °C for 6 h. NaCl-6h represents the seedlings that were treated with high salt stress for 6 h (100 mmol/L NaCl). PEG6000-6h represents the seedlings that were treated at 20% PEG stress for 6 h. The expression levels of *Hsps* were detected in the above treated seedlings and ZH11-CK. Data are means ± SD (n = 3). Different letters indicate significant differences at $p < 0.05$ by ANOVA and Duncan's test.

After high-temperature treatment of rice seedlings, the transcriptome analysis showed massive upregulation of thousands of genes, with *Hsp* genes most significantly enriched in the "protein processing in the endoplasmic reticulum" pathway (Figure S3A–F). At the same time, the expression of other *Hsp* genes was also upregulated in multiple KEGG enrichment pathways (Figure S3G), including 'plant and pathogen interaction', 'RNA degradation', 'Spliceosome', and 'Endocytosis'. It can be inferred that under heat stress, Hsp proteins may play an important role.

2.4. Creation of CRISPR Mutants

To identify the biological function of the 15 *Hsp* genes (1 *sHsp*, 4 *Hsp40*, 1 *Hsp60*, 7 *Hsp70*, and 2 *Hsp90* genes), their mutants were generated by the CRISPR/Cas9 gene editing system in a *japonica* rice cultivar ZH11. To completely turn off the gene function in the mutants, target sites for gene editing were located in the front exon region of each gene (Figure S4A). It was observed that most mutants of 14 genes exhibited early termination of the translation process; 1 gene exhibited the amino acid replacement in its two lines (Figure S4B). Just one line was obtained for genes *DjB6* and *Hsp90-4*; one or two lines of each gene were studied further. Among the 15 genes, the analogs of gene editing target sites of *Hsp16.9A*, *cHsp70-6*, and *cHsp70-7* also appear on other homologous genes, but the corresponding sites of these homologous genes were not mutated (Figure S5); the gene editing target sites of the other 12 genes have good specificity, and there are no analogs in their homologous genes. So, stable homozygous gene editing mutants were obtained for the 15 genes.

2.5. *Hsp* Genes Regulate Rice Yield

To investigate whether *Hsp* genes regulate rice yield, the mutants of 15 *Hsp* genes and the wild type were grown under natural high-temperature conditions. No significant differences in plant morphology, such as plant height and tiller number, were observed between mutants and wild-type plants (Figure S6A–C). The growth periods of the wild type and the mutants were also consistent (Figure S6D). Compared to the wild type, the grain shape, grain weight, and seed setting rate showed changes in the mutants. Among the 15 mutants, there were 7 mutants exhibiting changes in grain length, including *djc43*, *djc79*, *hsp60-11*, *mthsp70-1*, *mthsp70-3*, *chsp70-6*, and *hsp110-7* (Figure 6A); 5 mutants exhibited changes in grain width, such as *mthsp70-3*, *chsp70-6*, *chsp70-7*, *hsp110-7*, and *hsp90-1* (Figure 6B); there were 3 mutants exhibiting changes on grain thickness, for instance, *djc43*, *mthsp70-3*, and *hsp110-7* (Figure 6C); and 5 mutants exhibited changes on 1000-grain weight, including *djc43*, *djc79*, *mthsp70-3*, *chsp70-6*, and *hsp110-7* (Figure 6D). Except for the changes in grain size, the seed setting rate of nine mutants changed, with eight significantly decreased and one increased (Figure 6H). And finally, the smaller size and reduced seed setting rate of mutants *djc43*, *hsp60-11*, *mthsp70-3*, and *hsp90-1* led to a decreased yield per plant. For mutants *djc79* and *chsp70-7*, the grain size became larger, but the seed setting rate decreased, resulting in a decreased yield per plant. The yield per plant of *mthsp70-1* also reduced due to the smaller grain length, with no changes in seed setting rate. Only two mutants showed an increase in yield per plant (*hsp110-7* and *chsp70-6*); for mutant *hsp110-7*, the seed setting rate reduced from 82.13% to 77.00%, but grain length, width, thickness, and 1000-grain weight all increased significantly, resulting in an increased yield per plant. For mutant *chsp70-6*, the grain length, width, thickness, and 1000-grain weight all decreased, but the seed setting rate increased from 82.13% to 88.05%, resulting in an increased yield per plant. These results indicate that the loss-of-function of most *Hsp* genes leads to changes in grain size, reduced seed setting rate, and decreased yield per plant under natural high-temperature conditions.

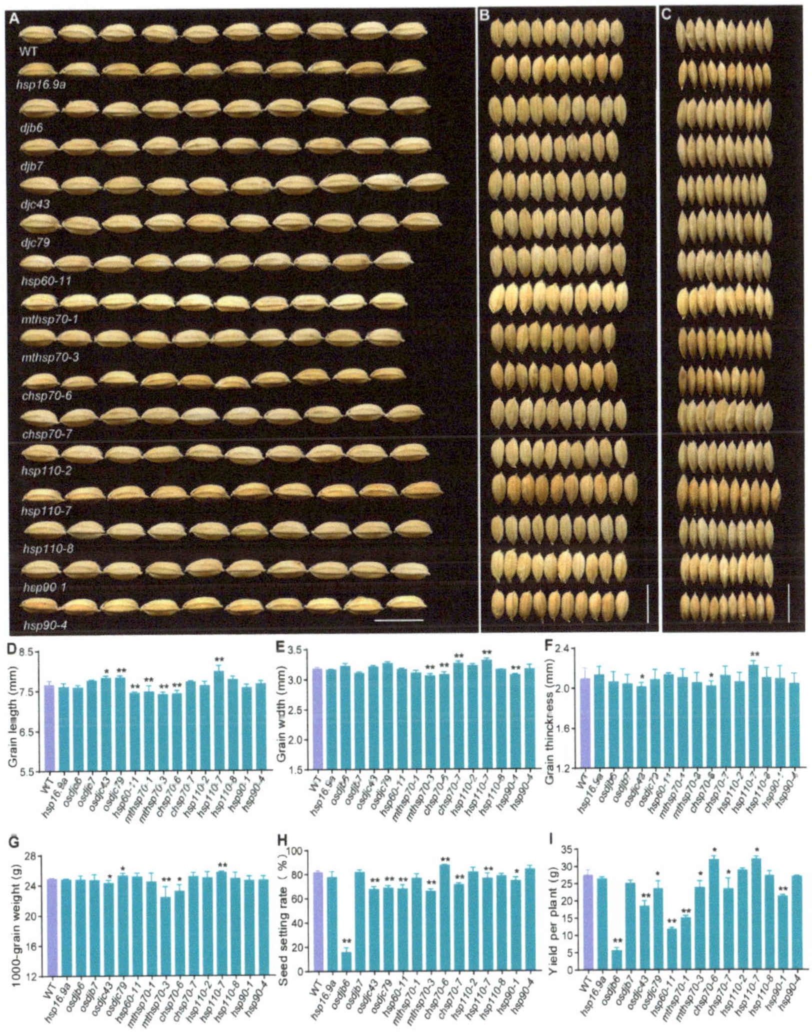

Figure 6. Grain size- and yield-related traits of ZH11 and the *hsp* mutants. (**A–C**) Comparison of the grain length (**A**), grain width (**B**), and grain thickness (**C**) of ZH11 and *hsp* mutants; scale bars = 1 cm. (**D–I**) Grain length (**D**), grain width (**E**), grain thickness (**F**), 1000-grain weight (**G**), seed setting rate (**H**), and yield per plant (**I**) of ZH11 and the mutants. The investigated plants were grown in natural high-temperature conditions in fields in Hangzhou in 2023. Data are means ± SD; n = 20 in (**D–F**), n = 3 in (**G**), and n = 10 in (**H**,**I**), and no less than 200 grains per replication in (**G**). Asterisks show statistical significance between the WT and the mutants, as determined by Student's *t*-test (* $p < 0.05$; ** $p < 0.01$).

2.6. Hsp Genes Regulate Rice Quality

We also compared the rice quality of the mutants of 15 *Hsps*. When grown under natural high-temperature conditions, mutants of eight *Hsps* (*hsp16.9a*, *djb7*, *djc79*, *chsp70-6*, *chsp70-7*, *hsp110-7*, *hsp110-8*, and *hsp90-4*) exhibited an increased chalkiness rate and degree, while two mutants (*hsp60-11* and *hsp90-1*) displayed a decreased chalkiness rate and degree (Figure 7A–C). Subsequently, the contents of storage substances in the mature endosperm were measured. Compared to the wild type, the total starch content decreased in mutants *djc79*, *chsp70-6*, *hsp110-2*, *hsp110-7*, *hsp110-8*, and *hsp90-4*, while it increased in mutants *djb7* and *mthsp70-3* (Figure 7D). Additionally, the amylose content decreased in 10 mutants (*hsp16.9a*, *djb6*, *djb7*, *djc79*, *hsp60-11*, *mthsp70-3*, *chsp70-6*, *chsp70-7*, *hsp110-7* and *hsp110-8*), whereas it increased in 3 mutants (*mthsp70-1*, *hsp110-2* and *hsp90-1*) (Figure 7E). Furthermore, the total protein content increased in 10 mutants (*hsp16.9a*, *djb6*, *djb7*, *djc79*, *hsp60-11*, *mthsp70-1*, *chsp70-6*, *chsp70-7*, *hsp110-8*, and *hsp90-1*) (Figure 7F). These results indicated that the mutation of these *Hsp* genes leads to an increased rate of chalkiness. In these mutants, at least one component of seed amylose or total starch content was reduced, while total protein content was elevated.

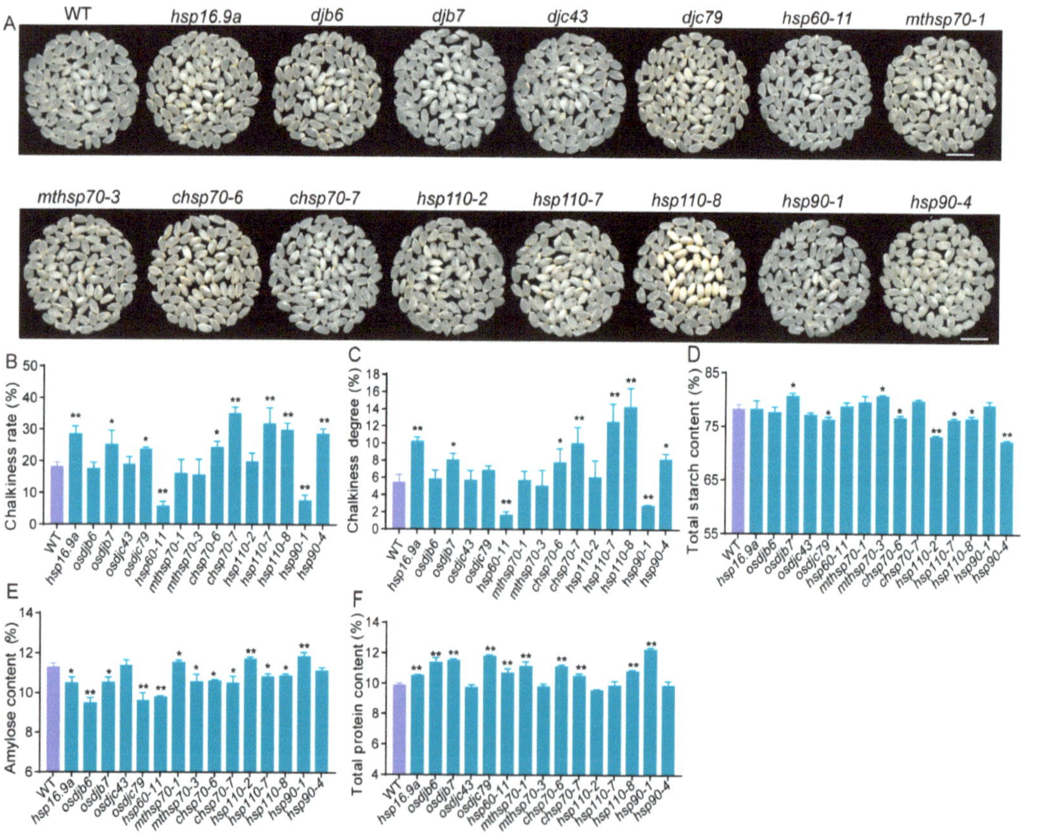

Figure 7. Grain quality of ZH11 and *hsp* mutants under natural high temperature in the Hangzhou field in 2023. (**A**) Appearance of mature grains of ZH11 and 15 *hsp* mutants. Scale bars = 1 cm. (**B**,**C**) Chalkiness rate and chalkiness degree of ZH11 and 15 *hsp* mutants grains. (**D**–**F**) Total starch, amylose, and total protein contents of ZH11 and the mutant grains. Data are means ± SD (n = 3); no less than 200 grains per replication in (**B**,**C**). Asterisks show the statistical significance between the WT and the mutants, as determined by Student's *t*-test (* $p < 0.05$; ** $p < 0.01$).

Furthermore, the plants of ZH11 and all mutants at the filling stage were subjected to high temperatures in the greenhouse. The high temperature was set at 28/34 °C for a duration of 15 days; the normal temperature was set at 22 °C/28 °C, and then the chalkiness phenotype was investigated. The results showed that under high temperatures, both ZH11 and all mutants showed increased chalkiness rate and chalkiness degree. Among these mutants, 13 mutants showed higher changes in chalkiness rate or/and chalkiness degree than ZH11 after high-temperature stress (Figure S8, Table S2). Overall, these *Hsp* genes play an important role in regulating rice quality under high-temperature conditions.

2.7. Hsp Genes Affect Starch Synthesis Pathway Under Naturally High Temperature

The starch in rice endosperm rapidly accumulates 5–10 days after flowering; the chalkiness phenotype of grains is closely related to the obstruction of starch synthesis [47]. Based on the observed chalkiness phenotype of the *hsp* mutants, we selected 29 major starch metabolism-related genes, including 13 starch synthesis genes (*OsAGPS1, OsAGPS2a, OsAGPS2b, OsAGPL1, OsAGPL2, OsAGPL4, OsSSI, OsSSIIa, OsSSIIIa, OsSSIIIb, OsSSIVb, OsGBSSI*, and *OsGBSSII*), eight starch degradation genes (*OsBEI, OsBEIIa, OsBEIIb, OsISA1-OsISA3, OsRAmy1A*, and *OsRAmy3D*), and eight sugar transport genes (*OsSUT1-OsSUT5, OsSWEET11, OsSWEET13*, and *OsSWEET14*); the expression of these genes was analyzed in 10d endosperm of ZH11 and seven mutants (*hsp16.9a, chsp70-6, chsp70-7, hsp110-7, hsp110-8, djc79*, and *hsp90-4*) exhibiting an increased chalkiness rate under high temperature (28/34 °C). These mutants were categorized into two groups according to the changes in the nutrient content in seeds (Figure 8).

The first category consists of mutants *hsp16.9a, chsp70-6, chsp70-7*, and *hsp110-7* (Figure 8A–D), with reduced amylose contents. The expression of *OsAGPS2b* decreased, the expression of *OsGBSSI* and *OsGBSSII* decreased, and the expression of *OsRAmy1A* and *OsRAmy3D* increased. The second category, including three mutants (Figure 8E–G) *hsp110-8, djc79*, and *hsp90-4*, exhibited decreased total starch content in their seeds. Compared with those in ZH11, the expression levels of *OsAGPL1, OsAGPL2*, and *OsAGPL4* all significantly decreased, and the expression of *OsRAmy1A* and *OsRAmy3D*, along with *OsSUT2* and *OsSWEET14*, increased in the three mutants. Furthermore, mutants *djc79* and *hsp110-8*, with reduced amylose contents, exhibited decreased expression of the *OsSSI* and *OsGBSSI* genes. Taken together, these mutants showed altered expression of key starch synthesis and degradation genes, confirming the impact of *hsp* mutations on starch accumulation and chalkiness formation.

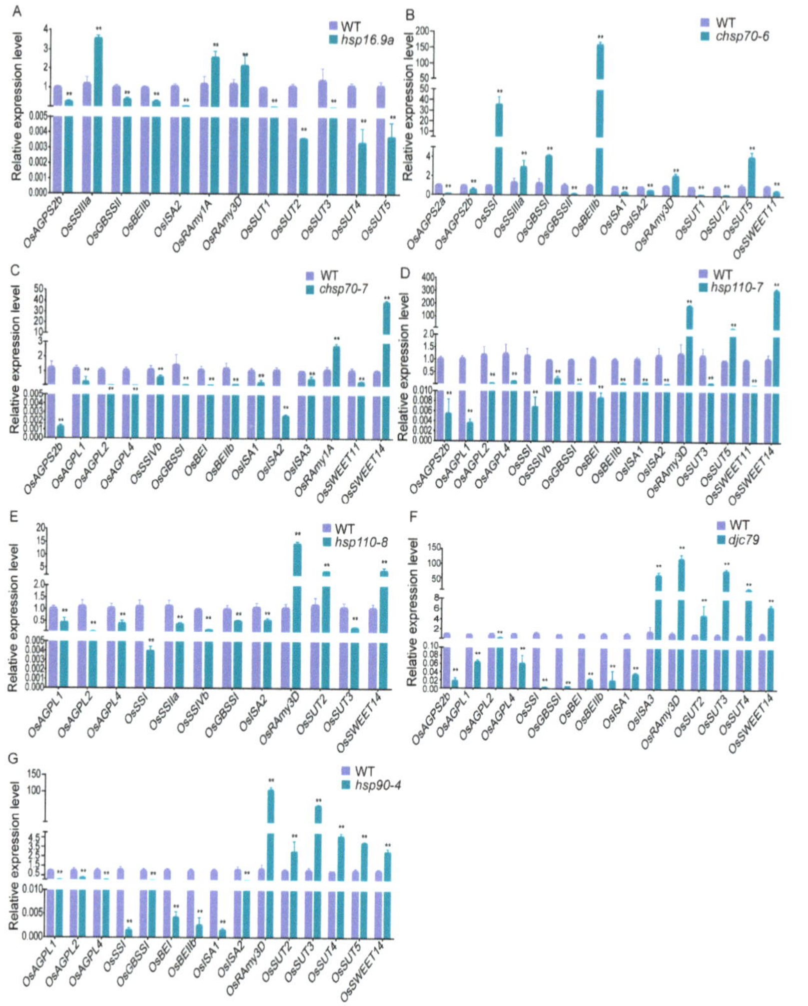

Figure 8. The expression levels of starch synthesis-related genes in developing endosperm of the wild type (WT) and 7 mutants (*hsp16.9a*, *chsp70-6*, *chsp70-7*, *hsp110-7*, *hsp110-8*, *djc79*, and *hsp90-4*). (**A–G**) The relative expression level of starch synthesis-related genes in endosperm at 10d after flowering under natural high temperature in the Hangzhou field in 2023. The data presented here are the relative expression levels of the genes that are expressed differently between mutants and the wild type. The rice UBIQUITIN gene was used as the internal control. Data are means ± SD of three individual replicates. Asterisks show the statistical significance between the WT and the mutants, as determined by Student's *t*-test (** $p < 0.01$).

3. Discussion

3.1. The Diversity and Abundance of Hsp Genes Are Notable Across Species

Hsp gene families have been extensively identified across various species, including both animals and plants. In animals, the bovine genome contains a total of 67 *Hsp* genes, comprising 10 *sHsp* genes, 43 *Hsp40* genes, 10 *Hsp70* genes, and 4 *Hsp90* genes [48]. A total of 47 *Hsp* genes have been identified in A. *glabripennis*, including 21 *sHsp* genes, 9 *Hsp60* genes, 14 *Hsp70* genes, and 3 *Hsp90* genes [49]. The whitefly has 26 *Hsp* genes, including 3 *Hsp90* genes, 17 *Hsp70* genes, 1 *Hsp60* gene, and 5 *sHsp* genes [50]. The apple snail possesses 42 *Hsp* genes, including 12 *sHsp* genes, 13 *Hsp40* and *Hsp70* genes, 1 *Hsp60* gene, and 3 *Hsp90* genes [51]. Additionally, the Hsp family genes has been analyzed in various plant species. In *Arabidopsis*, six *Hsp90* genes and three *Hsp100* genes have been identified [37,52]. In maize, 99 *DnaJ* and 22 *Hsp70* genes have been identified through whole-genome analysis [53,54]. Other crop species exhibit considerable variability in *Hsp* gene abundance: wheat has 7 *sHsp* genes, barley has 13 *sHsp* genes [55], and sorghum has 113 *Hsp40/DnaJ* genes [56]. In potatoes, 48 *sHsp* genes have been identified [57], while in soybeans, 51 *sHsp* genes are considered likely candidates [58]. In horticultural crops, the grape genome contains 48 *sHsp* genes and 78 *DnaJ* genes [59,60]. The *Ziziphus jujuba* has 21 *Hsp70* genes and 6 *Hsp100* genes [61]. In rice, four Hsp subfamilies have been identified, including 23 *sHsp* genes, 104 *Hsp40* genes, 32 *Hsp70* genes, and 3 *Hsp100* genes [9,10,20,26]. In this study, we identified 22 more rice *Hsp60* genes and 8 *Hsp90* members (Figure 1C,E). So far, the number of identified rice *Hsp* genes is up to 192. These findings demonstrate that *Hsp* genes are highly abundant and diverse across both animals and plants; among them, *Hsp* is the most abundant in rice. Given this diversity, *Hsps* play multifaceted roles across different organisms, warranting further investigation into their functional diversity.

3.2. The Functions of Hsps Are Crucial for Seed Biological Growth and Development

The Hsp family genes play a fundamental role in regulating plant growth and development, and the loss of these genes impacts both yield-related and quality-related traits [62–67]. Previous studies have shown that rice *Hsp* genes *HSP60-3B* can regulate seed setting by influencing starch biosynthesis in male gametophytes [43]. Rice *cHsp70-2* and *Hsp101* can positively regulate rice grain width and 1000-grain weight [44,47], but the underlying mechanisms remain unclear. HSP90.2 regulates wheat yield by affecting photosynthetic capacity [35,68]. Other studies showed that genes may influence cell number or size by influencing cytokinin metabolism, thereby regulating grain size [69–71]. In this study, five genes (*Hsp60-11*, *mtHsp70-1*, *mtHsp70-3*, *cHsp70-6*, and *Hsp90-1*) positively regulate rice grain size, while *DjC79* and *Hsp110-7* have a negative regulatory function (Figure 6D–F); eight genes (*DjB6*, *DjC43*, *DjC79*, *Hsp60-11*, *mtHsp70-1*, *mtHsp70-3*, *cHsp70-7*, and *Hsp90-1*) positively regulate rice seed setting rate and yield per plant, while *cHsp70-6* plays a negative regulatory role (Figure 6H–I). These results indicate that rice *Hsp* genes regulate grain size and seed setting rate, ultimately affecting yield.

Notably, the mutants of 15 *Hsp* genes not only changed the yield-related traits but also the quality-related indicators. Rice *OsbZIP60/O3* can regulate the chalkiness phenotype by affecting the expression of starch synthesis related genes [72,73]. In this study, we found that *Hsp* genes regulate rice quality in the same way. The chalkiness rate of eight mutants increased under naturally high temperatures (Figure 7), and the content of total starch and/or amylose in the mutant seeds decreased, while the content of total protein remained unchanged or increased (Figure 7D–F). Then, the expression of starch synthesis-related genes was assessed in seven mutants exhibiting an increased chalkiness rate under naturally high temperatures. The results showed that the expression of starch synthesis genes decreased in the seven mutants tested, while genes responsible for starch degradation and sugar transport increased (Figure 8 and Figure S9), thus resulting in abnormal starch accumulation, leading to chalky grains. The results indicate that *Hsp* genes are critical for balancing starch synthesis and metabolism. This underscores the pivotal role of *Hsp* genes in seed filling and highlights their importance in maintaining both yield and quality, providing valuable genetic resources for future breeding programs aimed at producing high-yield, high-quality rice.

3.3. Hsp Genes Located in ER May Participate in Endosperm Development Under High Temperature

Global climate change gradually raises temperatures, leading to more frequent extreme heat events [2,74]. When temperatures exceed 37 °C for more than three consecutive days during the reproductive growth stage, it adversely affects pollen development and fertilization, reducing both yield and quality [75]. High temperature during the grain-filling stage in rice can induce endoplasmic reticulum stress, resulting in a chalkiness phenotype of seeds [76,77]. In our study, transcriptomic analysis revealed that the Hsp proteins responsible for protein processing in the endoplasmic reticulum can quickly respond to high temperatures by upregulating the expression of corresponding genes (Figure S3D–G). Furthermore, an in-depth study of the 15 *Hsp* genes found that the loss-of-function of 8 genes exhibited a more pronounced chalkiness phenotype after high-temperature treatment (Figures 7 and S8, Table S2), suggesting that these *Hsp* genes are functionally involved in heat stress response, with seven proteins localized in the ER (Figure 4D). When plants are subjected to heat stress, many denatured proteins accumulate; the Hsp chaperone proteins are responsible for the folding, translocation, and degradation of denatured proteins [26,27]. The activation of the unfolded protein response (UPR) can be triggered by the upregulation of ER-localized Hsp70 members *Bip1-Bip5* [72,73]. It is speculated that in our study, the mutants with loss-of-function of the ER-localized Hsp proteins disrupt the activation of the UPR pathway, resulting in the accumulation of denatured proteins, and leading to the formation of chalky grains. The activation of the UPR can affect the storage of substances and the expression of starch synthesis-related genes [72,73]. In this study, the loss-of-function of the ER-localized proteins resulted in alterations in the content of stored substances (Figures 4D and 7D–F). And the expression of starch synthesis-related genes was affected in all the tested mutants (Figure 8). It is reasonably hypothesized that Hsp proteins located in the ER likely participate in the UPR pathway to maintain ER homeostasis, ensuring the normal development of rice endosperm under high temperatures.

The pathways through which Hsps respond to stress in nature are highly complex. Future research could delve into these interactions between Hsps to better understand their collective role. The functions of Hsps under other stresses can also be studied, leveraging the cross-protection effect of Hsps, which may enhance plant resistance to multiple stresses, thereby improving yield and quality.

4. Materials and Methods

4.1. Identification and Phylogenetic Analysis of Hsp Genes on Rice

In previous studies, four Hsp subfamilies have been reported, including 23 *sHsp* genes, 104 *Hsp40* genes, 32 *Hsp70* genes, and 3 *Hsp100* genes [9,10,20,26]. To identify all members of the Hsp60 and Hsp90 subfamilies, homology searches were conducted based on previously reported *Hsp* genes. Genomic and protein sequence data were downloaded from the National Center for Biotechnology Information (NCBI, https://www.ncbi.nlm.nih.gov (accessed on 11 April 2024)). The conserved domain data of sHsp, Hsp70, and Hsp90 were sourced from the Pfam database and visualized using the TBtools-II software. For Hsp40, Hsp60, and Hsp100, the conserved domain data were obtained from the NCBI Conserved Domain Database and visualized using the tool (https://www.chiplot.online/conserved_domains_plot.html (accessed on 12 April 2024)). A phylogenetic tree was constructed using the Neighbor-Joining method in MEGA 11 software, enabling insights into the evolutionary relationships among these gene families.

4.2. Chromosomal Location and Gene Structure Analysis

The chromosomal locations of the identified Hsp genes were determined using the MG2C website (http://mg2c.iask.in/mg2c_v2.0 (accessed on 17 April 2024)). The genomic and coding sequence (CDS) data were downloaded from the website of NCBI, and the gene structure was visualized through the GSDS website (http://gsds.gao-lab.org/index.php (accessed on 17 April 2024)), illustrating the exon–intron configurations and gene models.

4.3. Gene Expression Profile Analysis

4.3.1. Prediction of Expression Levels of 192 Genes Across Different Tissues

The expression patterns of 192 Hsp genes across various rice tissues were retrieved from the RiceData website (https://www.ricedata.cn/gene/index.htm (accessed on 25 April 2024)). These expression patterns were visualized using TBtools-II (accessed on 25

April 2024) software, providing a comprehensive overview of the tissue-specific expression of Hsp genes.

4.3.2. Expression Analysis Across Different Tissues

Total RNAs were extracted from roots, stems, leaves, panicles, and developing grains 5, 10, 15, 20, and 25 days after flowering using the RNAprep Pure Plant Kit (DP441, TIANGEN, Beijing, China). The extracted RNA was reverse-transcribed into complementary DNA (cDNA) using the ReverTra Ace qPCR RT kit (Toyobo, Osaka, Japan). The gene expression patterns were evaluated through quantitative real-time PCR (RT-qPCR) using SYBR Green Real-Time PCR Master Mix (Toyobo, Osaka, Japan), with each sample repeated three times to ensure accuracy. The rice *Ubiquitin* gene (Os03g0234200) was used as an internal control. The RT-qPCR reaction system is 10uL in total, including 1 uL of upstream primer, 1 uL of downstream primer, 2 uL of cDNA, 5 uL of mix, and 1 uL of ddH_2O. The qRT-PCR program and system were as follows: Step 1: 95 °C for 3 min; Step 2: 95 °C for 10 s, 58 °C for 15 s, 72 °C for 15 s, Step 2 for 45 cycles; Step 3: 72 °C for 2min. These expression patterns were visualized using TBtools-II software.

4.3.3. Induced Expressions Under Different Stress Conditions

Two-week-old ZH11 seedlings were subjected to various stress conditions, including heat stress (42 °C for 10 and 60 min, followed by heat recovery for 6 h), cold stress (5 °C for 6 h), high salt stress (100 mmol/L NaCl for 6 h), and drought stress (20% PEG for 6 h). Gene expression levels under these stress conditions were examined using RT-qPCR, offering insights into the stress-responsive behavior of the Hsp genes.

4.3.4. RNA Extraction and Transcriptome Analysis

For transcriptome analysis, two-week-old ZH11 seedlings were subjected to heat stress (42 °C for 10, 30, and 60 min). Total RNA was extracted and subjected to standard transcriptome sequencing procedures, including RNA purification, cDNA synthesis, adapter ligation, and PCR amplification. The raw sequence data were cleaned and mapped to the rice reference genome for further analysis.

4.4. Subcellular Localization Analysis

To examine the subcellular localization of the Hsps, their coding sequences without termination codons were cloned into the pYBA1132-GFP vector. Then, the fusion of GFP and Hsp proteins together with marker vectors (HDEL-mCherry) was co-transformed into the protoplast of rice. DAPI was used as the nuclear marker. The GFP control protein was evenly distributed in the cytoplasm and nucleus. Confocal laser scanning microscopy (Karl Zeiss, Jena, Germany) was employed to observe the fluorescence signals, enabling the determination of the precise subcellular localization of the Hsp proteins.

4.5. Plant Materials and Growth Conditions

The *japonica* rice cultivar Zhonghua 11 (ZH11) was used as the primary plant material in this study. ZH11 was cultivated by the Institute of Crops of the Chinese Academy of Agricultural Sciences in 1979 with Jingfeng No.5/Tetep/Fujin, and cultivated in 1984. Mutant plants were generated by the CRISPR/Cas9 gene editing system. The mutants and wild type (ZH11) were grown in the fields of Hangzhou in 2023; the average temperature outdoors in the daytime was 34 °C from 20 July 2023 to 20 August 2023, following standard agronomic practices. After flowering, some well-grown ZH11 and mutant plants were moved to plant growth chambers under normal-temperature (28 °C, 12 h light/22 °C, 12 h dark) and high-temperature (35 °C, 12 h light/28 °C, 12 h dark) conditions for 15 days, and then moved to normal-temperature conditions until the grains are ripe.

4.6. Methods of Phenotype Evaluation

Plant height is measured as the height of the highest part of the spike above the ground. The grain length, grain width, and grain thickness were measured by the Wanshen SC-G automatic seed testing and analysis system (Wseen, Hangzhou, China). Mature and plump grains were used to measure 1000-grain weight. The chalkiness rate and chalkiness degree were measured by the Wanshen SC-E Rice Quality Analyzer (Wseen, Hangzhou, China), with each sample repeated three times to ensure accuracy.

4.7. Determination of Total Starch, Amylose Starch, and Total Protein Contents in the Endosperm

The amylose starch and total starch contents of mature endosperm were measured using the Megazyme K-TSTA and K-AMYL starch assay kits. The total protein content was measured following previously established protocols [78].

4.8. Statistical Analysis

All statistical analyses were conducted using GraphPad Prism 9. Significant differences between the treatment and control groups were determined using t-tests and one-way ANOVA, followed by Duncan's multiple comparisons, with *p*-values less than 0.05 or 0.01 considered statistically significant.

5. Conclusions

In this study, the Hsp family in rice was systematically analyzed and classified into six subfamilies: sHsp, Hsp40, Hsp60, Hsp70, Hsp90, and Hsp100. Bioinformatic analyses revealed a high degree of conservation within each subfamily, suggesting functional significance. Structural analysis, gene expression profiling, and subcellular localization studies indicated that these characteristics are likely correlated with their specialized roles in cellular processes. Notably, Hsps were shown to play a pivotal role in the grain-filling process, and the loss-of-function of most Hsps led to a decline in rice yield and quality, particularly under high-temperature conditions. Therefore, by introducing powerful *Hsp* genes into rice varieties through genetic engineering or traditional breeding methods, it may be possible to improve the yield and quality of rice, which is significant for meeting the increasing demand for food. In conclusion, our findings provide new insights into the mechanism of heat protein regulation of rice yield and quality, and provide a new theoretical basis and gene resources for the genetic improvement of rice yield and quality.

Supplementary Materials: The following supporting information can be downloaded at https://www.mdpi.com/article/10.3390/ijms252211931/s1.

Author Contributions: Conceptualization, H.W. and X.W.; methodology, H.W., N.D., L.M., L.X. and G.J.; validation, H.W., L.C., F.Z. and G.S.; formal analysis, H.W. and N.D.; investigation, H.W., Y.W. and S.W.; resources, S.H., Z.S., S.T., F.L. and X.W.; data curation, H.W. and S.C.; writing—original draft preparation, H.W. and L.C.; writing—review and editing, H.W., R.C. and X.W.; supervision, P.H. and X.W. All authors have read and agreed to the published version of the manuscript.

Funding: This work was supported by grants from the National Key Research and Development Program of China (2023YFF1000500), the National Natural Science Foundation of China (32372099 and 32188102), and the Project of Laboratory of Advanced Agricultural Sciences, Heilongjiang Province (ZY04JD05-005).

Data Availability Statement: The datasets supporting the conclusions of this article are included within the article and Supplementary Materials.

Conflicts of Interest: The authors declare no conflicts of interest.

References

1. Lindquist, S. The heat-shock response. *Ann. Rev. Biochem.* **1986**, *55*, 1151–1191. [CrossRef] [PubMed]
2. Chen, B., Feder, M.E., Kang, L. Evolution of heat-shock protein expression underlying adaptive responses to environmental stress. *Mol. Ecol.* **2018**, *27*, 3040–3054. [CrossRef]

3. Ruan, W.J. Cloning and Functional Analysis of Heat Shock Protein 70 (HSP70) Gene from *Chimonanthus praecox*. Ph.D. Dissertation, Southwest University, Chongqing, China, 2015.
4. Zhang, J.G.; He, C.Y.; Duan, A.G. Research progress of plant heat shock proteins. *J. Fujian For. Univ.* **2005**, *25*, 187–192.
5. Evgen'ev, M.B.; Garbuz, D.G.; Zatsepina, O.G. *Heat Shock Proteins and Whole Body Adaptation to Extreme Environments*, 1st ed.; Springer: Berlin/Heidelberg, Germany, 2014; pp. 1–83.
6. Feder, M.E.; Hofmann, G.E. Heat-shock proteins, molecular chaperones, and the stress response: Evolutionary and ecological physiology. *Annu. Rev. Physiol.* **1999**, *61*, 243–282. [CrossRef]
7. Wallin, R.P.; Lundqvist, A.; More, S.H.; Bonin, A.; Kiessling, R.; Ljunggren, H.G. Heat-shock proteins as activators of the innate immune system. *Trends Immunol.* **2002**, *23*, 130–135. [CrossRef] [PubMed]
8. Arbeitman, M.N.; Hogness, D.S. Molecular chaperones activate the Drosophila ecdysone receptor, an RXR heterodimer. *Cell* **2000**, *101*, 67–77. [CrossRef]
9. Sarkar, N.K.; Kim, Y.K.; Grover, A. Rice sHsp genes: Genomic organization and expression profiling under stress and development. *BMC Genom.* **2009**, *10*, 393. [CrossRef] [PubMed]
10. Sarkar, N.K.; Thapar, U.; Kundnani, P.; Panwar, P.; Grover, A. Functional relevance of J-protein family of rice (*Oryza sativa*). *Cell Stress Chaperones* **2013**, *18*, 321–331. [CrossRef]
11. Huang, L.H.; Wang, H.S.; Kang, L. Different evolutionary lineages of large and small heat shock proteins in eukaryotes. *Cell Res.* **2008**, *18*, 1074–1076. [CrossRef]
12. Kim, K.K.; Kim, R.; Kim, S. Crystal structure of a small heat-shock protein. *Nature* **1998**, *39*, 595–599. [CrossRef]
13. Haslbeck, M.; Franzmann, T.; Weinfurtner, D.; Buchner, J. Some like it hot: The structure and function of small heat-shock proteins. *Nat. Struct. Mol. Biol.* **2005**, *12*, 842–846. [CrossRef] [PubMed]
14. Kampinga, H.H.; Hageman, J.; Vos, M.J.; Kubota, H.; Tanguay, R.M.; Bruford, E.A.; Cheetham, M.E.; Chen, B.; Hightower, L.E. Guidelines for the nomenclature of the human heat shock proteins. *Cell Stress Chaperones* **2009**, *14*, 105–111. [CrossRef] [PubMed]
15. Hennessy, F.; Nicoll, W.S.; Zimmermann, R.; Cheetham, M.E.; Blatch, G.L. Not all J domains are created equal: Implications for the specificity of Hsp40–Hsp70 interactions. *Protein Sci.* **2005**, *14*, 1697–1709. [CrossRef]
16. Li, Z.Y.; Long, R.C.; Zhang, T.J.; Yang, Q.C.; Kang, J.M. Research Progress on Plant Heat Shock Protein. *Biotechnol. Bulletin.* **2016**, *32*, 7–13.
17. Ke, X.L.; Zou, W.X.; Ren, Y.F.; Wang, Z.Q.; Lin, J.; Wu, X.; Zhao, J.; Kirsten, B. Functional divergence of chloroplast Cpn60α subunits during *Arabidopsis* embryo development. *PLoS Genet.* **2017**, *13*, e1007036. [CrossRef] [PubMed]
18. Jacob, P.; Hirt, H.; Bendahmane, A. The heat-shock protein/chaperone network and multiple stress resistance. *Plant Biotecnol. J.* **2016**, *15*, 405–414. [CrossRef] [PubMed]
19. Chaston, J.J.; Smits, C.; Aragão, D.; Wong, A.S.; Ahsan, B.; Sandin, S.; Molugu, S.K.; Molugu, S.K.; Bernal, R.A.; Stock, D.; et al. Structural and Functional Insights into the Evolution and Stress Adaptation of Type II Chaperonins. *Structure* **2016**, *24*, 364–374. [CrossRef]
20. Sarkar, N.K.; Kundnani, P.; Grover, A. Functional an-alysis of Hsp70 superfamily proteins of rice (*Oryza sativa* L.). *Cell Stress Chaperones* **2013**, *18*, 427–437. [CrossRef] [PubMed]
21. Rosenzweig, R.; Nillegoda, N.B.; Mayer, M.P.; Bukau, B. The Hsp70 chaperone network. *Nat. Rev. Mol. Cell Biol.* **2019**, *20*, 665–680. [CrossRef]
22. Lee-Yoon, D.; Easton, D.; Murawski, M.; Burd, R.; Subjeck, J.R. Identifi cation of a major subfamily of large hsp70-like proteins through the clonin g of the mammalian 110-kDa heat shock protein. *J. Biol. Chem.* **1995**, *270*, 15725–15733. [CrossRef]
23. Jackson, S.E.; Queitsch, C.; Toft, D. Hsp90: From structure to phenotype. *Nat. Struct. Mol. Biol.* **2004**, *11*, 1152–1155. [CrossRef] [PubMed]
24. Li, J.; Sha, B. Crystal structure of the E.cili HSP100 CIpb N-terminal dimain. *Structure* **2003**, *11*, 323–328. [CrossRef]
25. Mishra, R.; Grover, A. ClpB/Hsp100 proteins and heat stress tolerance in plants. *Crit. Rev. Biotechnol.* **2016**, *36*, 862–875. [CrossRef] [PubMed]
26. Singh, A.; Singh, U.; Mittal, D.; Grover, A. Genome-wide analysis of rice ClpB/HSP100, ClpC and ClpD genes. *BMC Genom.* **2010**, *11*, 95. [CrossRef]
27. Qiu, X.B.; Shao, Y.M.; Miao, S.; Wang, L. The diversity of the DnaJ/Hsp40 family, the crucial partners for Hsp70 chaperones. *Cell Mol. Life Sci.* **2006**, *63*, 2560–2570. [CrossRef] [PubMed]
28. Wehmeyer, N.; Hernandez, L.D.; Finkelstein, R.R.; Vierling, E. Synthesis of small heat-shock proteins is part of the developmental program of late seed maturation. *Plant Physiol.* **1996**, *112*, 747–757. [CrossRef]
29. Ma, W.; Guan, X.Y.; Li, J.; Pan, R.H.; Wang, L.Y.; Liu, F.J.; Ma, H.Y.; Zhu, S.J.; Hu, J.; Ruan, Y.L.; et al. Mitochondrial small heat shock protein mediates seed germination via thermal sensing. *Proc. Natl. Acad. Sci. USA* **2019**, *116*, 4716–4721. [CrossRef]
30. Yamamoto, M.; Maruyama, D.; Endo, T.; Nishikawa, S. *Arabidopsis thaliana* has a set of J proteins in the endoplasmic reticulum that are conserved from yeast to animals and plants. *Plant Cell Physiol.* **2008**, *49*, 1547–1562. [CrossRef]
31. Yang, K.Z.; Xia, C.; Liu, X.L.; Dou, X.Y.; Wang, W.; Chen, L.Q.; Zhang, X.Q.; Xie, L.F.; He, L.; Ma, X.; et al. A mutation in Thermosensitive Male Sterile 1, encoding a heat shock protein with DnaJ and PDI domains, leads to thermosensitive gametophytic male sterility in *Arabidopsis*. *Plant J.* **2009**, *57*, 870–882. [CrossRef] [PubMed]
32. Michael, W.; Jorg, S.; Les, F. Effects of near-lethal heat stress on bud break, heat-shock proteins and ubiquitin in dormant poplar. *Tree Physiol.* **1997**, *17*, 453–460.

33. Marocco, A.; Santucci, A.; Cerioli, S.; Motto, M.; Di Fonzo, N.; Thompson, R.; Salamini, F. Three High-Lysine Mutations Control the Level of ATP-Binding HSP70-like Proteins in the Maize Endosperm. *Plant Cell* **1991**, *3*, 507–515. [PubMed]
34. Takabatake, R.; Ando, Y.; Seo, S.; Katou, S.; Tsuda, S.; Ohashi, Y.; Mitsuhara, I. MAP kinases function downstream of HSP90 and upstream of mitochondaia in TMV resistance gene N-mediated hypersensitive cell death. *Plant Cell Physiol.* **2007**, *48*, 498–510. [CrossRef] [PubMed]
35. Yan, Y.; Wang, M.L.; Guo, Y.T.; Ding, C.H.; Niu, K.X.; Li, X.M.; Sun, C.; Dong, Z.; Cui, D.; Rasheed, A.; et al. HSP90.2 promotes CO_2 assimilation rate, grain weight and yield in wheat. *Plant Biotechnol. J.* **2023**, *21*, 1229–1239. [CrossRef] [PubMed]
36. Bao, F.; Huang, X.Z.; Zhu, C.P.; Zhang, X.Y.; Li, X.; Yang, S.H. *Arabidopsis* HSP90 protein modulates RPP4-mediated temperature-dependent cell death and defense responses. *New Phytol.* **2014**, *202*, 1320–1334. [CrossRef] [PubMed]
37. Lee, U.; Rioflorido, I.; Hong, S.W.; Larkindale, J.; Waters, E.R.; Vierling, E. The *Arabidopsis* ClpB/Hsp100 family of proteins: Chaperones for stress and chloroplast development. *Plant J.* **2007**, *49*, 115–127. [CrossRef] [PubMed]
38. Zhu, X.B.; Liang, S.H.; Yin, J.J.; Yuan, C.; Wang, J.; Li, W.T.; He, M.; Wang, J.C.; Chen, W.L.; Ma, B.T.; et al. The DnaJ DjA7/8 is essential for chloroplast development in rice (*Oryza sativa*). *Gene* **2015**, *574*, 11–19. [CrossRef] [PubMed]
39. Zhang, H. Effect of Heat Shock Protein gene *OsHSP40* on the seed setting rate of rice (*Oryza sativa*). Master's Dissertation, Nanchang University, Nanchang, China, 2018.
40. Zou, J.; Liu, C.F.; Liu, A.L.; Zou, D.; Chen, X.B. Over expression of OsHsp17.0 and OsHsp23.7 enhances drought and salt tolerance in rice. *J. Plant Physiol.* **2012**, *169*, 628–635. [CrossRef]
41. Yu, J.J.; Nou, I.S.; Kang, K.K. Overexpression of *Oshsp16.9* Gene Encoding Small Heat Shock Protein Enhances Tolerance to Abiotic Stresses in Rice. *Plant Breed. Biotechnol.* **2014**, *2*, 370–379.
42. Jiang, Q.; Mei, J.; Gong, X.D.; Xu, J.L.; Zhang, J.H.; Teng, S.; Lina, D.Z.; Dong, Y.H. Importance of the rice TCD9 encoding subunit of chaperonin protein60 (Cpn60) for the chloroplast development during the early leaf stage. *Plant Sci.* **2014**, *215–216*, 172–179. [CrossRef]
43. Lin, S.; Liu, Z.; Sun, S.Y.; Xue, F.; Li, H.; Tursun, A.; Cao, L.; Zhang, L.; Wilson, Z.A.; Zhang, D.; et al. Rice HEAT SHOCK PROTEIN60-3B maintains malefertility under high temperature by starch granule biogenesis. *Plant Physiol.* **2023**, *3*, 2301–2317. [CrossRef]
44. Zhe, X.P.; Teng, X.; Wang, Y.L.; Hao, Y.Y.; Jing, R.N.; Wang, Y.F.; Liu, Y.; Zhu, J.P.; Wu, M.M.; Zhong, M.S.; et al. FLOURY ENDOSPERM11 encoding a plastid heat shock protein 70 is essential for amyloplast development in rice. *Plant Sci.* **2018**, *277*, 89–99. [CrossRef] [PubMed]
45. Tabassum, R.; Dosaka, T.; Ichida, H.; Morita, R.; Ding, Y.; Abe, T.; Katsube-Tanaka, T. FLOURY ENDOSPERM11-2 encodes plastid HSP70-2 involved with the temperature-dependent chalkiness of rice (*Oryza sativa* L.) grains. *Plant J.* **2020**, *2*, 604–616. [CrossRef] [PubMed]
46. Qi, Y.; Wang, H.; Zou, Y.; Liu, C.; Liu, Y.; Wang, Y.; Zhang, W. Over-expression of mitochondrial heat shock protein 70 suppresses programmed cell death in rice. *FEBS Lett.* **2011**, *585*, 231–239. [CrossRef] [PubMed]
47. Wu, H.; Ren, Y.; Dong, H.; Xie, C.; Wang, Y.; Wang, Y.; Lei, C.; Cheng, Z.; Jiang, L.; Wan, J. FLOURY ENDOSPERM24, a heat shock protein 101 (HSP101), is required for starch biosynthesis and endosperm development in rice. *New Phytol.* **2024**, *242*, 2635–2651. [CrossRef] [PubMed]
48. Ajayi, O.O.; Peters, S.O.; De Donato, M.; Sowande, S.O.; Mujibi, D.N.; Morenikeji, O.B.; Thomas, B.N.; Adeleke, M.A.; Imumorin, I.G. Computational genome-wide identification of heat shock protein genes in the bovine genome. *F1000Research* **2018**, *20*, 1504. [CrossRef]
49. Xu, Y.; Shi, F.; Li, Y.; Zong, S.; Tao, J. Genome-wide identification and expression analysis of the Hsp gene superfamily in asian long-horned beetle (*Anoplophora glabripennis*). *Int. J. Biol. Macromol.* **2022**, *200*, 583–592. [CrossRef] [PubMed]
50. Wang, X.R.; Wang, C.; Xue, F.; Tong, D.; Sheng, S.; Wang, X.W. Genome-wide identification and characterization of HSP gene superfamily in whitefly (*Bemisia tabacl*) and expression profiling analysis under temperature stress. *Insect Sci.* **2019**, *1*, 44–57. [CrossRef]
51. Gao, Y.; Li, J.N.; Pu, J.J.; Tao, K.X.; Zhao, X.X.; Yang, Q.Q. Genome-wide identification and characterization of the HSP gene superfamily in apple snails (*Gastropoda: Ampullariidae*) and expression analysis under temperature stress. *Int. J. Biol. Macromol.* **2022**, *222*, 2545–2555. [CrossRef]
52. Milioni, D.; Hatzopoulos, P. Genomic organization of hsp90 gene family in *Arabidopsis*. *Plant Mol. Biol.* **1997**, *35*, 955–961. [CrossRef]
53. Cao, L.R.; Wang, G.R.; Fahim, A.M.; Pang, Y.Y.; Zhang, Q.J.; Zhang, X.; Wang, Z.H.; Lu, X.M. Comprehensive Analysis of the DnaJ/HSP40 Gene Family in Maize (*Zea mays* L.) Reveals that ZmDnaJ96 Enhances Abiotic Stress Tolerance. *J. Plant Growth Regul.* **2024**, *43*, 1548–1569. [CrossRef]
54. Jiang, L.; Hu, W.; Qian, Y.; Ren, Q.; Zhang, J. Genome-wide identification, classification and expression analysis of the Hsf and Hsp70 gene families in maize. *Gene* **2021**, *20*, 145348. [CrossRef] [PubMed]
55. Pandey, B.; Kaur, A.; Gupta, O.P.; Sharma, I.; Sharma, P. Identification of HSP20 Gene Family in Wheat and Barley and Their Differential Expression Profiling Under Heat Stress. *Appl. Biochem. Biotechnol.* **2015**, *175*, 2427–2446. [CrossRef] [PubMed]
56. Nagaraju, M.; Kumar, A.; Rajasheker, G.; Manohar Rao, D.; Kavi Kishor, P.B. DnaJs, the critical drivers of Hsp70s: Genome wide screening, characterization and expression of DnaJ family genes in *Sorghum bicolor*. *Mol. Biol. Rep.* **2020**, *47*, 7379–7390. [CrossRef] [PubMed]

57. Zhao, P.; Wang, D.; Wang, R.; Kong, N.; Zhang, C.; Yang, C.; Wu, W.; Ma, H.; Chen, Q. Genome-wide analysis of the potato Hsp20 gene family: Identification, genomic organization and expression profiles in response to heat stress. *BMC Genom.* **2018**, *19*, 61. [CrossRef]
58. Lopes-Caitar, V.S.; Carvalho, M.C.; Darben, L.M.; Kuwahara, M.K.; Nepomuceno, A.L.; Dias, W.P.; Abdelnoor, R.V.; Marcelino-Guimarães, F.C. Genome-wide analysis of the Hsp20 gene family in soybean: Comprehensive sequence, genomic organization andexpression profile analysis under abiotic and biotic stresses. *BMC Genom.* **2013**, *28*, 577.
59. Ji, X.R.; Yu, Y.H.; Ni, P.Y.; Zhang, G.H.; Guo, D.L. Genome-wide identification of small heat-shock protein (HSP20) gene family in grape and expression profile during berry development. *BMC Plant Biol.* **2019**, *19*, 433. [CrossRef]
60. Chen, T.; Xu, T.; Zhang, T.; Liu, T.; Shen, L.; Chen, Z.; Wu, Y.; Yang, J. Genome-Wide Identification and Characterization of DnaJ Gene Family in Grape (*Vitis vinifera* L.). *Horticulturae* **2021**, *7*, 589. [CrossRef]
61. Panzade, K.P.; Kale, S.S.; Chavan, N.R.; Hatzade, B. Genome-wide analysis of Hsp70 and Hsp100 gene families in *Ziziphus jujuba*. *Cell Stress Chaperones* **2021**, *26*, 341–353. [CrossRef] [PubMed]
62. Waters, E.R. The evolution, function, structure, and expression of the plant sHSPs. *J. Exp. Bot.* **2013**, *64*, 391–403. [CrossRef]
63. Apuya, N.R.; Yadegari, R.; Fischer, R.L.; Harada, J.J.; Zimmerman, J.L.; Goldberg, R.B. The *Arabidopsis* embryo mutant schlepperless has 42a defect in the chaperonin-60 alpha gene. *Plant Physiol.* **2001**, *126*, 717–730. [CrossRef]
64. Vierling, E. The Roles of Heat Shock Proteins in Plants. *Annu. Rev. Plant Biol.* **1991**, *42*, 579–620. [CrossRef]
65. Venetianer, A.; Pirity, M.; Hever-Szabo, A. The function of heat-shock proteins in stress tolerance. *Cell Biol. Int.* **1994**, *18*, 605–615. [CrossRef] [PubMed]
66. Chen, B.; Wagner, A. Hsp90 is important for fecundity, longevity, and buffering of cryptic deleterious variation in wild fly populations. *BMC Evol. Biol.* **2012**, *12*, 25. [CrossRef] [PubMed]
67. Lerman, D.N.; Feder, M.E. Naturally occurring transposable elements disrupt hsp70 promoter function in *Drosophila melanogaster*. *Mol. Biol. Evol.* **2005**, *22*, 776–783. [CrossRef] [PubMed]
68. Ding, C.Q.; Shao, Z.J.; Yan, Y.P.; Zhang, G.H.; Zeng, D.L.; Zhu, L.; Hu, J.; Gao, Z.Y.; Dong, G.J.; Qian, Q. Carotenoid isomerase regulates rice tillering and grain productivity by its biosynthesis pathway. *J. Integr. Plant Biol.* **2024**, *66*, 172–175. [CrossRef] [PubMed]
69. Xue, P.; Chen, Y.Y.; Wen, X.X.; Wang, B.F.; Yang, Q.Q.; Gong, K.; Kang, Y.W.; Sun, L.P.; Yu, P.; Cao, L.Y.; et al. Dissection of Closely Linked Quantitative Trait Locis Controlling Grain Size in Rice. *Front. Plant Sci.* **2022**, *12*, 804444. [CrossRef] [PubMed]
70. Xiong, D.; Wang, R.; Wang, Y.; Li, Y.; Sun, G.; Yao, S. *SLG2* specifically regulates grain width through WOX11-mediated cell expansion control in rice. *Plant Biotechnol. J.* **2023**, *21*, 1904–1918. [CrossRef]
71. Guo, T.; Lu, Z.Q.; Shan, J.X.; Ye, W.W.; Dong, N.Q.; Lin, H.X. ERECTA1 Acts Upstream of the OsMKKK10-OsMKK4-OsMPK6 Cascade to Control Spikelet Number by Regulating Cytokinin Metabolism in Rice. *Plant Cell* **2020**, *32*, 2763–2779. [CrossRef]
72. Cao, R.; Zhao, S.; Jiao, G.; Duan, Y.; Ma, L.; Dong, N.; Lu, F.; Zhu, M.; Shao, G.; Hu, S.; et al. OPAQUE3, encoding a transmembrane bZIP transcription factor, regulates endosperm storage protein and starch biosynthesis in rice. *Plant Commun.* **2022**, *6*, 100463. [CrossRef]
73. Yang, W.; Xu, P.; Zhang, J.; Zhang, S.; Li, Z.; Yang, K.; Chang, X.; Li, Y.B. OsbZIP60-mediated unfolded protein response regulates grain chalkiness in rice. *J. Genet. Genom.* **2022**, *49*, 414–426. [CrossRef]
74. Grover, A.; Aggarwal, P.K.; Kapoor, A.; Katiyar, A.S.; Agarwal, M.; Chandramouli, A. Addressing abiotic stresses in agriculture through transgenic technology. *Curr. Sci.* **2003**, *84*, 355–367.
75. Zou, J.; Liu, A.L.; Chen, X.B.; Zhou, X.Y.; Gao, G.F.; Wang, W.F.; Zhang, X.W. Expression analysis of nine rice heat shock protein genes under abiotic stresses and ABA treatment. *J. Plant Physiol.* **2009**, *166*, 851–861. [CrossRef] [PubMed]
76. Howell, S.H. Endoplasmic reticulum stress responses in plants. *Annu. Rev. Plant Biol.* **2013**, *64*, 477–499. [CrossRef] [PubMed]
77. Ren, Y.; Huang, Z.; Jiang, H.; Wang, Z.; Wu, F.; Xiong, Y.; Yao, J. A heat stress responsive NAC transcription factor heterodimer plays key roles in rice grain filling. *J. Exp. Bot.* **2021**, *72*, 2947–2964. [CrossRef] [PubMed]
78. Kang, H.G.; Park, S.; Matsuoka, M.; An, G. White-core endosperm floury endosperm-4 in rice is generated by knockout mutations in the C-type pyruvate orthophosphate dikinase gene (OsPPDKB). *Plant J.* **2005**, *42*, 901–911. [CrossRef] [PubMed]

International Journal of
Molecular Sciences

MDPI

Article

Identification and Characterization of *Shaker* Potassium Channel Gene Family and Response to Salt and Chilling Stress in Rice

Quanxiang Tian [1,2], Tongyuan Yu [1,†], Mengyuan Dong [1,†], Yue Hu [1], Xiaoguang Chen [1,2], Yuan Xue [1], Yunxia Fang [1,2], Jian Zhang [3], Xiaoqin Zhang [1,2,*] and Dawei Xue [1,2,*]

1 College of Life and Environmental Sciences, Hangzhou Normal University, Hangzhou 311121, China; quanxiang@hznu.edu.cn (Q.T.); 2022111010017@stu.hznu.edu.cn (M.D.); xg.chen@hotmail.com (X.C.); 2023111010055@stu.hznu.edu.cn (Y.X.); 20173008@hznu.edu.cn (Y.F.)

2 Zhejiang Provincial Key Laboratory for Genetic Improvement and Quality Control of Medicinal Plants, Hangzhou Normal University, Hangzhou 311121, China

3 State Key Laboratory of Rice Biology and Breeding, China National Rice Research Institute, Hangzhou 311400, China; zhangjian@caas.cn

* Correspondence: zxq@hznu.edu.cn (X.Z.); dwxue@hznu.edu.cn (D.X.)

† These authors have contributed equally to this work and share second authorship.

Abstract: *Shaker* potassium channel proteins are a class of voltage-gated ion channels responsible for K^+ uptake and translocation, playing a crucial role in plant growth and salt tolerance. In this study, bioinformatic analysis was performed to identify the members within the *Shaker* gene family. Moreover, the expression patterns of rice *Shaker(OsShaker)* K^+ channel genes were analyzed in different tissues and salt treatment by RT–qPCR. The results revealed that there were eight *OsShaker* K^+ channel genes distributed on chromosomes 1, 2, 5, 6 and 7 in rice, and their promoters contained a variety of cis-regulatory elements, including hormone-responsive, light-responsive, and stress-responsive elements, etc. Most of the *OsShaker* K^+ channel genes were expressed in all tissues of rice, but at different levels in different tissues. In addition, the expression of *OsShaker* K^+ channel genes differed in the timing, organization and intensity of response to salt and chilling stress. In conclusion, our findings provide a reference for the understanding of *OsShaker* K^+ channel genes, as well as their potential functions in response to salt and chilling stress in rice.

Keywords: rice; *Shaker* potassium channel; abiotic stress; bioinformatics analysis; expression patterns

Citation: Tian, Q.; Yu, T.; Dong, M.; Hu, Y.; Chen, X.; Xue, Y.; Fang, Y.; Zhang, J.; Zhang, X.; Xue, D. Identification and Characterization of *Shaker* Potassium Channel Gene Family and Response to Salt and Chilling Stress in Rice. *Int. J. Mol. Sci.* **2024**, 25, 9728. https://doi.org/10.3390/ijms25179728

Academic Editor: Daniela Trono

Received: 24 July 2024
Revised: 30 August 2024
Accepted: 5 September 2024
Published: 8 September 2024

1. Introduction

Soil salinity is a global environmental problem that restricts the growth and productivity of plants [1]. The area of saline soils in the world is about 950 million hectares [2]. It is estimated that, by the middle of the 21st century, about 50% of the world's arable land will be affected by soil salinization, and this proportion is expected to increase further due to global warming and irrational irrigation [3]. Rice is an important cereal crops for the world's population. However, it is one of the most salt-susceptible cereals, especially at the seedling stage. Therefore, it is crucial from the perspective of global food security to identify the key genes and breed new varieties of salt-tolerant rice.

Potassium (K^+), which accounts for 2–10% of plant dry weight, is essential not only for normal growth and development, but also for salt stress response [4–6]. For instance, salt stress normally leads to K^+ efflux accompanying the accumulation of Na^+ in the cytoplasm, which can competitively combine the K^+ binding sites of enzymes, as Na^+ and K^+ share similar chemical properties, resulting in cellular metabolic disorders. Therefore, maintaining cytoplasmic K^+ concentration is vital for plant salt tolerance [7,8]. K^+ exists in the soil in a variety of forms, However, due to the adsorption of silicates, the available K^+ content in the soil is usually very low [9,10]. Moreover, plant cells need a high concentration of K^+ to maintain normal metabolism. Cytoplasmic K^+ content is relatively stable, ranging from 60 to 200 mM, while vacuolar K^+ concentration is more variable, ranging from 20 to 500 mM [4,11,12]. Therefore, plants have evolved efficient K^+ absorption and transport systems to maintain optimal growth in low potassium environments or under abiotic stresses [13].

In plants, K^+ absorption and transport are executed mainly by K^+ transporters and channels. There are three families of K^+ transporters in plants [14]. HAK (high-affinity K^+)/KUP (K^+ uptake)/KT(K^+ transporter) is the largest family of K^+ transporters, only found in plants, with a K^+/H^+ symporter function [15]. The HKT (High-Affinity Potassium Transporter) family is mainly responsible for the transport of K^+/Na^+ cotransporters or Na^+ [16]. The CPA (Cation proton antiporters) family can be further divided into the CHX (Cation/H^+ exchangers), NHX (Na^+/H^+ exchangers), and KEA (K^+ efflux antiporters) subfamilies [14], and only some members have been reported to be involved in K^+ transport in *Arabidopsis* [17,18]. Four K^+ channel families exist in plants: *Shaker* (*Shaker*-type K^+ channels), TPK (Two-pore K^+-channels), Kir (K^+-inward rectifier)-like channels, and NSCC (non-selective cation channels). TPK members generally consists of two ion-conducting pores and four transmembrane domains, and this structure is thought to be similar to that of the KCNK (Two-pore-domain) family in animals [19]. Kir-like has only two transmembrane domains and one ion conducting pore, similar to animal Kir. TPK and Kir-like channels cannot perceive osmolarity changes due to their lack of osmotic pressure-sensing structures. However, many TPK and Kir-like channels have an EF-hand domain at the C-terminal that can sense changes in calcium ions, suggesting that they may be sensitive to intracellular or intercellular changes in Ca^{2+} [20]. TPK family members are mainly located in the vacuole membrane to mediate K^+ transport, and a few are localized in the plasma membrane [21]. NSCC probably plays the major role in K^+ uptake at high concentrations (>10 mM). *Shaker* channels were the first potassium channel proteins identified and are responsible for the uptake and translocation of K^+ [22]. The basic structure of the *Shaker* channel consists of four alpha-subunits that surround each other to form a central aqueous pore for the permeation of K^+ [22]. A typical α-subunit usually contains a short intracellular N-terminus of only about 60 amino acids, a C-terminus, six transmembrane segments (S1–S6) and a conserved ion-conducting pore region [23]. The fourth of these transmembrane structures (S4) senses and transmits voltage changes and controls the opening and closing of the channel [23]. The *Shaker* family can be divided into three functional subfamilies based on their voltage dependence: inwardly rectifying potassium channels (Kin), weakly rectifying potassium channels (K-weak), and outwardly rectifying potassium channels (K-out) [24].

The *Shaker* family is important in plant potassium ion uptake and transport under abiotic stresses. *Shaker* family member AtGORK is an outward K^+ channel, which negatively regulates salt tolerance in *Arabidopsis*, and lack of *AtGORK* in *Arabidopsis* could enhance salt tolerance [25]. In rice, *OsKAT1* can significantly increase the cellular K^+ content and K^+/Na^+ ratio of suspension cells and alleviate the inhibitory effect of salt stress on cell proliferation [26]. In addition, it was recently found that, under salt stress, the phloem-localized *OsAKT2* is responsible for re-transporting K^+ leaking from the phloem into the phloem and transporting it through the phloem cycle to critical sites, such as roots and young leaves, to maintain K^+ and Na^+ homeostasis at these sites, the mutation of *OsAKT2* resulted in significant sensitivity to salt stress in rice [27,28]. *OsK5.2* expression was significantly up-regulated in the shoot under salt stress, resulting in a large outflow of K^+ from the stomatal guard cells, stomatal closure, and a rapid decrease in the transpiration flow, reducing the transport of Na^+ to the above-ground part along with the transpiration flow, and improving the salt tolerance of rice [29]. In eggplant, expression of *Shaker* family genes *AKT1*, *KAT1* and *SOS1* was significantly up-regulated in leaves of the salt-tolerant variety ST118 [30]. In addition, *Shaker* family members have also been found to be involved in the regulation of salt tolerance in other plants. GmAKT1 mediates K^+ uptake and maintains Na^+/K^+ homeostasis under salt stress, and overexpression of *GmAKT1* enhances salt tolerance in *Arabidopsis* and soybean [31,32]. In *Zygophyllum xanthoxylum*, AKT1 enhances salt tolerance by regulating root cell Na^+/K^+ homeostasis [33]. PbrKAT1 exhibits typical inward rectifier currents in *Xenopus* oocytes, and its activity is inhibited by external Na^+ [34].

Rice is one of the most important food crops in the world, providing food for more than half of the global population. The *Shaker* family plays an important role in plant resistance to abiotic stress, However, this family has been less widely studied and lacks systematic bioinformatics analysis for identification and response to adversity stress in rice. In this study, we employed bioinformatics and publicly available data to identify the *Shaker* potassium channel in the rice genome and performed analysis of its gene structure, phylogenetic tree, conserved protein sequences, and promoter cis-acting elements, etc. In addition, we used reverse transcription quantitative polymerase chain reaction(RT–qPCR) to investigate the tissue expression pattern of the *Shaker* genes and their transcriptional response to salt and chilling stress. Our results provide a set of potential candidate *Shaker* K^+ channel genes for the future genetic modification of K^+ transport and salt and chilling tolerance in higher plants.

2. Results

2.1. Identification of the OsShaker K+ Channel Genes

Eight *OsShaker* K^+ channel genes were identified in the rice genome, which were distributed on chromosomes 1, 2, 5, 6, and 7 (Figure 1, Supplementary Table S1). The *OsShaker* K^+ channel genes are named according to the homologous relationships with *Arabidopsis*.

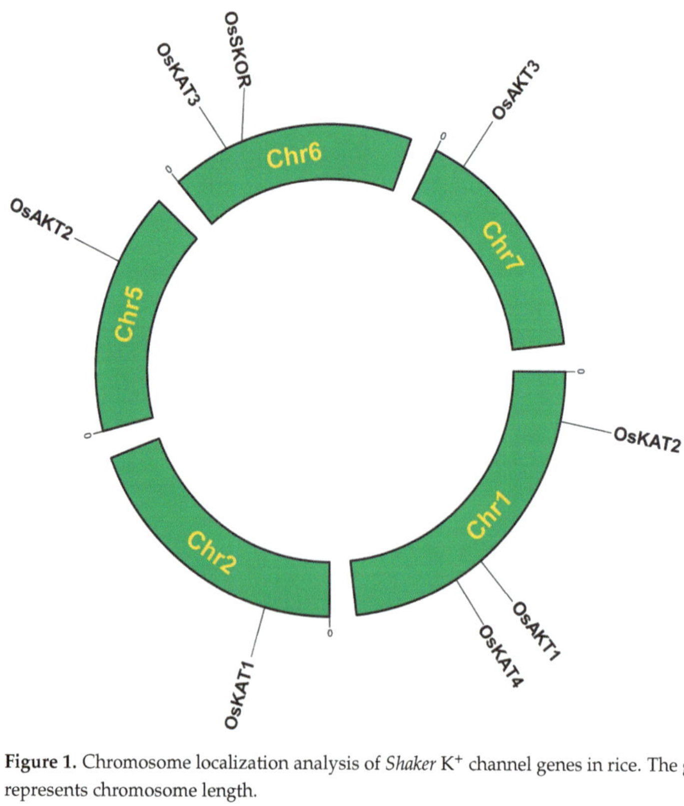

Figure 1. Chromosome localization analysis of *Shaker* K^+ channel genes in rice. The green arc length represents chromosome length.

The results of physicochemical property analysis showed that the amino acid lengths of the *Shaker* family proteins ranged from 373 to 935; the amino acid number of the OsKAT4 was the lowest, while the OsAKT1 has the longest peptide. The molecular masses of the proteins ranged from 41.67 to 102.02 kDa, with the smallest being the OsKAT4 and the largest being the OsAKT1, with 41.67 kDa and 102.02 kDa, respectively. The isoelectric points ranged from 5.66 to 10.9, with the smallest being the OsSKOR and the largest being the OsAKT3. The hydrophobicity values of the proteins ranged from −0.3 to 0.308, and the hydrophobicity values of the proteins were distributed in both positive and negative, indicating that the hydrophobicity of the rice *Shaker* family proteins is not fixed. The predicted subcellular localization showed that all *OsShaker* K^+ channel proteins were localized in the cytoplasmic membrane. In addition, OsAKT3 and OsKAT1 have extra nuclear localization signal peptides (Table 1, Supplementary Table S2).

Table 1. Physiochemical properties of the *OsShaker* K^+ channel genes.

Name	Gene ID	Accession Number	Chromosome Position	Length (aa)	MW (kDa)	PI	GRAVY	Localization	NLS Predicts
OsAKT1	*Os01g0648000*	*LOC_Os01g45990*	Chr1	935	102.02	6.78	−0.149	Plas	NA
OsAKT2	*Os05g0428700*	*LOC_Os05g35410*	Chr5	703	76.70	6.40	−0.071	Plas	NA
OsAKT3	*Os07g0175400*	*LOC_Os07g07910*	Chr7	711	77.46	10.9	−0.355	Plas	Positive
OsKAT1	*Os02g0245800*	*LOC_Os02g14840*	Chr2	718	81.03	6.76	−0.243	Plas	Positive
OsKAT2	*Os01g0210700*	*LOC_Os01g11250*	Chr1	601	68.17	8.57	−0.076	Plas	NA
OsKAT3	*Os06g0254200*	*LOC_Os06g14310*	Chr6	610	67.73	9.03	0.149	Plas	NA
OsKAT4	*Os01g0718700*	*LOC_Os01g52070*	Chr1	373	41.67	9.19	0.308	Plas	NA
OsSKOR	*Os06g0250600*	*LOC_Os06g14030*	Chr6	858	95.10	5.66	−0.211	Plas	NA

Note: MW, Molecular weight (kDa); pI, Isoelectric point; GRAVY, Grand average of hydropathicity; NLS predicts, Nuclear Localization Signal predicts. Plas, Plasma membrane. "NA" is an abbreviation for "Not Available" and indicates no NLS. "Positive" indicates the presence of NLS.

2.2. *OsShaker K^+ Channel Proteins' Sequence Alignment, Gene Structure and Conserved Motif Analysis*

Comparative analysis of the *OsShaker* K^+ channel protein sequences showed that they have four highly conserved domains: the Ion_trans_2 domain, the cNMP domain, the ANK domain and the KHA domain (Figure 2a). Structural analysis of the *OsShaker* K^+ channel genes demonstrated that the length of the family genes ranged from 1776 (OsKAT3) to 7499 (OsSKOR) bp. The number of exons in the *OsShaker* K^+ channel genes ranged from 1 to 11, except for *OsKAT3*, which contained one exon; the other members contained more than six exons, among which *OsAKT1*, *OsAKT2*, *OsKAT1*, *OsKAT2*, and *OsSKOR* contained 11 exons (Figure 2b).

Conserved motif analysis is important for exploring the structural composition of proteins, as well as their function. The MEME online tool was used to analyze the number and distribution of the Motifs of *OsShaker* K^+ channel proteins; a total of 14 Motifs were characterized, named Motif1–Motif14. In the conserved motif map, the relative degree of conservatism increased with the height of the letter and the frequency of the base corresponding to the letter. As shown in Figure 2c, OsAKT2 does not contain Motif7 and Motif9, while OsAKT1 and OsAKT3 contain 14 Motifs in the same number and order (Supplementary Table S3). The similarities and differences shown by these conserved motifs may be attributed to the adaptation of the functions of *OsShaker* K^+ channel proteins to the evolutionary direction during the evolutionary process.

2.3. *Phylogenetic and Gene Duplication Analysis of the OsShaker K^+ Channel Genes*

To further investigate the *OsShaker* K^+ channel genes' evolutionary relationship between rice and other species, such as *Arabidopsis thaliana*, *Hordeum vulgare*, *Zea mays*, and *Glycine max*, a phylogenetic tree was constructed. The results showed that *Shaker* K^+ channel proteins are divided into five subfamilies (Clusters I–V)., and the proteins of *OsShaker* K^+ channel were closely related to *Shaker* in monocotyledonous plants, such as barley and corn (Figure 3a), indicating that the *Shaker* K^+ channel gene family in monocotyledonous and dicotyledonous plants shows different evolutionary trends.

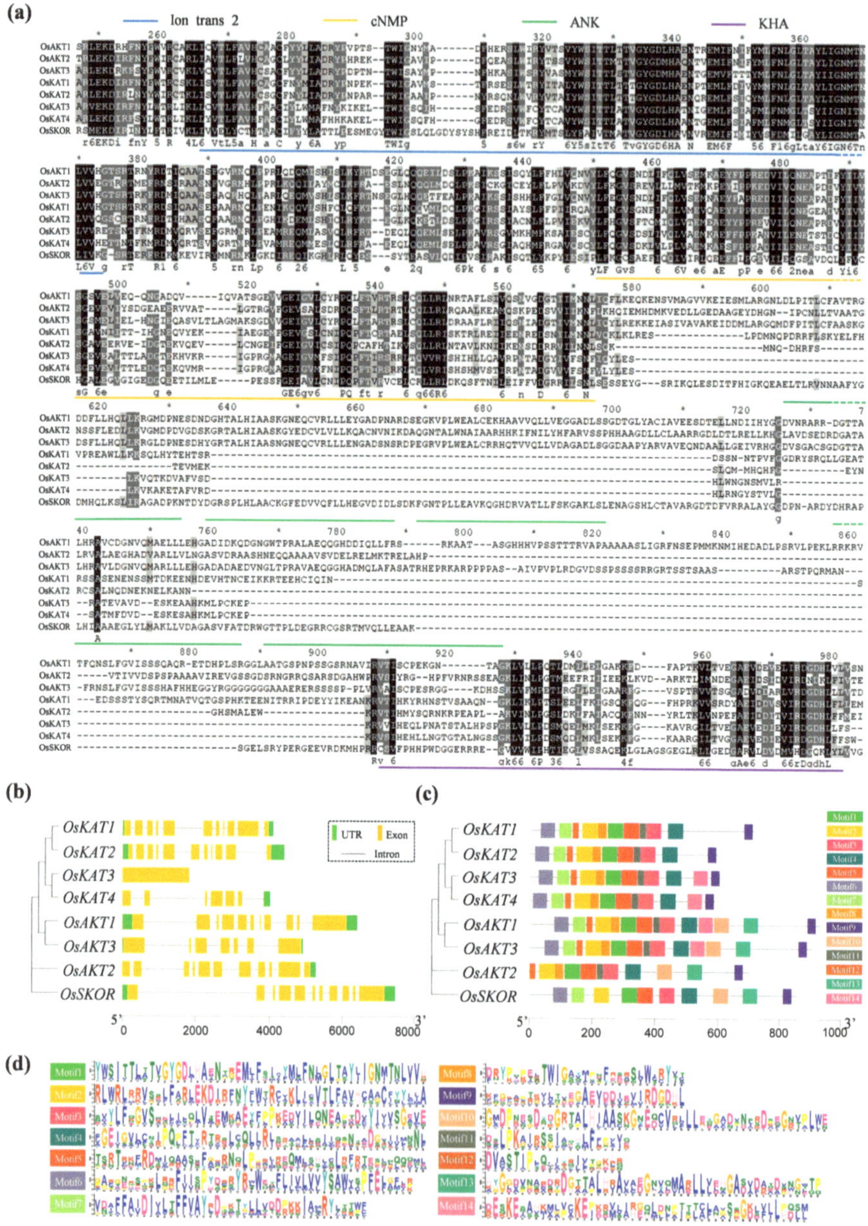

Figure 2. Analysis of protein sequence, gene structure and motifs of the *OsShaker* K^+ channel proteins. (**a**) Protein sequence comparison of the *OsShaker* K^+ channel, the underlining in different colors representing different structural domains. (**b**) Analysis of the gene structure of *OsShaker* K^+ channel genes, containing introns, exons, and UTR region, indicated by the black line, yellow and green rectangles, respectively. (**c**) Analysis of conserved motifs of the *OsShaker* K^+ channel proteins, different colored rectangles representing different motifs. (**d**) Sequence Logo of the *OsShaker* K^+ channel protein motifs. Different letters represent different amino acids.

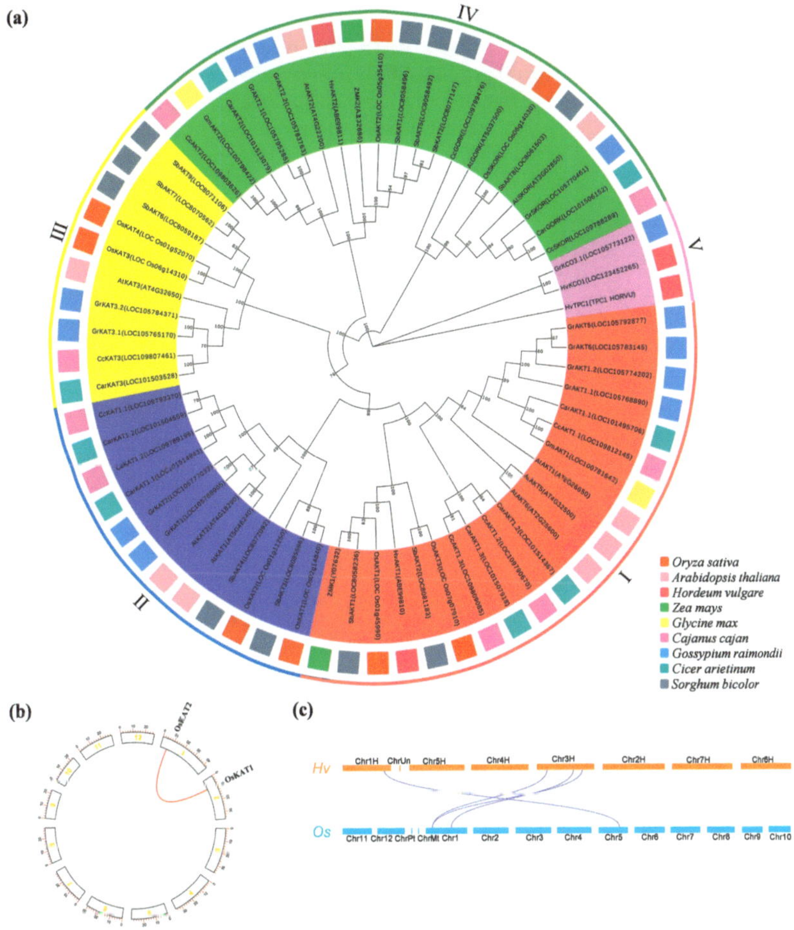

Figure 3. Phylogenetic, gene duplication and collinearity analysis of *OsShaker* K^+ channel genes. (**a**) Phylogenetic analysis of the *Shaker* gene in *Oryza sativa*, *Arabidopsis thaliana*, *Hordeum vulgare*, *Zea mays*, *Glycine max*, *Cajanus cajan*, *Gossypium raimondii*, *Cicer arietinum*, *Sorghum bicolor*. *Oryza sativa*, *Arabidopsis thaliana*, *Hordeum vulgare*, *Zea mays*, *Glycine max*, *Cajanus cajan*, *Gossypium raimondii*, *Cicer arietinum*, *Sorghum bicolor* are prefixed with Os, At, Hv, Zm, Gm, Cc, Gr, Car and Sb, respectively. (**b**) Gene segmental duplications of the *OsShaker* K^+ channel genes; the red curve represents one pair of segmental duplicated genes. The scale bar at the periphery of the chromosome represents the physical location (Kb). (**c**) Collinearity analysis of *OsShaker K*$^+$ *genes* in rice and barley.

Using TBtools, one pair of segmental duplication events was identified in the *OsShaker* K^+ channel family, including *OsKAT2* on chromosome 1 and *OsKAT1* on chromosome 2 (Figure 3b). In addition, interspecies collinearity analysis revealed that there are bonds of *Shaker* K^+ channel genes between rice and barley. Three *OsShaker* K^+ channel family members (*OsAKT1*, *OsAKT2*, and *OsKAT2*) and four barley *Shaker* gene family members had covariate duplications (Figure 3c, Supplementary Table S4), indicating that the *OsShaker* family and the barley *Shaker* family are closely related during the evolutionary process.

2.4. Protein Structures and Transmembrane Domain Analysis of OsShaker K$^+$ Channel Proteins

Four secondary structures of *OsShaker* K$^+$ channel proteins were characterized, i.e., alpha(α) helix, beta(β) turn, random coil and extended strand, and all were in the ratio of α-helix > random coil > extended strand > β-turn (Table 2). Accordingly, it can be noticed that α-helix and random coil are the main components of *OsShaker* K$^+$ channel proteins, while β-turn and extended strand are scattered throughout the protein sequence, playing an auxiliary role in modification. The three-dimensional structure of *OsShaker* K$^+$ channel proteins was analyzed, and the results are shown in Figure 4, indicating that the three-dimensional structures of *OsShaker* K$^+$ channel proteins show a similarity. As shown in Figure 5, the transmembrane domain analysis of *OsShaker* K$^+$ channel proteins showed that they all have transmembrane structural domains, and the predicted number of transmembrane helices is 3–6 (Figure 5, Supplementary Table S5). OsAKT2 contains three transmembrane domains, and OsKAT2, OsKAT3, OsKAT4 contain six transmembrane structural domains, while the remaining *OsShaker* K$^+$ channel proteins have five transmembrane structural domains (Figure 5).

Table 2. Secondary structure analysis of *OsShaker* K$^+$ channel proteins.

Protein Name	Percentage				Distribution of Secondary Structure Elements
	Ah	Bt	Rc	Es	
OsAKT1	40.43	7.38	37.86	14.33	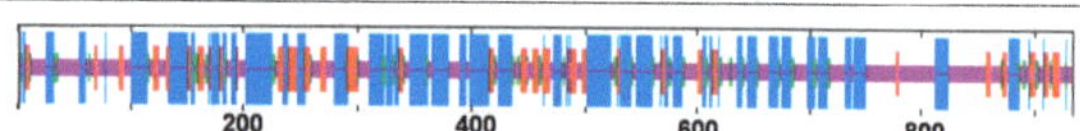
OsAKT2	43.10	7.11	33.71	16.07	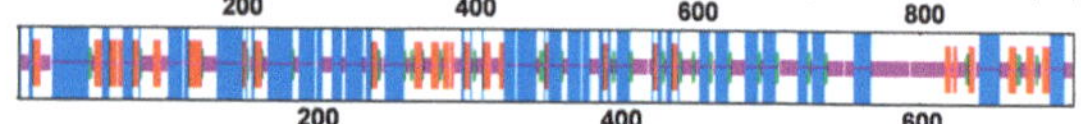
OsAKT3	40.79	7.50	35.50	16.21	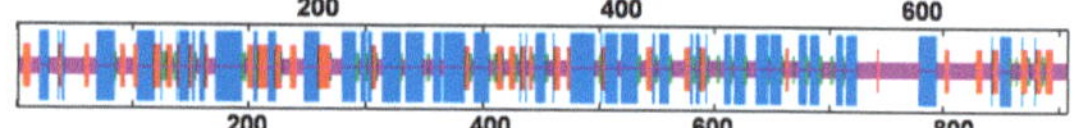
OsKAT1	41.36	4.32	38.02	16.30	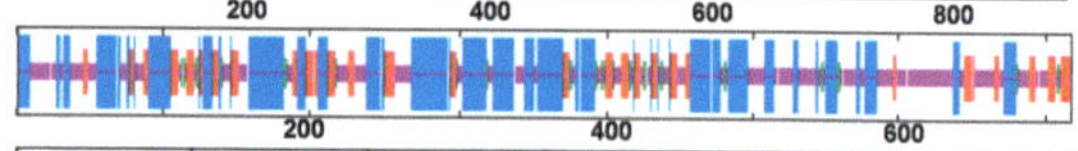
OsKAT2	47.92	4.99	28.79	18.30	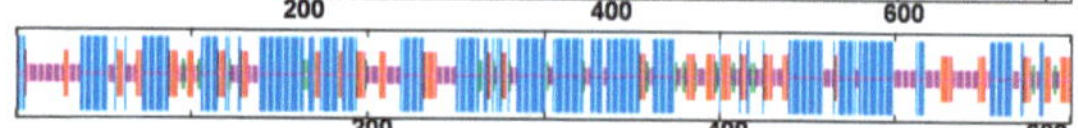
OsKAT3	46.39	4.75	30.00	18.85	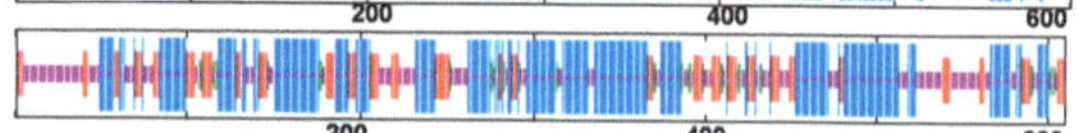
OsKAT4	45.36	5.90	29.51	19.22	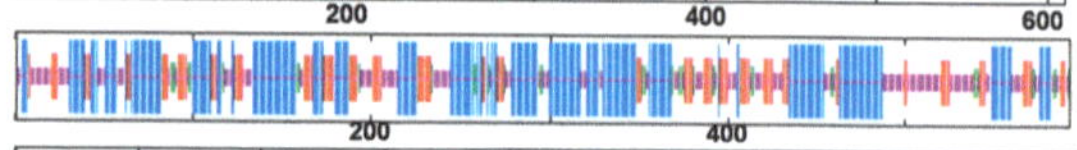
OsSKOR	44.76	8.28	30.89	16.08	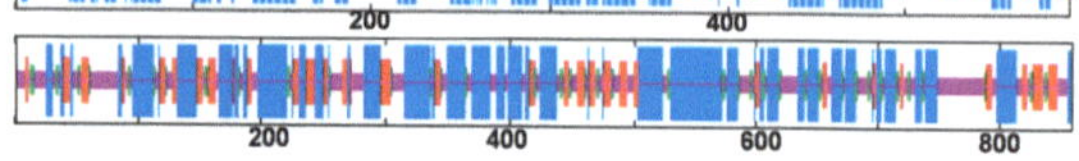

Note: In the secondary structure component distribution diagram, the blue structure represents Alpha helix (Ah), the green structure represents Beta turn (Bt), the yellow structure represents Random coil (Rc), and the red structure represents Extended strand (Es).

2.5. The Cis-Acting Element Analysis of OsShaker K$^+$ Channel Genes Promoter

The cis-acting elements in the promoter play a critical role in the regulation of gene expression. The analysis of cis-acting elements of *OsShaker* K$^+$ channel genes showed that the promoter sequences contained hormone-responsive cis-acting elements (eABRE, as-1, AuxRR-core, CGTCA-motif, ERE, GARE-motif, TATC-box, TGACG-box, TGACG-box, TGACG-box, TGACG-box, TGACG-box, TGACG-box, TGACG-box, TGACG-box, etc.), light-responsive cis-acting elements (AAGAA-motif, AE-box, Box4, G-Box, GATT-motif, ATC-motif, GT1-motif, Sp1, TCCC-motif, ACE, etc.), stress-responsive cis-acting elements (AP-1, ARE, DRE core, LTR, MYB, MYC, STRE, TC-rich repeats, W box, WRE3, etc.), promoter and site-binding-associated elements (CCAAT-box, MBS, MRE, CTAG-motif), and tissue and developmental elements (AAGAA-motif, ATC-motif, ATC-motif, ATC1-

motif, Sp1, TCCC-motif, ACE, etc.) (Figure 6, Supplementary Table S6). The results indicate that *OsShaker* K^+ channel genes may be regulated by hormones, light, and abiotic stresses. Among them, *OsAKT1* and *OsKAT2* contained AP-1 elements in response to cadmium stress, and *OsSKOR* contained LTR elements in response to stress. *OsKAT3* contained seven drought stress-responsive elements and *OsKAT4* contained eight cis-regulatory elements necessary for anaerobic induction.

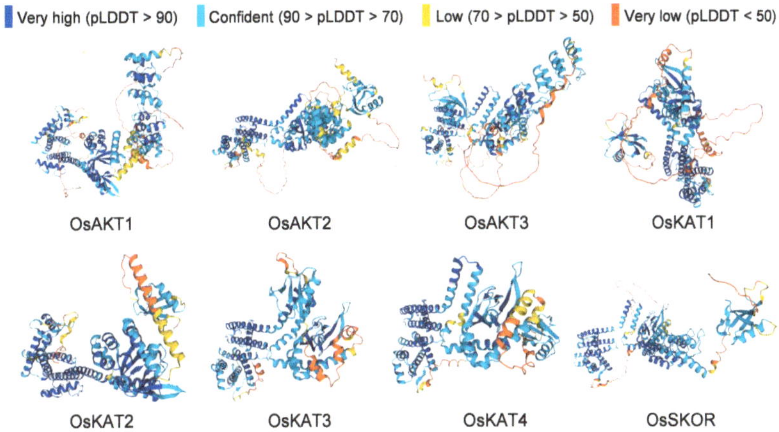

Figure 4. Protein 3D structure prediction model of *OsShaker* K^+ channel proteins. AlphaFold produces a per-residue confidence score (pLDDT) between 0 and 100.

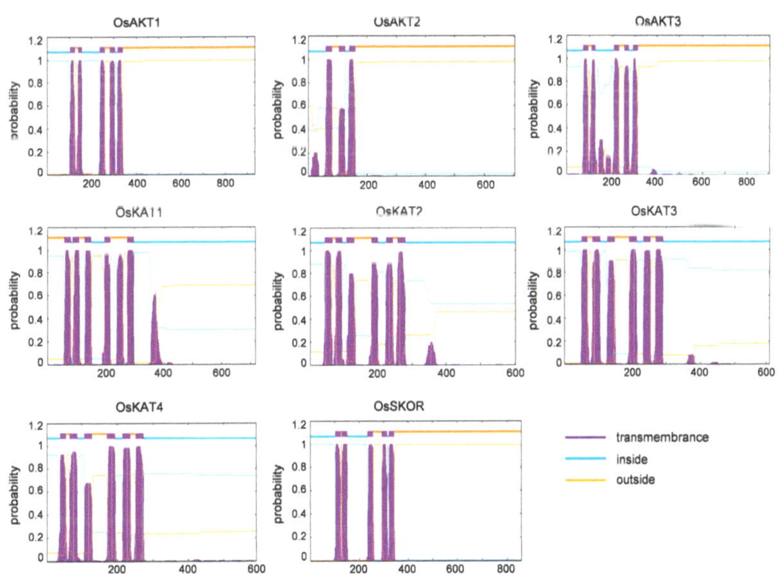

Figure 5. Transmembrane structure analysis of *OsShaker* K^+ channel proteins.

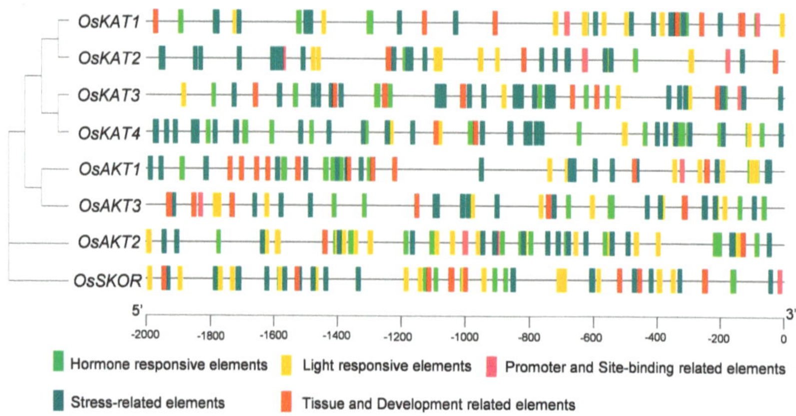

Figure 6. Prediction of cis-acting elements in *OsShaker* K^+ channel genes promoter. Promoter sequences (−2 Kb) of *OsShaker* K^+ channel genes were analyzed by PlantCARE. Different cis-elements are represented by different colors.

2.6. OsShaker K^+ Channel Proteins' PPI Network Analysis

The PPI network analysis of *OsShaker* K^+ channel proteins was performed using the Interacting Protein Prediction STRING website and Cytoscape software. As shown in Figure 7, OsAKT1, OsAKT2, OsAKT3 and OsSKOR proteins potentially interacted with OsCIPK23, which is the calcium-regulated phosphatase class B protein. OsCBL1 and OsCIPK23 have been identified as upstream regulators of OsAKT1. OsAKT2 interacts with OsAKT1, OsAKT3, OsKAT1, OsKAT2, OsKAT3, and OsKAT4, whereas the OsSKOR only interacted with OsCIPK23 and OsSAPK1 proteins. The results indicate that *OsShaker* K^+ channel proteins could perform different physiological functions by interacting with a variety of other proteins.

2.7. OsShaker K^+ Channel Gene Expression Profiling and Tissue Specificity Analysis

Hierarchical clustering of the expression of *OsShaker* K^+ channel genes in different organs and developmental periods obtained from the database on the RGAP website showed that *OsSKOR* had the highest expression in stamens and anthers, whereas *OsAKT1* and *OsAKT2* were found to be predominantly expressed in the shoots. *OsKAT3* and *OsKAT4* were poorly expressed in all tissues and at all developmental stages of rice. Tissue-specific analyses of roots and shoots at seedling stage, as well as roots, stems, leaves at tasseling stage of ZH11, showed that the expression of *OsAKT3* and *OsSKOR* was higher in roots at seedling stage, and the expression of *OsAKT1, OsAKT2, OsKAT1,OsKAT2,* and *OsKAT3* were higher in shoots at seedling stage. In conclusion the expression of *OsShaker* K^+ channel genes was spatio-temporally specific (Figure 8, Supplementary Table S7).

2.8. Expression Pattern of OsShaker K^+ Channel Genes under Salt Stress

In order to study the response of *OsShaker* K^+ channel genes to salt stress, the expression of *OsShaker* K^+ channel genes of rice seedlings was analyzed (Figure 9). In the root, *OsAKT1, OsAKT3, OsKAT1* and *OsKAT2* were up-regulated, while the expression level of *OsKAT4, OsSKOR* were down-regulated. The expression of *OsAKT1* was up-regulated by 2.11-fold compared with the control. In the shoot, the expression of *OsAKT2, OsKAT1, OsKAT2, OsKAT4* and *OsSKOR* was induced to be up-regulated, among which the expression of *OsSKOR* was up-regulated by 5.80-fold. In contrast, the *OsKAT3* expression was down-regulated by salt stress in shoot, with about 1.9-fold. In general, *OsShaker* K^+ channel genes' expression in response to salt stress suggests that *OsShaker* K^+ channel genes may have different functions under salt stress.

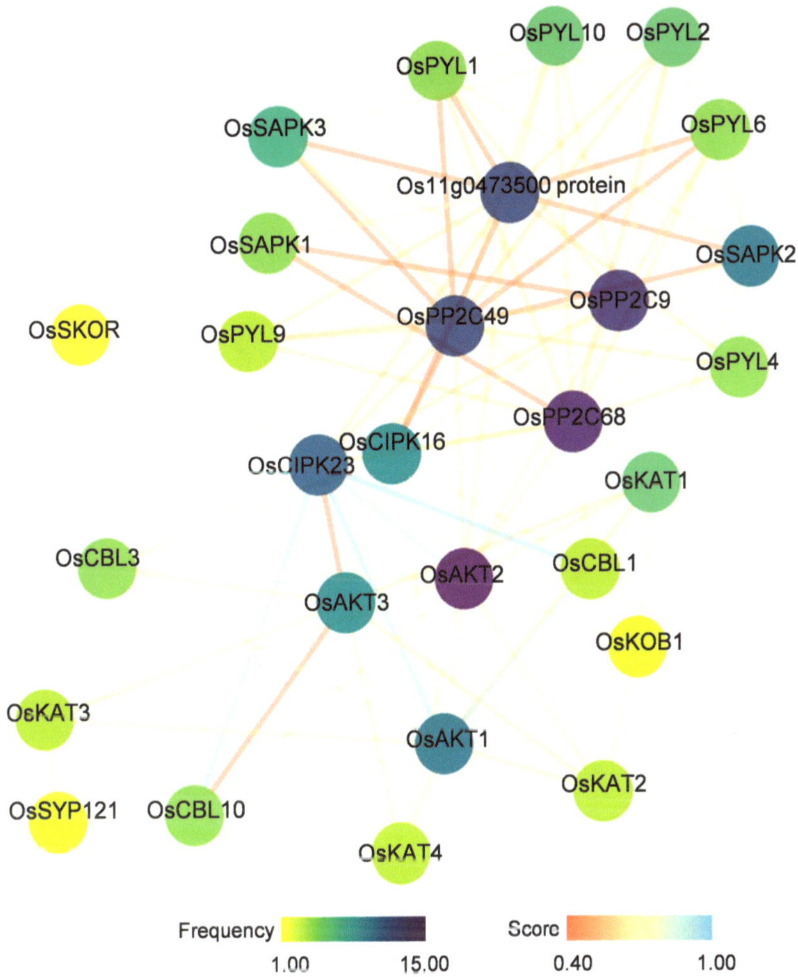

Figure 7. Protein–protein interaction (PPI) network analysis of *OsShaker* K^+ channel proteins. Nodes represent proteins with direct interactions, and colors represent interaction frequencies; edges represent different types of interactions, and colors represent the strength of interactions.

2.9. Expression Pattern of OsShaker K^+ Channel Genes under Chilling Treatment

To investigate the response of *OsShaker* K^+ channel genes to low temperature stress, after 12 hours of chilling stress treatment, the expression level of *OsShaker* K^+ channel genes in rice seedlings was analyzed by RT–qPCR. The results suggested that *OsAKT3*, *OsKAT1* and *OsKAT3* were induced, while *OsKAT2*, *OsKAT4*, *OsSKOR* were down-regulated in the shoot; however, *OsAKT1* and *OsAKT2* gene expression was not affected by cold stress. (Figure 10). Therefore, it can be concluded that *OsShaker* K^+ channel genes may also play an important role in response to low temperature.

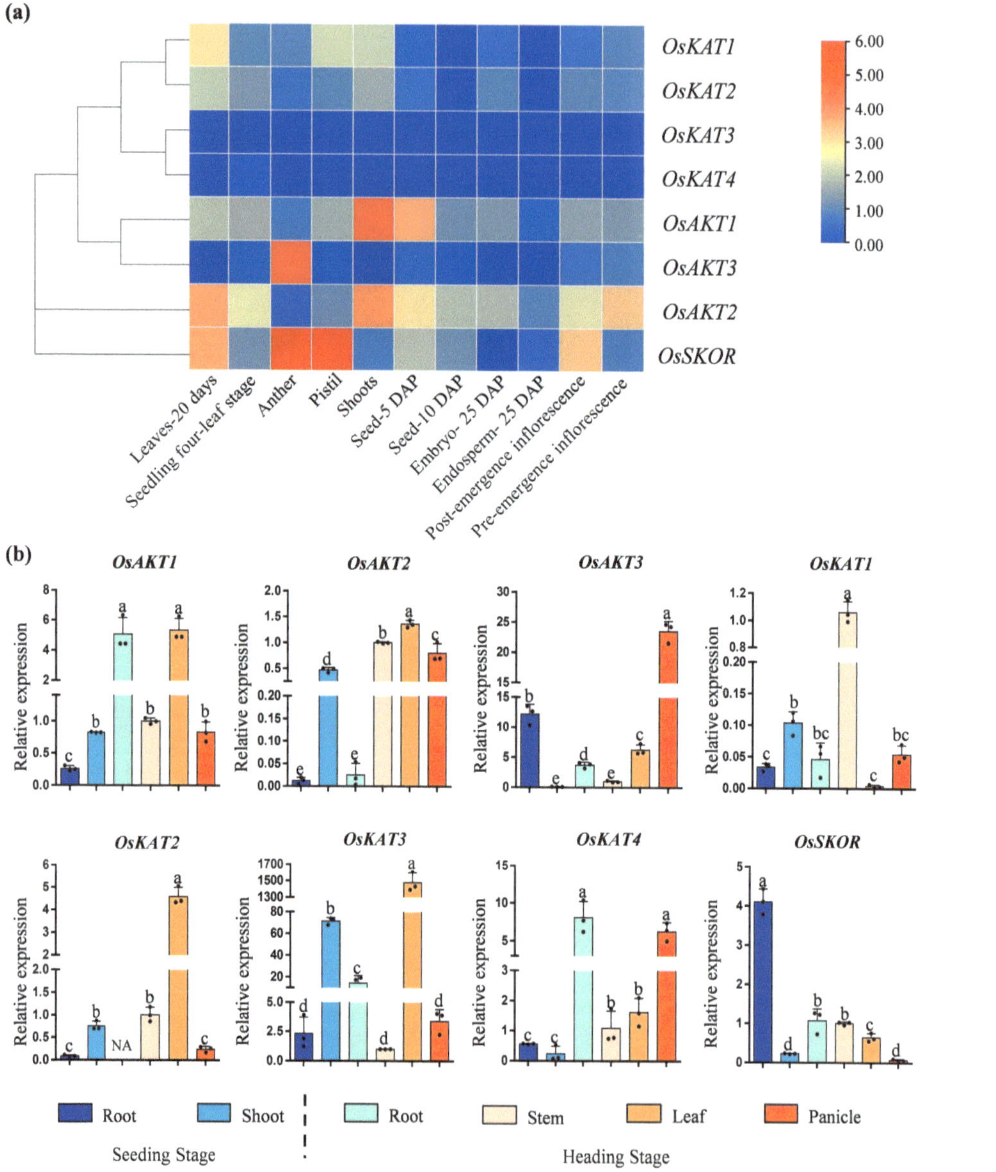

Figure 8. Tissue expression pattern of *OsShaker* K^+ channel genes. (**a**) The expression profiles of the *OsShaker* K^+ channel genes in the indica rice variety Minghui 63 obtained through the CREP database. The color scale represents relative expression levels from low (blue) to high (red). (**b**) The expression of *OsShaker* K^+ channel genes in root, shoot, stem, leaf, and panicle. *OsUBQ5* was used for internal parameter, and transcript abundance was normalized to Stem. The data represent means ± standard deviation (n = 3); different letters indicate significant differences.

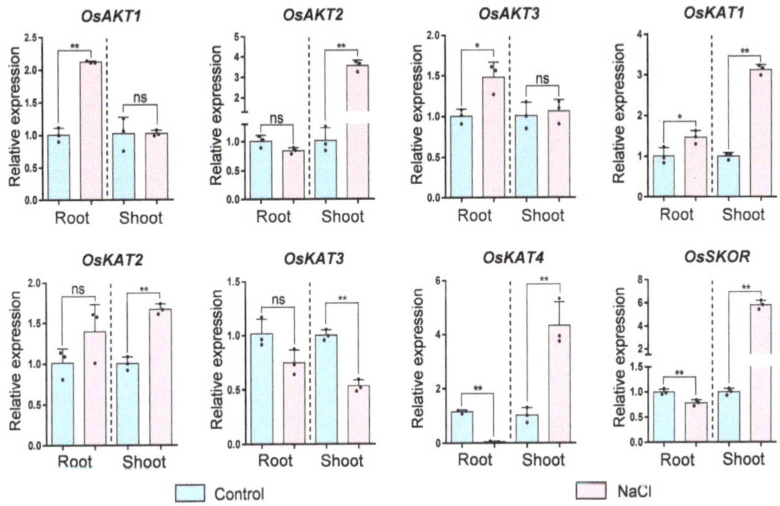

Figure 9. The expression analysis of *OsShaker* K^+ channel genes under salt stress. *OsUBQ5* was used for normalization, and transcript abundance was normalized to Control. The data represent means ± standard deviation ($n = 3$). Asterisks represents significant differences from Control (* $p < 0.05$; ** $p < 0.01$) by Student's *t*-test; ns indicates no significant difference.

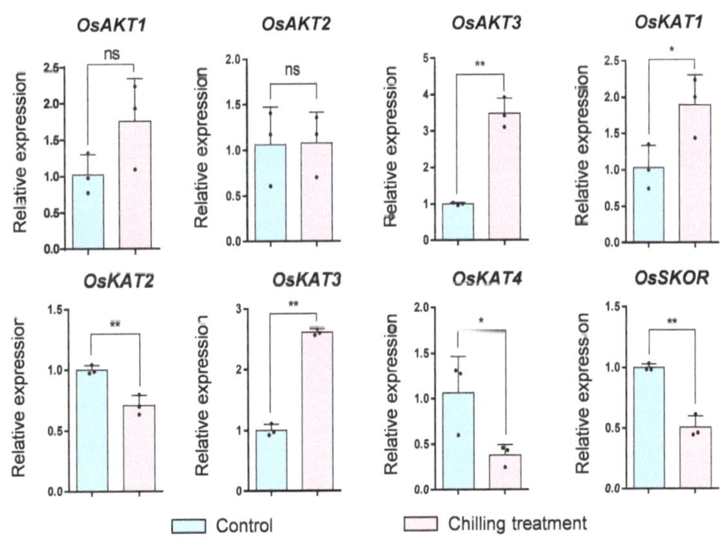

Figure 10. The expression analysis of *OsShaker* K^+ channel genes with chilling treatment. The data represent means ± standard deviation ($n = 3$). *OsUBQ5* was used for normalization, and transcript abundance was normalized to Control. Asterisks represents significant differences from Control (* $p < 0.05$; ** $p < 0.01$) by Student's *t*-test; ns indicates no significant difference.

3. Discussion

Soil salinity severely inhibits rice production. The intracellular Na^+/K^+ ratio is a key factor in determining the salt tolerance of plants [35]. Potassium channels are responsible for the uptake and transport of Na^+ and K^+ in plants, among which the *Shaker* family is the earliest identified potassium channel, which is involved in the regulation of plant salt tolerance and development [36]. The *Shaker* family has been systematically identified in

Arabidopsis thaliana, Hordeum vulgare, Cajanus cajan, Gossypium raimondii, etc. [31,37–42], but has been rarely studied in rice. In this study, 8 *OsShaker* K^+ channel genes were identified in the rice genome.

Gene structure and conserved motifs influence their function. In this study, most of the *OsShaker* K^+ channel genes were characterized as having multiple introns, but *OsKAT3* has no introns. Intronless genes are typically found in bacteria and primitive eukaryotes [43]. Thus, it is implied that *OsKAT3* is perhaps evolutionarily more conserved. In addition, the number and arrangement of motifs are the same in the same subfamily. For example, *OsAKT1* and *OsAKT3* have 15 motifs, and *OsAKT2* has only 7 motifs, suggesting that genes in the same subfamily may have similar functions. Repeatability and covariance analyses of the *OsShaker* K^+ channel genes demonstrated OsKAT1 and OsKAT2, a pair of fragment repeat events, both of which are potassium inward rectifier channel proteins. The predicted subcellular localization results showed that the *OsShaker* K^+ channel proteins were all localized in the cytoplasmic membrane, among which OsAKT2 was identified to be specifically localized in the plasma membrane [28,44].

Cis-acting elements in the promoter region are involved in the regulation of gene expression and can be used to predict biological functions, like stress response [45]. A variety of cis-acting elements, such as stress-responsive elements, light- and hormone-responsive elements, and tissue development-associated elements, were identified on the promoters of the *OsShaker* K^+ channel genes, suggesting that the *OsShaker* K^+ channel genes may be regulated by hormones, light, and adversity stresses, resulting in the regulation of rice growth, development, and response to adversity. Among them, *OsKAT3* contained seven drought stress-responsive elements, and *OsKAT4* contained eight cis-regulatory elements necessary for anaerobic induction. Analysis of interactions among *OsShaker* K^+ channel proteins showed that OsAKT2 interacts with six *OsShaker* K^+ channel members, except OsSKOR, which interacts with the ABA-responsive protein OsSAPK1. Moreover, the calmodulin phosphatase class B protein-interacting protein OsCIPK23, interacts with OsAKT1 [45]. In addition CIPK23 interactions with OsAKT1 have been experimentally demonstrated [46].

Potassium is one of the essential nutrients for plant growth and development, which can be taken up by plants from the environment through K^+ channels or transporter proteins in the root [47]. Transmembrane structure analysis of *OsShaker* K^+ channel proteins showed that they all have transmembrane structures, most of the proteins have 5–6 spans, and it is speculated that the *OsShaker* K^+ channel proteins may have the function of mediating the transmembrane transport of potassium ions. In rice, overexpression of *OsAKT1* increases potassium ion uptake in roots, whereas the mutant of the *OsAKT1* plant showed a significant reduction in potassium ion content in vivo and exhibited a potassium-deficient phenotype [48]. *OsAKT2* possesses the ability to mediate K^+ uptake in yeast and *Xenopus* oocytes [47], while AtAKT1 and AtHAK5 are responsible for K^+ uptake in *Arabidopsis* [49].

Shaker K^+ channel genes are widely involved in plant stomatal movement and drought stress response. In *Arabidopsis*, KAT1 and its homologous KAT2 are the primary inward rectifying channels in guard cells, mediating the influx of K^+ into these cells, leading to stomatal opening [50]. However, GORK is the only outward-rectifying K^+ channel in guard cells, which facilitate the efflux of K^+, resulting in stomatal closure [51,52]. AtAKT1 and AtAKT2 are involved in the drought stress response, and the *AKT1* mutant exhibits increased drought tolerance [53]. Additionally, HvAKT1 and HvAKT2 have also been identified as being involved in the regulation of drought tolerance in barley [54]. Here, four *KAT*, three *AKT* and one *SKOR* have been identified in the rice genome, among which *OsAKT1* has been reported to be involved in drought regulation [48] Gene expression under stress and the site of gene expression are closely related to gene function [45]. Tissue expression of *OsShaker* K^+ channel genes showed that *OsAKT2* and *OsKAT2* were mainly expressed in the shoot, whereas *OsAKT3* and *OsSKOR* were mainly expressed in the root. Among them, *OsAKT2* has been reported to mediate the distribution of K^+ among different leaves [27,28]. The expression of *OsShaker* K^+ channel genes was either up-regulated or down-regulated under salt stress, suggesting that these genes may be involved in the response process to salt stress in rice. The expression of *OsAKT1* was up-regulated in roots after salt stress treatment, as in the result of Golldack et al [55]. In this study, it was found that the expression of *OsAKT2* was significantly up-regulated in the shoot by salt stress, whereas the expression level in the roots was very low and no significant difference was obtained between normal condition and salt stress, which is consistent with the previous report [28]. Overexpression of *OsAKT2* improved salt tolerance in *Arabidopsis* by reabsorbing K^+ in the phloem sap and leaves [27]. In addition, in rice, *OsAKT2* promotes the reabsorption of K^+ from the phloem, controlling the distribution of shoot K^+ among different leaves, as well as increasing the K^+/Na^+ ratio in young leaves and other growth points to enhance the salt tolerance under salt stress [28]. The different expression sites and the responding path to salt stress of the *OsShaker* K^+ channel genes suggests that they may play different functions in different parts of the plant at different time points of the day during salt stress.

4. Materials and Methods

4.1. Gene Identification and Chromosomal Localization Analysis

The *Arabidopsis* Information Resource (TAIR) (http://www.arabidopsis.org/, accessed on 15 January 2024) was used to download the Arabidopsis *Shaker* protein sequences. The genome information of *OsShaker* K^+ channel genes was searched for in the InterPro database, based on the reported *Arabidopsis Shaker* family members. SMART(http://smart.embl-heidelberg.de/, accessed on 16 January 2024),an online tool, was used to identify the relevant domains (Ion_trans_2 structural domains, cNMP structural domains, ANK structural domains and KHA structural domains). *OsShaker* K^+ channel protein information wsd collected from Rice Date (https://www.ricedata.cn/gene/, accessed on 17 January 2024) and the Rice Genome Annotation Project (http://rice.plantbiology.msu.edu/index.shtml), and utilized TBtools II v2.031 [56] to create chromosomal localization maps.

4.2. Protein Sequence Comparison, Phylogenetic Tree and Gene Duplication Analysis

Sequence comparison and output of the conserved structural domains of *OsShaker* K^+ channel genes were accomplished by MEGA7.0 and geneDoc. The phylogenetic trees of rice, *Arabidopsis*, barley, soybean, and corn *Shaker* K^+ *channel* gene family were constructed in MEGA7.0 using the neighbor-joining method., and their amino acid sequences were download from the Ensembl Plants database (https://plants.ensembl.org/index.html, accessed on 22 January 2024) [57]. Analyses with 1000 replicates were conducted to evaluate the phylogenetic tree. The segmental duplication events of *Shaker* K^+ *channel* gene family proteins between rice and barley were analyzed using TBtools II v2.031. An all-vs-all BLASTP local search was performed among rice and barley to identify potential homologous gene pairs ($E < 1 \times 10^{-10}$). With homologous pairs as input, homologous chains were identified by MCScanX software. The results were presented in the form of Circos plots [56].

4.3. Gene Structure and Conserved Motif Analysis

The coding sequences (CDS) and their corresponding genomic DNA sequences of *OsShaker* K^+ channel genes were downloaded in FASTA format on the Ensembl Plants database (http://plants.ensembl.org/index.html, accessed on 2 February 2024) [57]. Search for the specified genotype number on the Home and click 'Export data', and then select GFF3 (Generic Feature Format Version 3) format Output. The distribution of introns and exons and non-coding regions of the genes was mapped using TBtools II v2.031. The online tool MEME (http://meme-suite.org/, accessed on 2 February 2024) was used to search for motifs (we found the most statistically significant (low E-value, E-value < 0.05) motifs), and the conserved motifs were beautified using TBtools II v2.031 [56].

4.4. Protein Physicochemical Properties and Subcellular Localization Analysis

Protein physicochemical properties were analyzed by ExPASy (https://web.expasy.org/protparam/, accessed on 23 January 2024). We perform the compute parameters after keying the protein sequence, then find the corresponding calculation result in the output interface. In addition, we made predictions about subcellular localization through the WoLF (https://wolfpsort.hgc.jp/, accessed on 23 January 2024), and it is important to note here that we first selected the plant option. Nuclear localization sequence analysis by NucPred (https://nucpred.bioinfo.se/nucpred/, accessed on 23 January 2024). The NucPred score was greater than 0.6, the nucleated localization sequence was indicated as "Positive".

4.5. Protein Structures and Transmembrane Structural Domain Analysis

The secondary and tertiary structures of the protein were analyzed through the SOPMA (https://npsa-prabi.ibcp.fr/cgi-bin/npsa_automat.pl?page=/NPSA/npsa_sopma_f.html, accessed on 12 February 2024) and the SWISS-MODEL (https://swissmodel.expasy.org/, accessed on 13 February 2024), respectively. Among them, the SOPMA website needs to enter the protein sequence, and the Swiss-model website can enter the protein sequence or protein abbreviation. The protein transmembrane structural domains were analysed by the online software tool TMHMM 2.0 (https://services.healthtech.dtu.dk/service.php?DeepTMHMM, accessed on 13 February 2024) [58]. It should be noted here that, after the input of the protein sequence, the following output format has to be selected: 'Extensive, with graphics'.

4.6. Cis-Acting Elements Analysis

Download 2000 bp upstream of the *OsShaker* K^+ channel genes on the Ensembl (http://plants.ensembl.org/index.html, accessed on 27 January 2024) as the promoter sequence. Select 'FASTA sequence' format Output in 'Export data'. The cis-acting elements predictions of the promoter were performed by PlantCare (http://bioinformatics.psb.ugent.be/webtools/plantcare/html/, accessed on 27 January 2024). After receiving the email from PlantCare, download the file in tab format for preliminary statistical analysis. Then the promoter cis-acting elements were mapped by TBtools II v2.031 software [56,59].

4.7. Plant Growth and Stress Treatments

The *Oryza sativa* cv. *japonica* cultivar Zhonghua 11 (ZH11) used in this study was obtained from the China National Rice Research Institute (Hangzhou, China). Seedlings were grown in an artificial climate chamber(Intelligent light incubator #GXM-508C-4, Ningbo Jiangnan Instrument Factory, China) under the following conditions: 14 h of light at 28 °C, 10 h of dark at 24 °C, 70% humidity, and a light intensity 18,000 lx.

Salt treatment: Two-week-old hydroponic seedlings were transferred to the nutrient solution with 140 mM NaCl for treatment, and the shoot and root were taken for gene expression analysis after 5 h of salt stress [28].

Chilling treatment: Two-week-old hydroponic seedlings were treated under cold stress (4 °C). The shoots of the seedlings were collected after 12 h of treatment for RT–qPCR [60].

4.8. RNA Extraction and RT–qPCR Analysis

The total RNA was extracted by using Plant Total RNA Kit (Tiangen Biochemical Technology Co., Ltd. Code: DP432, Beijing, China), and 1 µg RNA was reversed transcribed into cDNA by using HiScriptIII 1st Strand cDNA Synthesis Kit (Vazyme Code: R312-02). PCR amplification was performed using a CFX96 Real-Time PCR Detection System (Bio-Rad, Hercules, CA, USA). The reaction conditions were pre-denaturation at 95 °C for 5 min, denaturation at 94 °C for 30 s, annealing at 60 °C for 40 s, and extension at 72 °C for 20 s. Forty amplification cycles were performed, with three replicates, and the data were processed by $2^{-\Delta\Delta Ct}$ [61,62]. Primer Premier 5 was used to design primers for RT–qPCR, and *OsUBQ5* was selected as the internal reference gene.

4.9. Statistical Analysis

Microsoft Excel 2010 and GraphPad Prism 8.0 were used for data analysis and fig. construction. The general linear model procedure in SPSS 21.0 was used for analysis of variance (ANOVA). Mean values were compared using Duncan's multiple comparison procedure at the 1% or 5% level of probability and the single-sample *t*-test (* $p < 0.05$; ** $p < 0.01$).

5. Conclusions

In this study, genome-wide identification and expression pattern analysis of *OsShaker* K^+ channel genes were performed in rice. Eight *OsShaker* K^+ channel genes were identified in the rice genome and divided into five groups. Members of the same subfamily have similar genetic characteristics. All *OsShaker* K^+ channel proteins were predicted to localized in the cytoplasmic membrane. The promoters of *OsShaker* K^+ channel gene contain certain elements which respond to abiotic stresses. In addition, the RT–qPCR analysis suggested that the expression of *OsShaker* K^+ channel gene was significantly influenced by salt and chilling. According to the findings, it can be concluded that *OsShaker* K^+ channel genes may play an important role in salt stress and chilling stress response. Overall, the results of this study enriched the understanding of the *Shaker* K^+ channel family, providing a theoretical basis for in-depth research on their functions in the response to abiotic stress, as well as the cultivation of new abiotic-tolerant rice varieties.

Supplementary Materials: The following supporting information can be downloaded at: https://www.mdpi.com/article/10.3390/ijms25179728/s1.

Author Contributions: D.X. and Q.T. conceived and designed the project. T.Y., M.D. and Q.T. analyzed the data. Q.T., T.Y., M.D., Y.H. and Y.X., co-wrote the manuscript. D.X., X.Z., Q.T., J.Z., Y.F. and X.C. revised and edited the manuscript. All authors have read and agreed to the published version of the manuscript.

Funding: This study was supported by the China national key R&D program (2022YFE0125600, 2022YFE0139400), National Natural Science Foundation of China (Grant No 32301744), Hangzhou Scientific and Technological Major Project (202203A01), Zhejiang Provincial Natural Science Foundation of China (LY21C130007).

Institutional Review Board Statement: Not applicable.

Informed Consent Statement: Not applicable.

Data Availability Statement: Data are contained within the article and Supplementary Materials.

Conflicts of Interest: The authors declare no conflicts of interest.

References

1. Liu, C.; Mao, B.; Yuan, D.; Chu, C.; Duan, M. Salt tolerance in rice: Physiological responses and molecular mechanisms. *Crop J.* **2022**, *10*, 13–25. [CrossRef]
2. van Zelm, E.; Zhang, Y.; Testerink, C. Salt Tolerance Mechanisms of Plants. *Annu. Rev. Plant Biol.* **2020**, *71*, 403–433. [CrossRef] [PubMed]
3. Mukhopadhyay, R.; Sarkar, B.; Jat, H.S.; Sharma, P.C.; Bolan, N.S. Soil salinity under climate change: Challenges for sus-tainable agriculture and food security. *J. Environ. Manag.* **2021**, *280*, 111736. [CrossRef] [PubMed]
4. Anschütz, U.; Becker, D.; Shabala, S. Going beyond nutrition: Regulation of potassium homoeostasis as a common denomi-nator of plant adaptive responses to environment. *J. Plant Physiol.* **2014**, *171*, 670–687. [CrossRef]
5. Hasanuzzaman, M.; Bhuyan, M.H.M.B.; Nahar, K.; Hossain, M.S.; Mahmud, J.A.; Hossen, M.S.; Masud, A.A.C.; Moumita; Fujita, M. Potassium: A Vital Regulator of Plant Responses and Tolerance to Abiotic Stresses. *Agronomy* **2018**, *8*, 31. [CrossRef]
6. Wang, M.; Zheng, Q.; Shen, Q.; Guo, S. The Critical Role of Potassium in Plant Stress Response. *Int. J. Mol. Sci.* **2013**, *14*, 7370–7390. [CrossRef]
7. Ismail, A.M.; Horie, T. Genomics, Physiology, and Molecular Breeding Approaches for Improving Salt Tolerance. *Annu. Rev. Plant Biol.* **2017**, *68*, 405–434. [CrossRef] [PubMed]
8. Niu, X.; Bressan, R.A.; Hasegawa, P.M.; Pardo, J.M. Ion Homeostasis in NaCl Stress Environments. *Plant Physiol.* **1995**, *109*, 735–742. [CrossRef] [PubMed]

9. Fageria, N.K. *The Use of Nutrients in Crop Plants*; CRC Press: Boca Raton, FL, USA, 2016. [CrossRef]
10. Wang, C.; Xie, Y.; Tan, Z. Soil potassium depletion in global cereal croplands and its implications. *Sci. Total. Environ.* **2024**, *907*, 167875. [CrossRef]
11. MacRobbie, E.A. Control of Volume and Turgor in Stomatal Guard Cells. *J. Membr. Biol.* **2006**, *210*, 131–142. [CrossRef]
12. Walker, D.J.; Leigh, R.A.; Miller, A.J. Potassium homeostasis in vacuolate plant cells. *Proc. Natl. Acad. Sci. USA* **1996**, *93*, 10510–10514. [CrossRef] [PubMed]
13. Nieves, M.; Alemán, F.; Martínez, V.; Rubio, F. K+ uptake in plant roots. The systems involved, their regulation and paral-lels in other organisms. *J. Plant Physiol.* **2014**, *171*, 688–695. [CrossRef] [PubMed]
14. Gierth, M.; Mäser, P. Potassium transporters in plants–involvement in K+ acquisition, redistribution and homeostasis. *FEBS Lett.* **2007**, *581*, 2348–2356. [CrossRef] [PubMed]
15. Yang, T.; Zhang, S.; Hu, Y.; Wu, F.; Hu, Q.; Chen, G.; Cai, J.; Wu, T.; Moran, N.; Yu, L.; et al. The role of a potassium trans-porter OsHAK5 in potassium acquisition and transport from roots to shoots in rice at low potassium supply levels. *Plant Physiol.* **2014**, *166*, 945–959. [CrossRef] [PubMed]
16. Almeida, P.; Katschnig, D.; De Boer, A.H. HKT Transporters—State of the Art. *Int. J. Mol. Sci.* **2013**, *14*, 20359–20385. [CrossRef]
17. Cellier, F.; Conéjéro, G.; Ricaud, L.; Luu, D.T.; Lepetit, M.; Gosti, F.; Casse, F. Characterization of AtCHX17, a member of the cation/H+ exchangers, CHX family, from Arabidopsis thaliana suggests a role in K+ homeostasis. *Plant J.* **2004**, *39*, 834–846. [CrossRef]
18. Zhu, X.; Pan, T.; Zhang, X.; Fan, L.; Quintero, F.J.; Zhao, H.; Su, X.; Li, X.; Villalta, I.; Mendoza, I.; et al. K+ Efflux Antiporters 4, 5, and 6 Mediate pH and K+ Homeostasis in Endomembrane Compartments. *Plant Physiol.* **2018**, *178*, 1657–1678. [CrossRef]
19. Dabravolski, S.A.; Isayenkov, S.V. Recent updates on the physiology and evolution of plant TPK/KCO channels. *Funct. Plant Biol.* **2023**, *50*, 17–28. [CrossRef]
20. Lebaudy, A.; Véry, A.A.; Sentenac, H. K+ channel activity in plants: Genes, regulations and functions. *FEBS Lett.* **2007**, *581*, 2357–2366. [CrossRef]
21. Voelker, C.; Schmidt, D.; Mueller-Roeber, B.; Czempinski, K. Members of the Arabidopsis AtTPK/KCO family form homomeric vacuolar channels in planta. *Plant J.* **2006**, *48*, 296–306. [CrossRef]
22. Véry, A.-A.; Nieves-Cordones, M.; Daly, M.; Khan, I.; Fizames, C.; Sentenac, H. Molecular biology of K+ transport across the plant cell membrane: What do we learn from comparison between plant species? *J. Plant Physiol.* **2014**, *171*, 748–769. [CrossRef] [PubMed]
23. Ragel, P.; Raddatz, N.; Leidi, E.O.; Quintero, F.J.; Pardo, J.M. Regulation of K+ Nutrition in Plants. *Front. Plant Sci.* **2019**, *10*, 281. [CrossRef] [PubMed]
24. Véry, A.-A.; Sentenac, H. Molecular Mechanisms and Regulation of K^+ Transport in Higher Plants. *Annu. Rev. Plant Biol.* **2003**, *54*, 575–603. [CrossRef]
25. Lim, C.W.; Kim, S.H.; Choi, H.W.; Luan, S.; Lee, S.C. The Shaker Type Potassium Channel, GORK, Regulates Abscisic Acid Signaling in Arabidopsis. *Plant Pathol. J.* **2019**, *35*, 684–691. [CrossRef] [PubMed]
26. Obata, T.; Kitamoto, H.K.; Nakamura, A.; Fukuda, A.; Tanaka, Y. Rice Shaker Potassium Channel OsKAT1 Confers Tolerance to Salinity Stress on Yeast and Rice Cells. *Plant Physiol.* **2007**, *144*, 1978–1985. [CrossRef] [PubMed]
27. Huang, Y.N.; Yang, S.Y.; Li, J.L.; Wang, S.F.; Wang, J.J.; Hao, D.L.; Su, Y.H. The rectification control and physiological rele-vance of potassium channel OsAKT2. *Plant Physiol.* **2021**, *187*, 2296–2310. [CrossRef]
28. Tian, Q.; Shen, L.; Luan, J.; Zhou, Z.; Guo, D.; Shen, Y.; Jing, W.; Zhang, B.; Zhang, Q.; Zhang, W. Rice shaker potassium channel OsAKT2 positively regulates salt tolerance and grain yield by mediating K^+ redistribution. *Plant Cell Environ.* **2021**, *44*, 2951–2965. [CrossRef]
29. Zhou, J.; Nguyen, T.H.; Hmidi, D.; Luu, D.T.; Sentenac, H.; Véry, A.A. The outward shaker channel OsK5.2 improves plant salt tolerance by contributing to control of both leaf transpiration and K+ secretion into xylem sap. *Plant Cell Environ.* **2022**, *45*, 1734–1748. [CrossRef]
30. Li, J.; Gao, Z.; Zhou, L.; Li, L.; Zhang, J.; Liu, Y.; Chen, H. Comparative transcriptome analysis reveals K+ transporter gene contributing to salt tolerance in eggplant. *BMC Plant Biol.* **2019**, *19*, 1–18. [CrossRef]
31. Feng, C.; He, C.; Wang, Y.; Xu, H.; Xu, K.; Zhao, Y.; Yao, B.; Zhang, Y.; Zhao, Y.; Carther, K.F.I.; et al. Genome-wide identification of soybean Shaker K+ channel gene family and functional characterization of GmAKT1 in transgenic Arabidopsis thaliana under salt and drought stress. *J. Plant Physiol.* **2021**, *266*, 153529. [CrossRef]
32. Wang, X.; Zhao, J.; Fang, Q.; Chang, X.; Sun, M.; Li, W.; Li, Y. GmAKT1 is involved in K+ uptake and Na+/K+ homeostasis in Arabidopsis and soybean plants. *Plant Sci.* **2020**, *304*, 110736. [CrossRef]
33. Ma, Q.; Hu, J.; Zhou, X.R.; Yuan, H.J.; Kumar, T.; Luan, S.; Wang, S.M. ZxAKT1 is essential for K+ uptake and K+/Na+ homeo-stasis in the succulent xerophyte Zygophyllum xanthoxylum. *Plant J.* **2017**, *90*, 48–60. [CrossRef]
34. Chen, G.; Chen, Q.; Qi, K.; Xie, Z.; Yin, H.; Wang, P.; Wang, R.; Huang, Z.; Zhang, S.; Wang, L.; et al. Identification of Shaker K+ channel family members in Rosaceae and a functional exploration of PbrKAT1. *Planta.* **2019**, *250*, 1911–1925. [CrossRef]
35. Assaha, D.V.M.; Ueda, A.; Saneoka, H.; Al-Yahyai, R.; Yaish, M.W. The role of Na+ and K+ transporters in salt stress adaptation in glycophytes. *Front. Physiol.* **2017**, *8*, 509. [CrossRef] [PubMed]

36. Hedrich, R. Ion Channels in Plants. *Physiol. Rev.* **2012**, *92*, 1777–1811. [CrossRef]
37. Siddique, M.H.; Babar, N.I.; Zameer, R.; Muzammil, S.; Nahid, N.; Ijaz, U.; Masroor, A.; Nadeem, M.; Rashid, M.A.R.; Hashem, A.; et al. Genome-Wide Identification, Genomic Organization, and Characterization of Potassium Transport-Related Genes in *Cajanus cajan* and Their Role in Abiotic Stress. *Plants* **2021**, *10*, 2238. [CrossRef] [PubMed]
38. Azeem, F.; Zameer, R.; Rehman Rashid, M.A.; Rasul, I.; Ul-Allah, S.; Siddique, M.H.; Fiaz, S.; Raza, A.; Younas, A.; Rasool, A.; et al. Genome-wide analysis of potassium transport genes in Gossypium rai-mondii suggest a role of GrHAK/KUP/KT8, GrAKT2.1 and GrAKT1.1 in response to abiotic stress. *Plant Physiol. Biochem.* **2022**, *170*, 110–122. [CrossRef] [PubMed]
39. Azeem, F.; Ahmad, B.; Atif, R.M.; Ali, M.A.; Nadeem, H.; Hussain, S.; Manzoor, H.; Azeem, M.; Afzal, M. Genome-Wide Analysis of Potassium Transport-Related Genes in Chickpea (Cicer arietinum L.) and Their Role in Abiotic Stress Responses. *Plant Mol. Biol. Rep.* **2018**, *36*, 451–468. [CrossRef]
40. Kumar, S.A.; Kumari, P.H.; Nagaraju, M.; Reddy, P.S.; Dheeraj, T.D.; Mack, A.; Katam, R.; Kishor, P.B.K. Genome-wide identification and multiple abiotic stress transcript profiling of potassium transport gene homologs in Sorghum bicolor. *Front. Plant Sci.* **2022**, *13*, 965530. [CrossRef]
41. Philippar, K.; Büchsenschütz, K.; Abshagen, M.; Fuchs, I.; Geiger, D.; Lacombe, B.; Hedrich, R. The K+ Channel KZM1 Mediates Potassium Uptake into the Phloem and Guard Cells of the C4 Grass Zea mays. *J. Biol. Chem.* **2003**, *278*, 16973–16981. [CrossRef]
42. Boscari, A.; Clément, M.; Volkov, V.; Golldack, D.; Hybiak, J.; Miller, A.J.; Amtmann, A.; Fricke, W. Potassium channels in barley: Cloning, functional characterization and expression analyses in relation to leaf growth and development. *Plant Cell Environ.* **2009**, *32*, 1761–1777. [CrossRef] [PubMed]
43. Hausner, G.; Hafez, M.; Edgell, D.R. Bacterial group I introns: Mobile RNA catalysts. *Mob. DNA* **2014**, *5*, 8. [CrossRef] [PubMed]
44. Shen, L.; Tian, Q.; Yang, L.; Zhang, H.; Shi, Y.; Shen, Y.; Zhou, Z.; Wu, Q.; Zhang, Q.; Zhang, W. Phosphatidic acid directly binds with rice potassium channel OsAKT2 to inhibit its activity. *Plant J.* **2020**, *102*, 649–665. [CrossRef]
45. Yu, T.; Cen, Q.; Kang, L.; Mou, W.; Zhang, X.; Fang, Y.; Zhang, X.; Tian, Q.; Xue, D. Identification and expression pattern analysis of the OsSnRK2 gene family in rice. *Front. Plant Sci.* **2022**, *13*, 1088281. [CrossRef] [PubMed]
46. Sánchez-Barrena, M.J.; Chaves-Sanjuan, A.; Raddatz, N.; Mendoza, I.; Cortés, Á.; Gago, F.; González-Rubio, J.M.; Benavente, J.L.; Quintero, F.J.; Pardo, J.M.; et al. Recognition and Activation of the Plant AKT1 Potassium Channel by the Kinase CIPK23. *Plant Physiol.* **2020**, *182*, 2143–2153. [CrossRef]
47. Li, J.; Long, Y.; Qi, G.-N.; Li, J.; Xu, Z.-J.; Wu, W.-H.; Wang, Y. The Os-AKT1 Channel Is Critical for K+ Uptake in Rice Roots and Is Modulated by the Rice CBL1-CIPK23 Complex. *Plant Cell* **2014**, *26*, 3387–3402. [CrossRef]
48. Ahmad, I.; Mian, A.; Maathuis, F.J.M. Overexpression of the rice AKT1 potassium channel affects potassium nutrition and rice drought tolerance. *J. Exp. Bot.* **2016**, *67*, 2689–2698. [CrossRef]
49. Gierth, M.; Mäser, P.; Schroeder, J.I. The Potassium Transporter *AtHAK5* Functions in K+ Deprivation-Induced High-Affinity K+ Uptake and *AKT1* K+ Channel Contribution to K+ Uptake Kinetics in Arabidopsis Roots. *Plant Physiol.* **2005**, *137*, 1105–1114. [CrossRef]
50. Szyroki, A.; Ivashikina, N.; Dietrich, P.; Roelfsema, M.R.G.; Ache, P.; Reintanz, B.; Deeken, R.; Godde, M.; Felle, H.; Steinmeyer, R.; et al. KAT1 is not essential for stomatal opening. *Proc. Natl. Acad. Sci. USA* **2001**, *98*, 2917–2921. [CrossRef]
51. Adem, G.D.; Chen, G.; Shabala, L.; Chen, Z.H.; Shabala, S. GORK Channel: A Master Switch of Plant Metabolism? *Trends Plant Sci.* **2020**, *25*, 434–445. [CrossRef]
52. Eisenach, C.; Papanatsiou, M.; Hillert, E.K.; Blatt, M.R. Clustering of the K+ channel GORK of Arabidopsis parallels its gat-ing by extracellular K+. *Plant J.* **2014**, *78*, 203–214. [CrossRef]
53. Nieves-Cordones, M.; Caballero, F.; Martínez, V.; Rubio, F. Disruption of the Arabidopsis thaliana Inward-Rectifier K+ Channel AKT1 Improves Plant Responses to Water Stress. *Plant Cell Physiol.* **2011**, *53*, 423–432. [CrossRef] [PubMed]
54. Feng, X.; Liu, W.; Qiu, C.W.; Zeng, F.; Wang, Y.; Zhang, G.; Chen, Z.H.; Wu, F. HvAKT2 and HvHAK1 confer drought toler-ance in barley through enhanced leaf mesophyll H+ homoeostasis. *Plant Biotechnol. J.* **2020**, *18*, 1683–1696. [CrossRef] [PubMed]
55. Golldack, D.; Quigley, F.; Michalowski, C.B.; Kamasani, U.R.; Bohnert, H.J. Salinity stress-tolerant and -sensitive rice (Oryza sativa L.) regulate AKT1-type potassium channel transcripts differently. *Plant Mol. Biol.* **2003**, *51*, 71–81. [CrossRef] [PubMed]
56. Chen, C.; Chen, H.; Zhang, Y.; Thomas, H.R.; Frank, M.H.; He, Y.; Xia, R. TBtools: An Integrative Toolkit Developed for Inter-active Analyses of Big Biological Data. *Mol. Plant.* **2020**, *13*, 1194–1202. [CrossRef]
57. Kinsella, R.J.; Kähäri, A.; Haider, S.; Zamora, J.; Proctor, G.; Spudich, G.; Almeida, J.; Staines, D.; Derwent, P.; Kerhornou, A.; et al. Ensembl BioMarts: A hub for data retrieval across taxonomic space. *Database* **2011**, *2011*, bar030. [CrossRef]
58. Krogh, A.; Larsson, B.; von Heijne, G.; Sonnhammer, E.L. Predicting transmembrane protein topology with a hidden Mar-kov model: Application to complete genomes. *J. Mol. Biol.* **2001**, *305*, 567–580. [CrossRef]
59. Lescot, M.; Déhais, P.; Thijs, G.; Marchal, K.; Moreau, Y.; Van de Peer, Y.; Rouzé, P.; Rombauts, S. PlantCARE, a database of plant cis-acting regulatory elements and a portal to tools for in silico analysis of promoter sequences. *Nucleic Acids Res.* **2002**, *30*, 325–327. [CrossRef]
60. Zhang, M.; Zhao, R.; Huang, K.; Huang, S.; Wang, H.; Wei, Z.; Li, Z.; Bian, M.; Jiang, W.; Wu, T.; et al. The OsWRKY63–OsWRKY76–OsDREB1B module regulates chilling tolerance in rice. *Plant J.* **2022**, *112*, 383–398. [CrossRef]

61. Livak, K.J.; Schmittgen, T.D. Analysis of relative gene expression data using real-time quantitative PCR and the 2(-Delta Delta C(T)) Method. *Methods.* **2001**, *25*, 402–408. [CrossRef]
62. Rathor, P.; Borza, T.; Stone, S.; Tonon, T.; Yurgel, S.; Potin, P.; Prithiviraj, B. A Novel Protein from *Ectocarpus* sp. Improves Salinity and High Temperature Stress Tolerance in *Arabidopsis thaliana*. *Int. J. Mol. Sci.* **2021**, *22*, 1971. [CrossRef] [PubMed]

International Journal of
Molecular Sciences

MDPI

Article

The Waxy Gene Has Pleiotropic Effects on Hot Water-Soluble and -Insoluble Amylose Contents in Rice (*Oryza sativa*) Grains

Hongkai Wu *, Siyuan Wang and Min Wu

College of Advanced Agricultural Sciences, Zhejiang A&F University, Hangzhou 311300, China; wangsy@stu.zafu.edu.cn (S.W.); rice@zafu.edu.cn (M.W.)
* Correspondence: wuhk@zafu.edu.cn

Abstract: Rice (*Oryza sativa*) is a cereal crop with a starchy endosperm. Starch is composed of amylose and amylopectin. Amylose content (AC) is the principal determinant of rice quality, but varieties with similar ACs can still vary substantially in their quality. In this study, we analyzed the total AC (TAC) and its constituent fractions, the hot water-soluble amylose content (SAC) and hot water-insoluble amylose content (IAC), in two sets of related chromosome segment substitution lines of rice with a common genetic background grown in two years. We searched for quantitative trait loci (QTLs) associated with SAC, IAC, and TAC and identified one common QTL (*qSAC–6*, *qIAC–6*, and *qTAC–6*) on chromosome 6. Map-based cloning revealed that the gene underlying the trait associated with this common QTL is Waxy (*Wx*). An analysis of the colors of soluble and insoluble starch–iodine complexes and their λ_{max} values (wavelengths at the positions of their peak absorbance values) as well as gel permeation chromatography revealed that *Wx* is responsible for the biosynthesis of amylose, comprising a large proportion of the soluble fractions of the SAC. *Wx* is also involved in the biosynthesis of long chains of amylopectin, comprising the hot water-insoluble fractions of the IAC. These findings highlight the pleiotropic effects of *Wx* on the SAC and IAC. This pleiotropy indicates that these traits have a positive genetic correlation. Therefore, further studies of rice quality should use rice varieties with the same *Wx* genotype to eliminate the pleiotropic effects of this gene, allowing the independent relationship between the SAC or IAC and rice quality to be elucidated through a multiple correlation analysis. These findings are applicable to other valuable cereal crops as well.

Keywords: rice (*Oryza sativa*) quality; amylose content; amylopectin fine structure; quantitative trait locus (QTL) mapping; pleiotropy

Citation: Wu, H.; Wang, S.; Wu, M. The Waxy Gene Has Pleiotropic Effects on Hot Water-Soluble and -Insoluble Amylose Contents in Rice (*Oryza sativa*) Grains. *Int. J. Mol. Sci.* **2024**, *25*, 6561. https://doi.org/10.3390/ijms25126561

Academic Editor: Micaela Carvajal

Received: 30 April 2024
Revised: 9 June 2024
Accepted: 10 June 2024
Published: 14 June 2024

1. Introduction

Improving rice (*Oryza sativa*) quality is a principal aim of rice breeding. Starch accounts for approximately 90% of the dry weight of the grain. The type of starch present in rice grain strongly influences quality [1]. The starch in rice endosperm is composed of two types of glucose polymers: the linear, slightly branched polymer amylose and the highly branched polymer amylopectin (AP) [2]. Amylose content (AC), as a percentage of flour weight rather than starch itself, is widely used to evaluate rice quality. Based on extensive studies, the AC is generally considered to be the principal determinant of rice quality [3]; however, AC alone cannot account for the difference in quality between varieties with similar ACs [4].

Bhattacharya and colleagues observed that the hot water-insoluble amylose content (IAC) differs markedly among different rice varieties, even when the total AC (TAC) is in the same range [4–6]. Juliano and colleagues observed that low amylose solubility was associated with a hard gel consistency and low breakdown [7] and very high setback values and faster retrogradation when measured with a Brabender viscograph [8].

The relationship between IAC and rice quality (texture and various other properties) has been thoroughly studied. The stickiness and consistency of cooked rice are strongly correlated with the IAC rather than the TAC [9], as are the viscograph characteristics [10].

The textural properties of cooked rice cannot be explained by the TAC alone, but they are well correlated with the IAC [11]. In addition, the retrogradation degree (%R) was found to be correlated with the IAC [12]. Thus, the IAC rather than the TAC is the key factor determining rice quality [9,10].

In recent decades, many studies have focused on the genetic control of AC. The Waxy (*Wx*) gene, which underlies a major quantitative trait locus (QTL) for AC on chromosome 6 [13,14], encodes granule-bound starch synthase I (GBSSI), which is responsible for amylose biosynthesis in the endosperm [15]. Several genes that give a dull grain phenotype were found to affect the splicing efficiency of the Wx^b allele, which results in low GBSSI activity [16–19]. An additional 43 QTLs nonallelic to the *Wx* locus have been detected throughout the rice genome (https://archive.gramene.org/db/qtl, accessed on 1 December 2023). To date, however, no QTL or gene controlling the IAC has been reported.

In this study, we aimed to identify QTLs for IAC in rice. We leached starch granules from rice flour in a boiling-water bath and separated the sample flour into the hot water-soluble (HWS) and hot water-insoluble (HWI) fractions, allowing us to separate the TAC into the hot water-soluble AC (SAC) and the IAC for further study. We conducted QTL analysis for TAC, SAC, and IAC using two sets of related chromosome segment substitution lines (CSSLs) of rice with a japonica rice cv. Koshihikari (Kosh) as a common genetic background and an indica rice cv. Kasalath or NonaBokra as a donor parent grown in two years. Our findings highlight the pleiotropic effects of *Wx* on SAC and IAC and suggest future avenues of research to further explore and improve grain quality in rice and other crops with starchy endosperm.

2. Results

2.1. Variation in the TAC, SAC, and IAC in Two Sets of CSSLs

We investigated the TAC, SAC, and IAC in two sets of CSSLs grown in two years: 2016 and 2017. Among the Kosh–Kasalath CSSL (KK–CSSL, chromosome segment substitution lines with Koshihikari as the background parent and Kasalath as the donor parent) plants (Figure 1), SL216 had the highest mean TAC, SAC, and IAC values in both 2016 and 2017: TAC (22.19%), SAC (17.39%), and IAC (10.14%) in 2016 and TAC (22.46%), SAC (19.88%), and IAC (10.85%) in 2017. These values were significantly higher than the values of the receptor parent Kosh ($p < 0.01$), although the mean values of TAC, SAC, and IAC in the other lines were similar to those of Kosh, ranging from 11.6% to 16.75%, 8.17% to 12.3%, and from 3.43% to 3.68%, respectively, in 2016, and 10.61% to 16.98%, 7.65% to 13.35%, and 4.06% to 5.84%, respectively, in 2017.

Among the Kosh–NonaBokra CSSL (KN–CSSL, chromosome segment substitution lines with Koshihikari as the background parent and NonaBokra as the donor parent) plants (Figure 2), SL519 and SL520 had significantly higher TAC, SAC, and IAC values than Kosh ($p < 0.01$), with mean values of 22.27%, 19.95%, and 8.29%, respectively, in 2016, and 23.63%, 20.29%, and 9.82%, respectively, in 2017. However, the mean values of TAC, SAC, and IAC in the other lines were similar to those of Kosh, ranging from 10.58% to 16.98%, 7.22% to 13.42%, and 3.68% to 6.84%, respectively, in 2016, and from 10.99% to 16.65%, 7.17% to 13.7%, and 3.8% to 6.71%, respectively, in 2017. These results indicate that QTLs for TAC, SAC, and IAC likely exist in the substituted segment of SL216, SL519, and SL520.

2.2. Mapping and Identifying QTLs for TAC, SAC, and IAC

In the KK–CSSL set, one common QTL was found to be associated with all three traits (*qTAC–6–1*, *qSAC–6–1*, and *qIAC–6–1* for TAC, SAC, and IAC, respectively). This QTL is located in the interval of the distal end–G200 on chromosome 6. By contrast, no QTL controlling any of these traits was detected in other genomic regions in either year of the study (Table 1). Similarly, in the KN–CSSL set, one common QTL (*qTAC–6–2*, *qSAC–6–2*, and *qIAC–6–2*) was detected in the interval of the distal end–RM5963 on chromosome 6, while no QTL controlling the TAC, SAC, or IAC was detected in other genomic regions in either year (Table 1). The shared genomic regions of these QTLs for each CSSL set suggest that the same genomic region controls the TAC, SAC, and IAC.

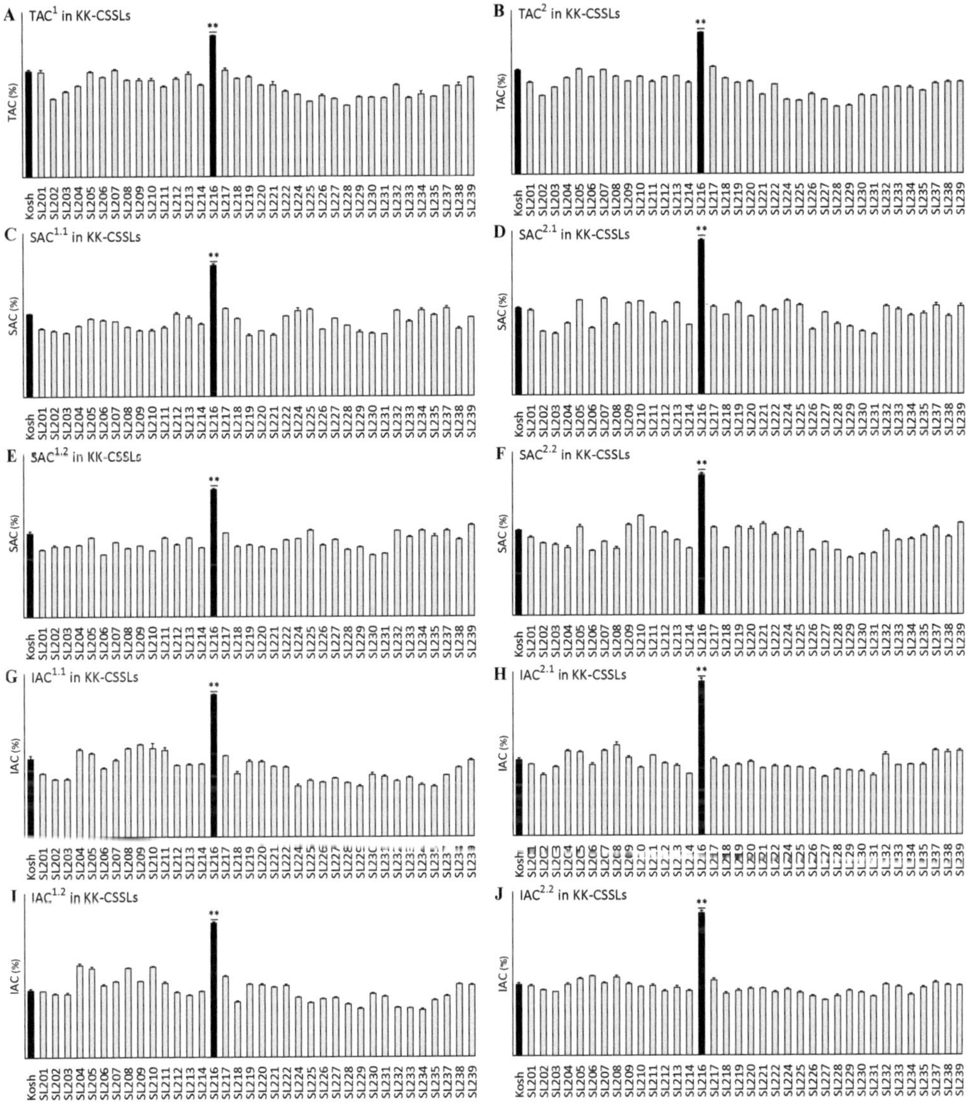

Figure 1. Variation in the total amylose content (TAC), hot water-soluble amylose content (SAC), and hot water-insoluble amylose content (IAC) in the KK-CSSLs including Kosh (Koshihikari, background parent). (**A**,**C**,**E**,**G**,**I**) The KK-CSSLs plants were grown in 2016 and the mean values of TAC, SAC, and IAC exhibited abundant variation, ranging from 11.6% to 16.75%, 8.17% to 12.3%, and 3.43% to 3.68%, respectively. (**B**,**D**,**F**,**H**,**J**) The KK-CSSLs plants were grown in 2017 and the mean values of TAC, SAC, and IAC exhibited abundant variation, ranging from 10.61% to 16.98%, 7.65% to 13.35%, and 4.06% to 5.84%, respectively. The superscript numbers for TAC, SAC, and IAC indicate 2016 or 2017 ("1" or "2" before the decimal point, respectively) and the first or second measurements ("1" or "2" after the decimal point, respectively). For instance, $SAC^{1.2}$ represents the second-measured SAC in 2016. Black bars represent the background parent (Kosh) or line with significantly higher TAC, SAC, or IAC than Kosh using an analysis of variance and the least significance difference test (**, $p < 0.01$). KK-CSSLs, chromosome segment substitution lines with Koshi as the background parent and Kasalath as the donor parent.

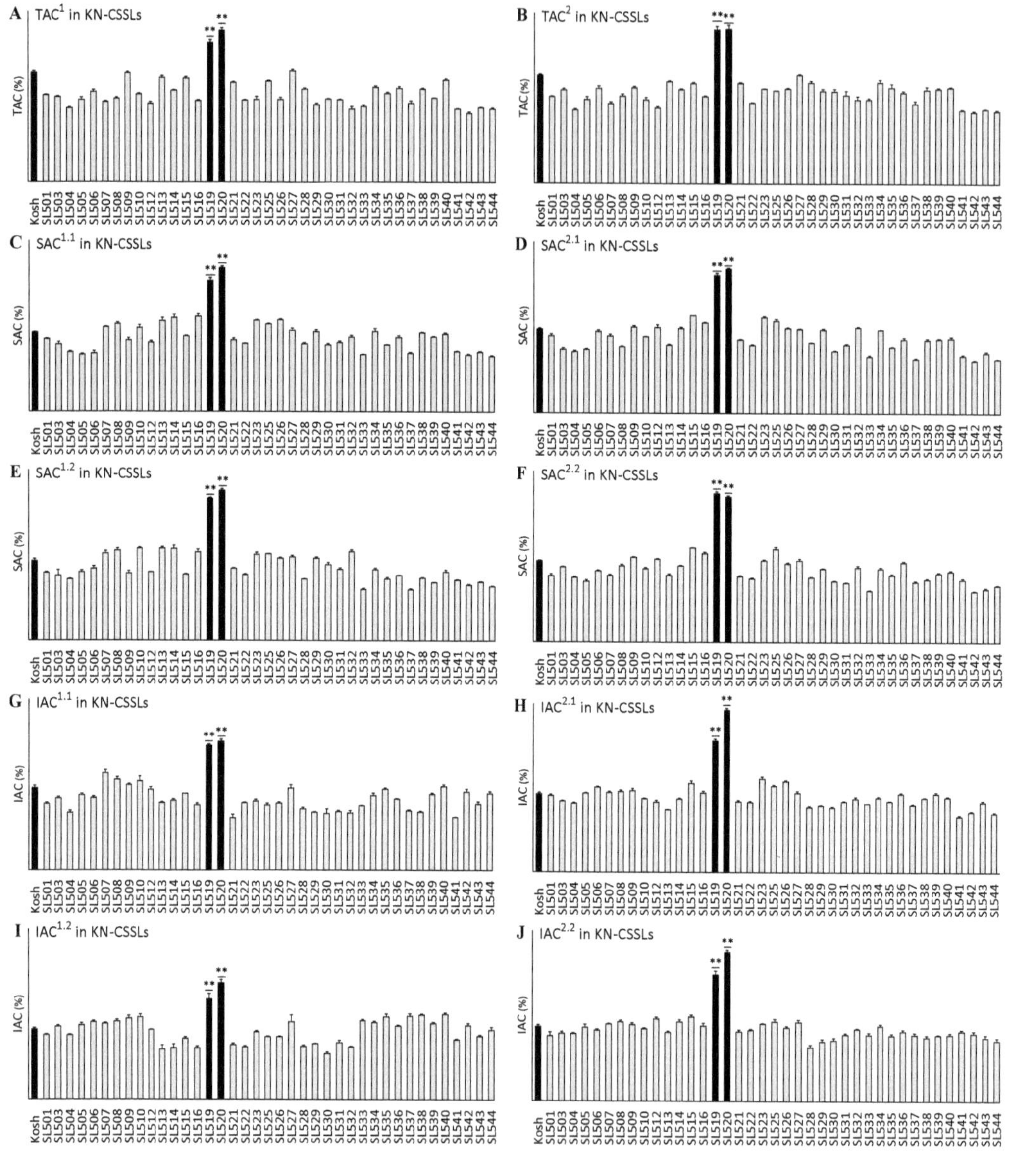

Figure 2. Variation in total amylose content (TAC), hot water-soluble amylose content (SAC), and hot water-insoluble amylose content (IAC) in the KN-CSSLs including Kosh (Koshihikari, background parent). (**A**,**C**,**E**,**G**,**I**) The KN-CSSLs plants were grown in 2016 and the mean values of TAC, SAC, and IAC exhibited abundant variation, ranging from 10.58% to 16.98%, 7.22% to 13.42%, and 3.68% to 6.84%, respectively. (**B**,**D**,**F**,**H**,**J**) The KN-CSSLs plants were grown in 2017 and the mean values of TAC, SAC, and IAC exhibited abundant variation, ranging from 10.99% to 16.65%, 7.17% to 13.7%, and 3.8% to 6.71%, respectively. The superscript numbers for TAC, SAC, and IAC are described in the Figure 1 legend. Black bars represent the background parent (Kosh) or lines with significantly higher TAC, SAC, and IAC than Kosh using an analysis of variance and the least significance difference test (**, $p < 0.01$). KN-CSSLs, chromosome segment substitution lines with Kosh as the background parent and NonaBokra as the donor parent.

Table 1. Detection of quantitative trait loci (QTLs) associated with total amylose content (TAC), hot water-soluble amylose content (SAC), and hot water-insoluble amylose content (IAC). Here, 2016.1 and 2016.2 are the first and second detection in plants grown in 2016, respectively; here, 2017.1 and 2017.2 are the first and second detection in plants grown in 2017, respectively. LOD, logarithm of odds; ADD, denotes the additive effect of substituting a donor (Kasalath or NonaBokra) allele for the corresponding receptor (Kosh) allele; M(QQ) and M(qq), mean values of the QQ genotype from the donor and qq genotype from the receptor at the QTL; PVE, phenotype variation explained by the QTL.

Populations	Traits	Years	QTLs	Chr.	Intervals	LOD	M(QQ) (%)	M(qq) (%)	ADD (%)	PVE (%)
KK-CSSL	TAC	2016	*qTAC-6-1*	6	end-G200	3.04	22.19	14.04	4.07	41.7
		2017	*qTAC-6-1*	6	end-G200	3.78	22.46	13.70	4.38	48.8
	SAC	2016.1	*qSAC-6-1*	6	end-G200	3.61	17.63	9.95	3.84	47.2
		2016.2	*qSAC-6-1*	6	end-G200	3.15	17.14	9.74	3.70	42.7
		2017.1	*qSAC-6-1*	6	end-G200	4.69	20.77	10.59	5.09	45.1
		2017.2	*qSAC-6-1*	6	end-G200	3.71	18.99	10.07	4.46	37.8
	IAC	2016.1	*qIAC-6-1*	6	end-G200	4.77	10.65	5.74	2.45	41.0
		2016.2	*qIAC-6-1*	6	end-G200	3.80	9.84	4.60	2.62	49.0
		2017.1	*qIAC-6-1*	6	end-G200	10.57	11.98	5.91	3.03	66.0
		2017.2	*qIAC-6-1*	6	end-G200	14.35	10.43	4.91	2.76	84.1
KN-CSSL	TAC	2016	*qTAC-6-2*	6	end-RM5963	6.37	22.27	13.02	4.62	57.8
		2017	*qTAC-6-2*	6	end-RM5963	8.64	23.63	13.99	4.82	70.1
	SAC	2016.1	*qSAC-6-2*	6	end-RM5963	4.32	19.30	10.18	4.56	44.3
		2016.2	*qSAC-6-2*	6	end-RM5963	4.56	20.59	10.02	5.29	46.1
		2017.2	*qSAC-6-2*	6	end-RM5963	7.00	20.71	9.45	5.63	56.3
		2017.1	*qSAC-6-2*	6	end-RM5963	5.34	19.87	9.93	4.97	46.8
	IAC	2016.1	*qIAC-6-2*	6	end-RM5963	5.33	8.92	4.89	2.01	44.1
		2016.2	*qIAC-6-2*	6	end-RM5963	3.57	7.66	4.82	1.42	38.5
		2017.1	*qIAC-6-2*	6	end-RM5963	14.28	11.15	5.80	2.67	75.5
		2017.2	*qIAC-6-2*	6	end-RM5963	14.70	9.72	4.93	2.39	82.4

To finely map the QTL on chromosome 6, we used polymorphic simple sequence repeat (SSR) markers to construct linkage maps for the KK–CSSLs (Figure 3A) and KN–CSSLs (Figure 3B). Using substitution mapping, *qTAC–6*, *qSAC–6*, and *qIAC–6* were mapped to the interval of the distal end–RM7088 (0–6,649,590 bp) on chromosome 6 in the KK–CSSLs (Figure 3A) and of the interval at the distal end-RM587 (0–2,292,076 bp) on this chromosome in the KN–CSSLs (Figure 3B). These regions overlap with the known *Wx* locus (1,765,622–1,770,653 bp) (https://rapdb.dna.affrc.go.jp/, accessed on 15 December 2023); therefore, we reasoned that the common QTL might be the *Wx* locus.

Three major alleles of the *Wx* gene are present in cultivated rice: Wx^a, Wx^b, and *wx*. Wx^a, which confers high AC, is mainly distributed in *indica* rice. This allele contains a G base at the 5′ first splice donor site of the first intron. Wx^b, which confers low AC, is mainly distributed in *japonica* rice. This allele has a G-to-T mutation at the 5′ first splice donor site of the first intron, which decreases its expression relative to Wx^a [20,21]. *wx* is a loss-of-function allele that produces waxy (opaque) endosperm lacking amylose, which is distinct from the semitransparent endosperm of rice containing Wx^a or Wx^b. The G/T polymorphism can be genotyped using the cleaved amplified polymorphic sequence marker PCR–AccI [22]. We used the PCR–AccI marker to genotype Kosh and the CSSLs (SL216, SL519, and SL520, each harboring the common QTL), which revealed that Kosh contains the +1T nucleotide while SL216, SL519, and SL520 each contain the +1G nucleotide. The G/T polymorphism suggests that the common QTL is identical to the *Wx* locus.

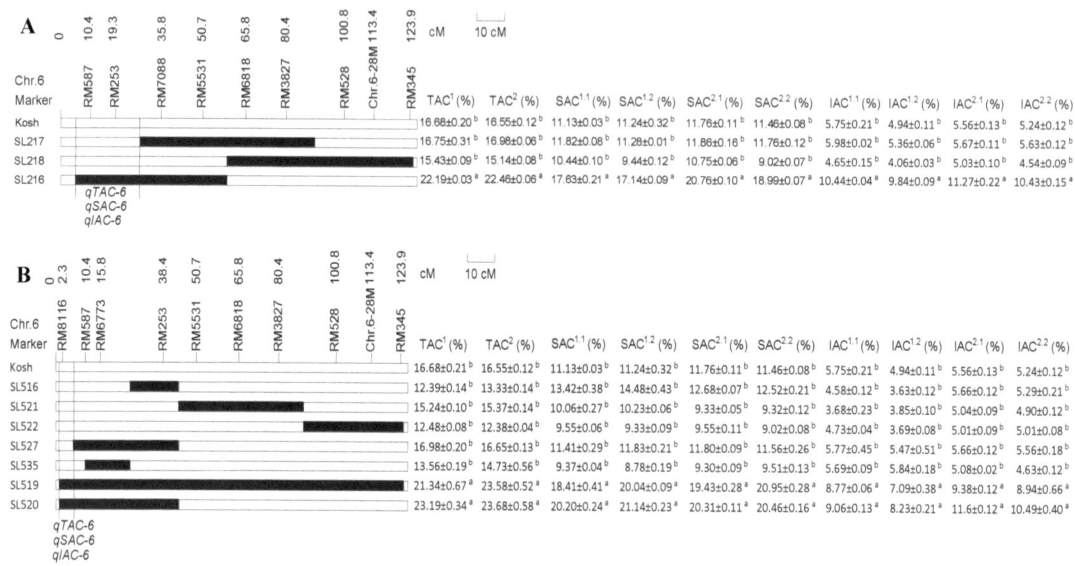

Figure 3. Substitution mapping of the quantitative trait loci (QTLs) *qTAC-6*, *qSAC-6*, and *qIAC-6* on chromosome 6. (**A**) The *qTAC–6*, *qSAC–6*, and *qIAC–6* were mapped to the interval of the distal end–RM7088 on chromosome 6 in the KK–CSSLs. (**B**) The *qTAC–6*, *qSAC–6*, and *qIAC–6* were mapped to the interval of the distal end–RM587 on chromosome 6 in the KN–CSSLs. The solid bars represent chromosome segments from the donor parent; the open bars represent chromosome segments from the background parent (Kosh). Different lowercase letters indicate that the mean values significantly differ ($p < 0.01$), while the same letter indicates no significant difference at $p > 0.1$ using an analysis of variance and the least significance difference test. The superscript numbers for TAC, SAC, and IAC are described in the Figure 1 legend.

To validate this notion, we developed an F_2 subpopulation by crossing SL520 and Kosh and assessed three markers, 90E3H (1,760,297–1,760,569 bp, upstream of the *Wx* locus), RM190 (1,765,625–1,765,746 bp, inside the *Wx* locus), and L17A (1,770,738–1,770,976 bp, downstream of the *Wx* locus) [23], all of which were polymorphic between SL520 and Kosh. Subsequently, we screened this population for recombinants between the three polymorphic marker loci and obtained two homozygous recombinants (SL520–219 and SL520–873) (Figure 4A). We phenotyped SL520–219 and SL520–873, along with SL520 and Kosh, for SAC and IAC, revealing no significant differences in these traits among SL520, SL520–219, and SL520–873, although all lines had significantly higher SAC and IAC than Kosh ($p < 0.001$). These results indicate that SL520–219 and SL520–873 contain a QTL in their overlapping segments (10.13 kbp genomic region) defined by 90E3H and L17A, in which only the *Wx* locus is present [23]. Therefore, the common QTL (*qSAC–6* and *qIAC–6*) is identical to the *Wx* locus, i.e., *Wx* has pleiotropic effects on the SAC and IAC of rice grains.

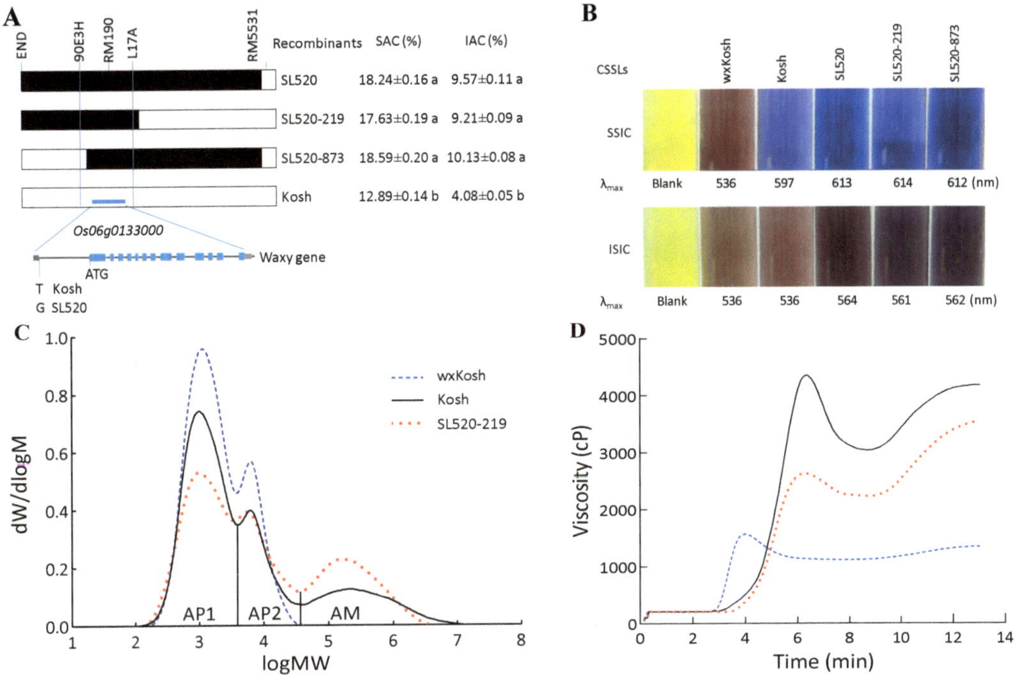

Figure 4. Identification of quantitative trait loci (QTLs) for hot water-soluble amylose content (SAC) and hot water-insoluble amylose content (IAC). (**A**) Fine-mapping of *qSAC-6* and *qIAC-6* using a set of smaller chromosome segment substitution lines (CSSLs) with their genotypes and phenotypes. The CSSLs were constituted by the recombinants (SL520-219 and SL520-873) from a cross between SL520 and Kosh. The common QTL (*qSAC-6* and *qIAC-6*) labeled from END and RM5531 was verified as the Waxy gene. The structural diagram of *Wx* was downloaded from https://rapdb.dna.affrc.go.jp/locus/?name=Os06g0133000, accessed on 30 December 2023. The different letters (a and b) mean that their genotypic means significantly differ at $p < 0.01$. (**B**) A composition analysis of the hot water-soluble (HWS) and hot water-insoluble (HWI) fractions of starch. The HWS fractions are composed of amylose and a small quantity of short-chain AP while the HWI fractions are composed of long-chain AP. SSIC, soluble starch–iodine complex; ISIC, insoluble starch–iodine complex. (**C**) Gel permeation chromatograms of isoamylase-debranched starch from wxKosh, Kosh, and SL520–219, which compose a set of near-isogenic lines (NILs) of the *Wx* gene corresponding to the *wx*, Wx^b, and Wx^a alleles, respectively, in the Kosh background. AP1, AP2, and AM correspond to short branch chains of amylopectin, long branch chains of amylopectin, and amylose, respectively. (**D**) Rapid viscosity profiles of flour from the set of NILs exhibited distinct pasting properties at 12% solids content. The legends of lines are the same as (**C**).

2.3. Compositions of the HWS and HWI Fractions of Starch

To uncover the mechanism underlying the pleiotropy associated with SAC and IAC, we analyzed the compositions of the HWS and HWI fractions of extracted starch. Amylose–iodine complexes appear blue, while AP–iodine complexes appear brown; thus, we examined the colors of the soluble starch–iodine complex (SSIC) and insoluble starch–iodine complex (ISIC) in wxKosh, Kosh, SL520, SL520–219, and SL520–873 (Figure 4B).

For wxKosh, both the iodine-stained SSIC and ISIC appeared brown, indicating that their HWS and HWI fractions both consisted of AP, with no amylose, which was validated by our GPC results (Figure 4B and Table 2). The iodine-stained SSICs of Kosh, SL520, SL520–219, and SL520–873 were all pure blue, indicating that their HWS fractions contained amylose. The ISIC of Kosh was brown, similar to that of wxKosh, suggesting that its HWI fraction may comprise AP. The ISICs of SL520, SL520–219, and SL520–873 were all dark brown rather than blue, suggesting that their HWI fractions also consisted of AP. Nevertheless, their color (dark brown) was not exactly the same as that of Kosh (brown).

Table 2. Gel permeation chromatography parameters of the starches of wxKosh, Kosh, and SL520-219. Data are the means ± standard deviations (SD) (n = 2). AP1, low-molecular-weight molecules, including A and short B amylopectin (AP) chains; AP2, high-molecular-weight AP molecules, including long B chains; AP1/AP2, the ratio of AP2 to AP1; AM, a broad range of amyloses. Further details are provided in the Figures 3 and 4C legends. The different letters (a, b and c) mean that their genotypic means significantly differ at $p < 0.01$.

NILs	AP1	AP2	AM	AP2/AP1
wxKosh	72.44 ± 0.48 a	27.56 ± 0.48 a	0	0.3804 ± 0.009 b
Kosh	59.68 ± 0.37 b	22.61 ± 0.43 c	17.71 ± 0.06 b	0.3788 ± 0.010 b
SL520-219	45.52 ± 0.46 c	26.40 ± 0.20 b	28.08 ± 0.96 a	0.5799 ± 0.005 a

Moreover, we investigated the λ_{max} values of the SSIC and ISIC of wxKosh, Kosh, SL520, SL520–219, and SL520–873 (Figure 4B). For wxKosh, its λ_{max} values for the SSIC and ISIC were both 536 nm. Its HWS fractions accounted for 8.34% of the total OD_{536} value, indicating that the AP of wxKosh is slightly soluble in hot water.

For Kosh, SL520, SL520–219, and SL520–873, their λ_{max} values for SSIC were 597, 614, 613, and 612 nm, respectively, which were slightly lower than that (650 nm) of a purified amylose–iodine complex [24] and much higher than that of AP (536 nm), indicating that, besides amylose, their HWS fractions contained a small amount of AP. Their λ_{max} values for ISIC were 536, 562, 561, and 564 nm, respectively. These results are in agreement with the results reported by Takeda et al. (1987), who showed that the λ_{max} values of purified AP–iodine complexes were ~535 nm from Wx^b varieties and ~565 nm from Wx^a varieties [25]. These results suggest that the HWI fractions of Kosh, SL520, SL520–219, and SL520–873 may consist of AP. The λ_{max} values for the ISIC of SL520, SL520–219, and SL520–873 were higher than that of Kosh. Coupled with the above finding that the ISIC of SL520, SL520–219, and SL520–873 was a darker brown than that of Kosh, these findings highlight the differences in the fine structures of AP between genotypes.

We analyzed the fine structure of starch by determining the relative molecular weight distributions of isoamylase-debranched starch in a set of NILs (wxKosh, Kosh, and SL520–219) using GPC (Figure 4C). The completely debranched starch was fractionated into three components: AP1, consisting of low-molecular-weight molecules, including A and short B chains (A + B1 chains); AP2, consisting of high-molecular-weight molecules, including long B chains; and AM, consisting of a broad range of amyloses [26]. In general, the AM represents the true AC [27].

GPC revealed that wxKosh lacks amylose, while Kosh and SL520–219 have low and high AM (true AC), respectively (Figure 4C; Table 2). These results indicate that *Wx* is responsible for the biosynthesis of amylose. Nevertheless, SL520–219 contained lower levels of long amylose chains than Kosh (Figure 4C), which is in agreement with the findings of Zhang et al. (2012; 2019) [23,28]. This is because amylose contains more branch linkages in the Wx^a vs. Wx^b varieties [29,30].

In terms of the fine structure of AP, the proportion of AP1 decreased, consequently increasing the proportion of AP2 and AM in wxKosh, Kosh, and SL520–219. AP2 and AM govern the texture of cooked rice [31]. The ratio of AP2 to AP1 (AP2/AP1) showed no significant difference between Kosh and wxKosh, whereas this ratio was significantly greater in SL520–219 than in Kosh and wxKosh. This may explain why the color and λ_{max} values of the ISIC of Kosh were the same as those of wxKosh, while the ISIC of SL520, SL520–219, and SL520–873 were darker and their λ_{max} values were higher than those of Kosh and wxKosh (Figure 4C). Thus, the color and λ_{max} value of the ISIC may be regarded as characteristics of the fine structure of AP [25,32]. These results indicate that the Wx^a

allele (SL520–219) is more strongly involved in the biosynthesis of long-branch AP chains than Wx^b (Kosh). The APs from the high-AC varieties (carrying the Wx^a allele) contain a high proportion of long B chains [33], which are known to be present in all high-AC cereal starches [34]. In addition, Wx^a is responsible for the biosynthesis of extra-long unit chains of AP. These differences resulted in the structural differences in the AP of the HWI fractions between the Wx^a and Wx^b varieties, and the structural differences contributed to the variation in their IAC (Figure 4B,C).

In conclusion, the HWS fractions are composed of amylose and a small quantity of short-chain AP, while the HWI fractions are composed of long-chain AP. These observations are in agreement with the findings of Bhattacharya et al. (2011) [3] and Butardo et al. (2019) [35]. These findings are also in accordance with the finding by Cuevas et al. (2010) that amylose readily dissolves in hot water while AP is largely insoluble [36]. The *Wx* gene is not only responsible for biosynthesizing amylose, which comprises a large proportion of the HWS fractions that were used to calculate the SAC, but it is also involved in the biosynthesis of long chains of AP, which comprises the HWI fractions that were used to calculate the IAC. As a result, *Wx* has pleiotropic effects on SAC and IAC.

2.4. Relationship between SAC or IAC and Quality

The RVA mimics the cooking process of rice. In the RVA, a flour–water suspension is subjected to a heat–hold–cool–hold temperature cycle [37], and the measured values reflect the texture of cooked rice [27]. To elucidate the relationships between SAC or IAC and cooked-rice quality, we measured the flour pasting properties of the lines in an RVA. wxKosh, Kosh, and SL520–219 exhibited distinct pasting properties at 12% solids content (Figure 4D). However, due to the pleiotropic effects of *Wx* on SAC and IAC, these values were positively correlated with each other: the higher the SAC, the higher the IAC, and vice versa (Figure 4A). As a result, the individual relationships between SAC or IAC and the pasting properties of the flour could not be independently determined.

3. Discussion

3.1. Amylose Content (AC) and Its Relationship with Rice Quality

The AC is the most important physicochemical index of rice quality. Estimating the AC relies on the binding of iodine to amylose to produce a blue amylose–iodine complex, whose absorbance is quantified at 620 nm using a spectrophotometer [38]. This absorbance values also contain the binding of iodine to amylopectin (AP); hence, the AC was designated as the 'apparent amylose content (AAC)' [25]. That is, the AC measured based on iodine coloration is a mixture of amylose and AP. In this way, the AC of milled rice grains can be ranked as waxy (0–2% of the flour weight), very low (3–9%), low (10–19%), intermediate (20–25%), or high (>25%) [39].

In this study, by leaching starch granules in a boiling-water bath, we separated the starch samples into the HWS and the HWI fractions, dividing the AC into the SAC and IAC. The HWS fractions are composed of amylose and a small quantity of short-chain AP, and the HWI fractions consist of long-chain AP; thus, the SAC roughly represents the "actual AC", and the IAC roughly represents the "AP content at 620 nm". The effect of the AC on rice quality included amylose and AP.

3.2. AP Fine Structure and Its Relationship to Quality

Starch granules are composed of amylose and AP [2]. AP is the more abundant and complex of the two components, with both influencing rice quality. In waxy rice, a longer average AP chain length and longer exterior AP chain length result in higher gelatinization temperatures, higher pasting consistencies, greater retrogradation, and a firmer texture of cooked rice [26]. In a study of 11 non-waxy rice cultivars, hard-cooked rice (which was hard after being cooked) tended to have a higher AC and more longer-chain AP than soft-cooked rice [40]. In a study of seven varieties of rice with cooked-rice textures ranging from very soft to very hard, the content of long B chains of AP correlated well with the sensory tenderness of the cooked rice [31]. Thus, variation in the fine structure of AP appears to be at the root of variation in rice texture [31,33,40].

Bhattacharya and colleagues investigated the water solubility of amylose, revealing that the IAC is a key factor in rice quality [9,10]; further research has widely confirmed this finding [12]. In the present study, we determined that the HWI fractions of starch consist of long-chain AP and that structural differences in the AP in the HWI fraction contribute to the observed variation in IAC (Figure 4A–C); therefore, the IAC (AP content at 620 nm) represents variation in the fine structure of AP. That is, IAC can be substituted for AP fine structure to study the impact of the latter on quality.

3.3. Pleiotropy of *Wx*

The starch biosynthesis pathway is a complex network, which requires the involvement of multiple isoforms of some enzymes, including granule-bound starch synthase (GBSS) which is responsible for amylose biosynthesis [15], and soluble starch synthase (SSS), starch-branching enzyme (SBE), and starch-debranching enzyme (DBE) which are together responsible for AP biosynthesis [41,42]. Moreover, ADP-glucose pyrophosphorylase (AGPase) is a rate-limiting step in starch biosynthesis [42]. Each enzyme plays a distinct role, but as parts of this biosynthesis pathway, enzymes (isoforms) involved in AP biosynthesis presumably have a function in amylose biosynthesis [43]. SSSIIIa plays an important part in the elongation of AP chains, and deficiency in the SSSIIIa caused an increase in the amounts of GBSSI which increased amylose content in rice grain [44]. BEIIb is capable of forming the branch linkages in AP biosynthesis, and the knockout of BEIIb led to increased amylose content in rice grain [45,46]. In this study, we showed that GBSSI encoded by *Wx* gene is not only responsible for biosynthesizing amylose but is also involved in the biosynthesis of long chains of AP. These relationships result in pleiotropic effects of starch synthesis-related genes.

As described above, the IAC is a key determinant of cooked-rice quality, which is particularly driven by variation in the fine structure of AP. However, all these findings were based on a small number of research materials, including two types of varieties, high AC (Wx^a) and low AC (Wx^b) varieties, meaning that the effect of *Wx* contributed to the findings. Here, we revealed that *Wx* has pleiotropic effects on the SAC and IAC and that this pleiotropy means that these traits are genetically correlated with each other; thus, the role of IAC in rice quality cannot be determined independently. We, therefore, suggest that further studies of rice quality should use rice varieties with the same *Wx* genotype to eliminate the effect of this gene, enabling the independent relationship between the SAC or IAC and rice quality to be elucidated through a multiple correlation analysis. We believe that this should be a major focus of research because many of the varieties currently cultivated in places such as China have similar TACs but show a huge variation in quality, with no clear pattern of quality characteristics. The current findings could also be applicable to other cereal crops.

4. Materials and Methods

4.1. Plant Materials

Two sets of related CSSLs obtained from the Rice Genome Resource Center, Japan (http://www.rgrc.dna.affrc.go.jp/stock.html, accessed on 15 September 2015) were used in this study. For both CSSLs, Koshihikari (Kosh), a *japonica* rice (*Oryza sativa*) variety with low AC, was used as the common receptor parent, while high-AC *indica* rice variety Kasalath or NonaBokra was used as the donor parent. The Kosh–Kasalath CSSL (KK–CSSL) set comprises 36 lines, although three were lacking in this study (SL215, SL223, and SL236), bringing about a gap of 13.2 cM between the distal end and C390 on chromosome 8. The Kosh–NonaBokra (KN–CSSL) set comprises 39 lines, although five were lacking (SL502, SL511, SL517, SL518, and SL524), bringing about a gap of 19.2 cM between the distal end and RM3308 on chromosome 4. The two sets of CSSLs were grown in Lingshui, Hainan province,

China (18° N, 109° E) in two years: during the winter season in November 2016–May 2017 (referred to as 2016 hereafter) and November 2017–May 2018 (referred to as 2017 hereafter).

To fine-map the common QTL that was identified (*qAC–6*, *qAC–6*, and *qAC–6*), an F_2 subpopulation was developed from a cross between SL520 and Kosh. Its two derived recombinants, SL520–219 and SL520–873 (in the common Kosh background), together with Kosh and waxy Kosh (wxKosh; developed by backcrossing waxy variety Suyuno as the donor parent six times to Kosh as the recurrent parent), possessed the Wx^a, Wx^a, Wx^b, and *wx* alleles, respectively. These lines constituted a set of near-isogenic lines (NILs), which were planted in Linan, Zhejiang province, China (30.16° N, 120.12° E) during the summer season (May–October 2021) for gel permeation chromatography (GPC) and analysis of pasting properties.

All the above plant materials were harvested ~45 days after heading when the seeds had turned yellow. The seeds were air-dried in ambient conditions to ~13% moisture content, stored at room temperature (~25 °C) for three months for after-ripening, and stored in the cold room at 16 °C and <65% relative humidity until use. The seeds were dehulled, milled, ground into flour, and passed through a 0.5 mm sieve before analysis.

4.2. Separation of the HWS and HWI Fractions

The HWS and HWI fractions of rice flour were separated as described by Shanthy et al. (1980) [47], with some modifications. Briefly, 0.1000 g milled rice flour was accurately weighed into a 50 mL centrifuge tube. The sample was thoroughly moistened in 1 mL of 95% ethanol, and 20 mL of distilled water was added instead of dilute alkali as a solvent. The samples were vortexed and heated in a boiling-water bath for 30 min with intermittent mixing to fully separate the soluble and insoluble starch. The samples were cooled to room temperature and centrifuged at 5000× *g* for 10 min, and the supernatant was collected in a 100 mL volumetric flask. To fully separate the soluble and insoluble fractions, the separation procedure was repeated one more time with 20 mL of distilled water added to the precipitate. Both supernatants were pooled into the same 100 mL volumetric flask. The supernatant is referred to as the HWS fraction and the precipitate as the HWI fraction hereafter. The flour sample was thus used to calculate the TAC, the HWS fraction was used to calculate the SAC, and the HWI fraction was used to calculate the IAC. Each separation experiment was performed in three biological replications.

4.3. Determination of the TAC, SAC, and IAC

The TAC was determined using the iodine colorimetric method, according to EN ISO 6647–1.2020 [48], with minor modifications. Briefly, 0.1000 g of the test sample was mixed with a sodium hydroxide solution, an iodine solution was added for color development, and the absorbance at 620 nm was measured against the blank solution using a spectrophotometer. The SAC and IAC were determined using the method described for the TAC, except that the iodine solution was directly added to the soluble starch fraction for color development. Each sample was determinated in three biological replications.

A calibration line was generated using a concentration series of purified amylose from potato (MilliporeSigma, Burlington, MA, USA) and defatted AP from wxKosh. The calibration line was used to determine the TAC. Another calibration line was generated using a set of standard samples (0.4%, 10.3%, 16.6%, and 26.6%) obtained from the China National Rice Research Institute and used to determine the SAC and IAC. The sample was scanned at a wavelength of 800–400 nm using a UV–visible spectrophotometer (UV2600; Shimadzu, Kyoto, Japan). The λ_{max} represents the wavelength at the position of the peak absorbance value of the iodine–starch complex.

4.4. QTL Analysis

QTL analysis was performed using the stepwise regression-based likelihood ratio tests for additive QTL (RSTEP–LRT–ADD) method for CSL (chromosome substitution line) functionality provided by the QTL IciMapping Version 4.1 software package [49], as described by Xu et al. (2023) [50]. A standard threshold logarithm of odds (LOD) score of 3.0 was used to suggest the presence of a putative QTL.

4.5. Molecular Weight Distribution of Starch

The molecular weight distribution of starch was determined using an Agilent PL–gel permeation chromatography (GPC) 220 system (Agilent Technologies, Santa Clara, CA, USA), as described by Zhang et al. (2021) [51].

4.6. Pasting Properties of Rice Flour

The pasting properties of the samples were determined using a Rapid Viscosity Analyzer (RVA; Model 3D; Newport Scientific, Warriewood, Australia) according to the approved method 61–02 (AACC 2000) described by Wu et al. (2020) [52].

4.7. Data Analysis

Analysis of variance (ANOVA) was performed to compare the genotypic mean TAC, SAC, and IAC values using the least significance difference (LSD) test. Data analysis was performed using SPSS Statistics version 17.0 software for Windows (IBM, Armonk, NY, USA).

5. Conclusions

Amylose content (AC) is the principal determinant of rice quality, but varieties with similar ACs still vary in their palatability. In this study, by analyzing the ACs and their constituents, hot water-soluble AC (SAC) and hot water-insoluble AC (IAC), a common QTL (*qAC–6*, *qSAC–6*, and *qIAC–6*) on chromosome 6 associated with all three AC components was revealed to be Waxy (*Wx*) gene. The finding highlights the pleiotropic effects of *Wx* on the SAC and IAC. It implies that further studies of rice quality should be performed using rice varieties with the same *Wx* genotype in order to eliminate the pleiotropic effects of this gene, which could enable the elucidation of the independent relationship between the SAC or IAC and rice quality via a multiple correlation analysis.

Author Contributions: Conceptualization, H.W.; data curation, S.W. and M.W.; formal analysis, S.W. and M.W.; investigation, M.W. and S.W.; methodology, M.W. and S.W.; project administration, H.W.; resources, H.W.; supervision, H.W.; validation, S.W.; visualization, S.W. and M.W.; writing—original draft, S.W.; writing—review and editing, H.W. All authors have read and agreed to the published version of the manuscript.

Funding: This work was funded by the Key Research and Development Program of Zhejiang province (2021C02063-6-5).

Institutional Review Board Statement: Not applicable.

Informed Consent Statement: Not applicable.

Data Availability Statement: The data presented in this study are available on request from the corresponding author.

Acknowledgments: We would like to thank Changquan Zhang, at Yangzhou University, for determining molecular weight distribution of starch using gel permeation chromatography.

Conflicts of Interest: The authors declare that they have no conflicts of interest (including financial and non-financial interests).

References

1. Ramesh, M.; Bhattacharya, K.R.; Mitchell, J.R. Developments in Understanding the Basis of Cooked-Rice Texture. *Crit. Rev. Food Sci. Nutr.* **2000**, *40*, 449–460. [CrossRef] [PubMed]
2. Kong, X. Fine Structure of Amylose and Amylopectin. In *Starch Structure, Functionality and Application in Foods*; Wang, S., Ed.; Springer Nature Singapore Pte Ltd.: Singapore, 2020; pp. 29–39.
3. Bhattacharya, K.R. Eating quality of rice. In *Rice Quality*; Bhattacharya, K.R., Ed.; Woodhead Publishing Limited: Sawston, UK, 2011; pp. 193–246.

4. Bhattacharya, K.R.; Sowbhagya, C.M.; Swamy, Y.M.I. Interrelationship between Certain Physicochemical Properties of Rice. *J. Food Sci.* **1972**, *37*, 733–735. [CrossRef]
5. Bhattacharya, K.R.; Sowbhagya, C.M.; Swamy, Y.M.I. Quality Profiles of Rice: A Tentative Scheme for Classification. *J. Food Sci.* **1982**, *47*, 564–569. [CrossRef]
6. Sowbhagya, C.M.; Ramesh, B.S.; Bhattacharya, K.R. The relationship between cooked-rice texture and the physicochemical characteristics of rice. *J. Cereal Sci.* **1987**, *5*, 287–297. [CrossRef]
7. Juliano, B.O.; Perdon, A.A. Gel and Molecular Properties of Waxy Rice Starch. *Starch-Stärke* **1975**, *27*, 69–71. [CrossRef]
8. Kongseree, N.; Juliano, B.O. Physicochemical properties of rice grain and starch from lines differing in amylose content and gelatinization temperature. *J. Agric. Food Chem.* **1972**, *20*, 714–718. [CrossRef]
9. Kumar, B.M.; Upadhyay, J.K.; Bhattacharya, K.R. Objective Tests for the Stickiness of Cooked Rice. *J. Texture Stud.* **1976**, *7*, 271–278. [CrossRef]
10. Indudhara Swamy, Y.M.; Sowbhagya, C.M.; Bhattacharya, K.R. Changes in the physicochemical properties of rice with ageing. *J. Sci. Food Agric.* **1978**, *29*, 627–639. [CrossRef]
11. Bhattacharya, K.R.; Sowbhagya, C.M.; Indudhara Swamy, Y.M. Importance of insoluble amylose as a determinant of rice quality. *J. Sci. Food Agric.* **1978**, *29*, 359–364. [CrossRef]
12. Iturriaga, L.B.; Lopez de Mishima, B.; Añon, M.C. A study of the retrogradation process in five argentine rice starches. *LWT—Food Sci. Technol.* **2010**, *43*, 670–674. [CrossRef]
13. He, P.; Li, S.G.; Qian, Q.; Ma, Y.Q.; Li, J.Z.; Wang, W.M.; Chen, Y.; Zhu, L.H. Genetic analysis of rice grain quality. *Theor. Appl. Genet.* **1999**, *98*, 502–508. [CrossRef]
14. Tan, Y.F.; Li, J.X.; Yu, S.B.; Xing, Y.Z.; Xu, C.G.; Zhang, Q. The three important traits for cooking and eating quality of rice grains are controlled by a single locus in an elite rice hybrid, Shanyou 63. *Theor. Appl. Genet.* **1999**, *99*, 642–648. [CrossRef] [PubMed]
15. Wang, Z.Y.; Zheng, F.Q.; Shen, G.Z.; Gao, J.P.; Snustad, D.P.; Li, M.G.; Zhang, J.L.; Hong, M.M. The amylose content in rice endosperm is related to the post-transcriptional regulation of the waxy gene. *Plant J.* **1995**, *7*, 613–622. [CrossRef] [PubMed]
16. Isshiki, M.; Nakajima, M.; Satoh, H.; Shimamoto, K. dull: Rice mutants with tissue-specific effects on the splicing of the waxy pre-mRNA. *Plant J.* **2001**, *23*, 451–460. [CrossRef] [PubMed]
17. Zeng, D.; Yan, M.; Wang, Y.; Liu, X.; Qian, Q.; Li, J. Du1, encoding a novel Prp1 protein, regulates starch biosynthesis through affecting the splicing of Wx b pre-mRNAs in rice (*Oryza sativa* L.). *Plant Mol. Biol.* **2007**, *65*, 501–509. [CrossRef] [PubMed]
18. Isshiki, M.; Matsuda, Y.; Takasaki, A.; Wong, H.L.; Satoh, H.; Shimamoto, K. Du3, a mRNA cap-binding protein gene, regulates amylose content in Japonica rice seeds. *Plant Biotechnol.* **2008**, *25*, 483–487. [CrossRef]
19. Cai, Y.; Zhang, W.; Fu, Y.; Shan, Z.; Xu, J.; Wang, P.; Kong, F.; Jin, J.; Yan, H.; Ge, X.; et al. Du13 encodes a C2H2 zinc finger protein that regulates Wxb pre-mRNA splicing and microRNA biogenesis in rice endosperm. *Plant Biotechnol. J.* **2022**, *20*, 1387–1401. [CrossRef] [PubMed]
20. Cai, X.L.; Wang, Z.Y.; Xing, Y.Y.; Zhang, J.L.; Hong, M.M. Aberrant splicing of intron 1 leads to the heterogeneous 5′ UTR and decreased expression of waxy gene in rice cultivars of intermediate amylose content. *Plant J.* **1998**, *14*, 459–465. [CrossRef] [PubMed]
21. Isshiki, M.; Morino, K.; Nakajima, M.; Okagaki, R.J.; Wessler, S.R.; Izawa, T.; Shimamoto, K. A naturally occurring functional allele of the rice waxy locus has a GT to TT mutation at the 5′ splice site of the first intron. *Plant J.* **1998**, *15*, 133–138. [CrossRef]
22. Cai, X.L.; Liu, Q.Q.; Tang, S.Z.; Gu, M.H.; Wang, Z.Y. Development of a molecular marker for screening the rice cultivars with intermediate amylose content in *Oryza sativa* subsp. indica. *J. Plant Physiol. Mol. Biol.* **2002**, *28*, 137–144.
23. Zhang, C.; Zhu, J.; Chen, S.; Fan, X.; Li, Q.; Lu, Y.; Wang, M.; Yu, H.; Yi, C.; Tang, S.; et al. Wx, the Ancestral Allele of Rice Waxy Gene. *Mol. Plant* **2019**, *12*, 1157–1166. [CrossRef] [PubMed]
24. Takeda, Y.; Hizukuri, S.; Juliano, B.O. Purification and structure of amylose from rice starch. *Carbohydr. Res.* **1986**, *148*, 299–308. [CrossRef]
25. Takeda, Y.; Hizukuri, S.; Juliano, B.O. Structures of rice amylopectins with low and high affinities for iodine. *Carbohydr. Res.* **1987**, *168*, 79–88. [CrossRef]
26. Zhu, L.-J.; Liu, Q.-Q.; Sang, Y.; Gu, M.-H.; Shi, Y.-C. Underlying reasons for waxy rice flours having different pasting properties. *Food Chem.* **2010**, *120*, 94–100. [CrossRef]
27. Zhang, C.; Chen, S.; Ren, X.; Lu, Y.; Liu, D.; Cai, X.; Li, Q.; Gao, J.; Liu, Q. Molecular Structure and Physicochemical Properties of Starches from Rice with Different Amylose Contents Resulting from Modification of OsGBSSI Activity. *J. Agric. Food Chem.* **2017**, *65*, 2222–2232. [CrossRef]
28. Zhang, C.; Zhu, L.; Shao, K.; Gu, M.; Liu, Q. Toward underlying reasons for rice starches having low viscosity and high amylose: Physiochemical and structural characteristics. *J. Sci. Food Agric.* **2012**, *93*, 1543–1551. [CrossRef]
29. Mizukami, H.; Takeda, Y.; Hizukuri, S. The structure of the hot-water soluble components in the starch granules of new Japanese rice cultivars. *Carbohydr. Polym.* **1999**, *38*, 329–336. [CrossRef]
30. Hanashiro, I.; Itoh, K.; Kuratomi, Y.; Yamazaki, M.; Igarashi, T.; Matsugasako, J.i.; Takeda, Y. Granule-Bound Starch Synthase I is Responsible for Biosynthesis of Extra-Long Unit Chains of Amylopectin in Rice. *Plant Cell Physiol.* **2008**, *49*, 925–933. [CrossRef] [PubMed]
31. Ramesh, M.; Zakiuddin Ali, S.; Bhattacharya, K.R. Structure of rice starch and its relation to cooked-rice texture. *Carbohydr. Polym.* **1999**, *38*, 337–347. [CrossRef]

32. Takeda, Y.; Maruta, N.; Hizukuri, S.; Juliano, B.O. Structures of indica rice starches (IR48 and IR64) having intermediate affinities for iodine. *Carbohydr. Res.* **1989**, *187*, 287–294. [CrossRef]
33. Radhika Reddy, K.; Zakiuddin Ali, S.; Bhattacharya, K.R. The fine structure of rice-starch amylopectin and its relation to the texture of cooked rice. *Carbohydr. Polym.* **1993**, *22*, 267–275. [CrossRef]
34. Hanashiro, I.; Matsugasako, J.-i.; Egashira, T.; Takeda, Y. Structural Characterization of Long Unit-Chains of Amylopectin. *J. Appl. Glycosci.* **2005**, *52*, 233–237. [CrossRef]
35. Butardo, V.M.; Sreenivasulu, N.; Juliano, B.O. Improving Rice Grain Quality: State-of-the-Art and Future Prospects. In *Rice Grain Quality: Methods and Protocols*; Sreenivasulu, N., Ed.; Springer Science+Business Media, LLC; Springer Nature: New York, NY, USA, 2019; Volume 1892, pp. 19–55.
36. Cuevas, R.P.; Gilbert, R.G.; Fitzgerald, M.A. Structural differences between hot-water-soluble and hot-water-insoluble fractions of starch in waxy rice (*Oryza sativa* L.). *Carbohydr. Polym.* **2010**, *81*, 524–532. [CrossRef]
37. Chen, M.-H.; Bergman, C.J.; Pinson, S.R.M.; Fjellstrom, R.G. Waxy gene haplotypes: Associations with pasting properties in an international rice germplasm collection. *J. Cereal Sci.* **2008**, *48*, 781–788. [CrossRef]
38. Williams, V.R.; Wu, W.-T.; Tsai, H.Y.; Bates, H.G. Rice Starch, Varietal Differences in Amylose Content of Rice Starch. *J. Agric. Food Chem.* **1958**, *6*, 47–48. [CrossRef]
39. Kumar, I.; Khush, G.S. Genetics of amylose content in rice (*Oryza sativa* L). *J. Genet.* **1986**, *65*, 1–11. [CrossRef]
40. Ong, M.H.; Blanshard, J.M.V. Texture determinants in cooked, parboiled rice. I: Rice starch amylose and the fine stucture of amylopectin. *J. Cereal Sci.* **1995**, *21*, 251–260. [CrossRef]
41. James, M.G.; Denyer, K.; Myers, A.M. Starch synthesis in the cereal endosperm. *Curr. Opin. Plant Biol.* **2003**, *6*, 215–222. [CrossRef] [PubMed]
42. Hannah, L.C.; Greene, T. The Complexities of Starch Biosynthesis in Cereal Endosperms. In *Molecular Genetic Approaches to Maize Improvement*; Springer: Berlin/Heidelberg, Germany, 2009; Volume 63, pp. 287–301. [CrossRef]
43. Tian, Z.; Qian, Q.; Liu, Q.; Yan, M.; Liu, X.; Yan, C.; Liu, G.; Gao, Z.; Tang, S.; Zeng, D.; et al. Allelic diversities in rice starch biosynthesis lead to a diverse array of rice eating and cooking qualities. *Proc. Natl. Acad. Sci. USA* **2009**, *106*, 21760–21765. [CrossRef]
44. Fujita, N.; Yoshida, M.; Kondo, T.; Saito, K.; Utsumi, Y.; Tokunaga, T.; Nishi, A.; Satoh, H.; Park, J.-H.; Jane, J.-L.; et al. Characterization of SSIIIa-Deficient Mutants of Rice: The Function of SSIIIa and Pleiotropic Effects by SSIIIa Deficiency in the Rice Endosperm. *Plant Physiol.* **2007**, *144*, 2009–2023. [CrossRef]
45. Nishi, A.; Nakamura, Y.; Tanaka, N.; Satoh, H. Biochemical and Genetic Analysis of the Effects ofAmylose-Extender Mutation in Rice Endosperm. *Plant Physiol.* **2001**, *127*, 459–472. [CrossRef]
46. Zhou, L.; Zhang, C.; Zhang, Y.; Wang, C.; Liu, Q. Genetic manipulation of endosperm amylose for designing superior quality rice to meet the demands in the 21st century. *J. Cereal Sci.* **2022**, *105*, 103481. [CrossRef]
47. Shanthy, A.P.; Sowbhagya, C.M.; Bhattacharya, K.R. Simplified Determination of Water-Insoluble Amylose Content of Rice. *Starch-Stärke* **1980**, *32*, 409–411. [CrossRef]
48. *EN ISO 6647-1:2020*; Rice—Determination of Amylose Content—Part 1: Spectrophotometric Method with a Defatting Procedure by Methanol and with Calibration Solutions of Potato Amylose and Waxy Rice Amylopectin. European Standards Coordinating Committee: Brussels, Belgium, 2020.
49. Meng, L.; Li, H.; Zhang, L.; Wang, J. QTL IciMapping: Integrated software for genetic linkage map construction and quantitative trait locus mapping in biparental populations. *Crop J.* **2015**, *3*, 269–283. [CrossRef]
50. Xu, S.; Fei, Y.; Wang, Y.; Zhao, W.; Hou, L.; Cao, Y.; Wu, M.; Wu, H. Identification of a Seed Vigor–Related QTL Cluster Associated with Weed Competitive Ability in Direct–Seeded Rice (*Oryza sativa* L.). *Rice* **2023**, *16*, 45. [CrossRef]
51. Feng, L.; Lu, C.; Yang, Y.; Lu, Y.; Li, Q.; Huang, L.; Fan, X.; Liu, Q.; Zhang, C. The Physicochemical Properties of Starch Are Affected by Wxlv in Indica Rice. *Foods* **2021**, *10*, 3089. [CrossRef]
52. Wu, M.; Xu, S.; Fei, Y.; Cao, Y.; Wu, H.; Yamamoto, T. The rice (*Oryza sativa*) Rc gene improves resistance to preharvest sprouting and retains seed and milled rice quality. *Plant Breed.* **2020**, *140*, 53–64. [CrossRef]

International Journal of
Molecular Sciences

Article

Genome-Wide Association Study Identifies Rice Panicle Blast-Resistant Gene *Pb4* Encoding a Wall-Associated Kinase

Yunxin Fan †, Lu Ma †, Xiaoqian Pan, Pujiang Tian, Wei Wang, Kunquan Liu, Ziwei Xiong, Changqing Li, Zhixue Wang, Jianfei Wang, Hongsheng Zhang and Yongmei Bao *

State Key Laboratory of Crop Genetics and Germplasm Enhancement, College of Agriculture, Jiangsu Collaborative Innovation Center for Modern Crop Production, Nanjing Agricultural University, Nanjing 210095, China; 15966697937@163.com (X.P.); 2021101060@stu.njau.edu.cn (P.T.); 2023201042@stu.njau.edu.cn (C.L.); hszhang@njau.edu.cn (H.Z.)

* Correspondence: yongmeibao@njau.edu.cn

† These authors contributed equally to this work.

Abstract: Rice blast is one of the most devastating diseases, causing a significant reduction in global rice production. Developing and utilizing resistant varieties has proven to be the most efficient and cost-effective approach to control blasts. However, due to environmental pressure and intense pathogenic selection, resistance has rapidly broken down, and more durable resistance genes are being discovered. In this paper, a novel wall-associated kinase (WAK) gene, *Pb4*, which confers resistance to rice blast, was identified through a genome-wide association study (GWAS) utilizing 249 rice accessions. Pb4 comprises an N-terminal signal peptide, extracellular GUB domain, EGF domain, EGF-Ca^{2+} domain, and intracellular Ser/Thr protein kinase domain. The extracellular domain (GUB domain, EGF domain, and EGF-Ca^{2+} domain) of Pb4 can interact with the extracellular domain of CEBiP. Additionally, its expression is induced by chitin and polygalacturonic acid. Furthermore, transgenic plants overexpressing *Pb4* enhance resistance to rice blast. In summary, this study identified a novel rice blast-resistant gene, *Pb4*, and provides a theoretical basis for understanding the role of WAKs in mediating rice resistance against rice blast disease.

Keywords: rice blast; genome-wide association study; Pb4; WAK

check for updates

Citation: Fan, Y.; Ma, L.; Pan, X.; Tian, P.; Wang, W.; Liu, K.; Xiong, Z.; Li, C.; Wang, Z.; Wang, J.; et al. Genome-Wide Association Study Identifies Rice Panicle Blast-Resistant Gene *Pb4* Encoding a Wall-Associated Kinase. *Int. J. Mol. Sci.* **2024**, *25*, 830. https://doi.org/10.3390/ijms25020830

Academic Editor: Abir U. Igamberdiev

Received: 22 November 2023
Revised: 27 December 2023
Accepted: 3 January 2024
Published: 9 January 2024

1. Introduction

Rice blast, caused by the pathogen *Magnaporthe oryzae* (*M. oryzae*), is one of the most devastating diseases that occurs throughout all growth stages of rice plants and seriously affects both the yield and grain quality of rice crops [1]. Using resistance genes from disease-resistant varieties for breeders is a more cost-effective and environmentally friendly approach. However, most resistance (R) genes conform to the gene-for-gene theory and exhibit race-specific resistance. Due to the high variability of rice blast fungus, the majority of resistance genes will become ineffective within 3–5 years [2]. Thus, more R genes and defense regulator (DR) genes that confer broad-spectrum resistance should be discovered.

To date, over 100 genes conferring resistance to rice blast have been identified, and at least 30 genes have been cloned in rice [2]. However, the majority of these genes are resistant to seedling blast, and only a few, such as *Pb1*, *Pb2*, *Pb3*, *Pi25*, *Pi64*, and *Pi68(t)*, confer resistance to panicle blast [3–8]. Among the cloned R genes, the majority encode a nucleotide-binding leucine-rich repeat (NLR) protein, except for *bsr-d1*, *bsr-k1*, *rod1*, *pi21*, and *Pi-d2*, where bsr-d1 functions as a C_2H_2-type transcription factor, while *Bsr-k1* encodes a protein containing tetratricopeptide repeats (TPRs) [9]. ROD1 acts as a calcium ion (Ca^{2+}) sensor and plays a crucial role in promoting the scavenging of reactive oxygen species (ROS) [10]. *Pi21* encodes a proline-rich protein containing a metal-binding domain [11]. *Pi-d2* encodes a receptor-like kinase protein with a predicted extracellular domain containing the bulb-type mannose-specific binding lectin (B-lectin), as well as an intracellular serine-threonine kinase domain [12]. There are also some receptor-like kinase (RLK) genes conferring resistance to other pathogens, such as *Xa21* and *Xa26*, both encoding an extracellular leucine-rich repeat (LRR) and intracellular kinase domain, conferring resistance to *Xanthomonas oryzae* pv. *oryzae* [13]. Further, *Xa4*, encoding a wall-associated kinase (WAK) protein, provides durable resistance against bacterial leaf blight [14].

WAKs are closely involved in the plant immune process, functioning as receptors of damage-associated molecular patterns (DAMPs). WAKs participate in various physiological processes, including responses to pathogens and metal stress and the regulation of cell elongation. In the context of pathogen response, WAKs are involved in recognizing and transmitting signals from the cell wall to the intracellular components, thereby activating defense responses. They can detect changes in the cell wall caused by pathogen invasion or other environmental stresses, initiating downstream signaling pathways that regulate the expression of defense-related genes and the production of defense compounds [15]. Structurally, WAKs consist of several conserved domains: an amino-terminal (N-terminal) signal peptide, an extracellular WAK galacturonan-binding (GUB) domain, an epidermal growth factor (EGF) domain, an EGF-Ca^{2+} domain, and a carboxy-terminal (C-terminal) cytoplasmic Ser/Thr protein kinase domain [16]. In *Arabidopsis*, the AtGRP3 protein, pectin, and oligogalacturonides (OGs) can bind to the extracellular domain of AtWAK1 and AtWAK2 [17–20]. WAKs play a crucial role in the plant defense response against pathogens. The transcription factor PIBP1 interacts with PigmR and activates the expression of *OsWAK14*, thereby enhancing the PigmR-mediated defense against rice blast fungus [21]. *OsWAK25* is induced by salicylic acid (SA) and mechanical injury, and overexpression of *OsWAK25* enhances rice blast resistance [22]. The resistance gene *Xa4* provides durable resistance against bacterial leaf blight by strengthening the cell wall by promoting cellulose synthesis and inhibiting cell wall loosening [14].

Genome-wide association studies (GWAS) have become a widely used tool for mapping resistance loci, analyzing the association between a specific resistance phenotype and natural genetic variations, such as single-nucleotide polymorphisms, indels, or the copy number of variations. GWAS is a highly efficient approach that conducts whole-genome scans to identify genomic regions linked to the phenotype of interest. This method has greatly benefited from the rapid advancements in high-throughput sequencing technologies, such as second-generation sequencing, SMART sequencing, and nanopore sequencing [23]. To date, researchers have gained valuable loci through GWAS assays by analyzing the genetic variation, population structure, and diversity in various crop species. For instance, 12 loci associated with bacterial blight resistance were identified through GWAS within the 3000 Rice Genomes Project [24]. A total of 27 rice blast-resistant loci were identified through the GWAS using the Rice Diversity Panel II (C-RDP-II). The rice blast resistance genes *PiPR1*, *RNG1*, and *RNG2* were successfully cloned through GWAS [25,26].

In this study, we conducted a GWAS involving 249 rice cultivars to analyze blast resistance loci and *Pb4* was identified to positively regulate rice resistance. Pb4 is a typical WAK protein, which consists of an N-terminal signal peptide, extracellular GUB domain, EGF domain, EGF-Ca^{2+} domain, TM domain, and intracellular Ser/Thr protein kinase domain. The extracellular domain of Pb4 can interact with the extracellular domain of CEBiP, and its expression is induced by chitin and polygalacturonic acid. These findings provide a theoretical basis for understanding the role of WAKs in mediating rice resistance against rice blast disease.

2. Results

2.1. Structure and Phenotype of the Population

To map the rice blast-resistant genes, 249 cultivars with relatively uniform growth periods were selected from the 3K rice accessions. The population structure was analyzed using the 3K core SNP, with k (the number of groups) ranging from 2 to 12. The cross-validation error was minimized when k = 7 (Figure 1A,B). Therefore, 249 rice varieties were divided into seven subgroups: 26 GJ-adm, 12 GJ-sbtrp, 30 GJ-tmp, 32 GJ-trpA, 20 GJ-trpB, 32 XI-1A, 49 XI-1B, 20 XI-adm, 14 cA (Aus), and 14 admix individuals (Figure 1C). A maximum likelihood (ML) phylogenetic tree was constructed using the 3K core SNP, showing that individuals with the same group structure were more likely to aggregate together and form a clade (Figure 1D). To investigate the panicle blast resistance of the population, 249 cultivars were inoculated with the blast strain Hoku1 at the booting stage in the field (Nanjing, China) over different years. The percentage of diseased grains varied greatly from 0 to 100%, indicating a broad range of panicle blast resistance levels (Figure 2A,B). The average percentage of diseased grains in the GJ subgroups was 19% and 22% in 2020 and 2021, respectively, whereas, in the XI subgroups, it was 9% and 8%, showing a higher susceptibility compared to the XI subgroups (Figure 2C,D). A moderate positive linear correlation ($r = 0.62$, p-value $< 2.2 \times 10^{-16}$) was calculated between phenotypes in two different years, indicating that the data can be treated as replicated data for GWAS analysis.

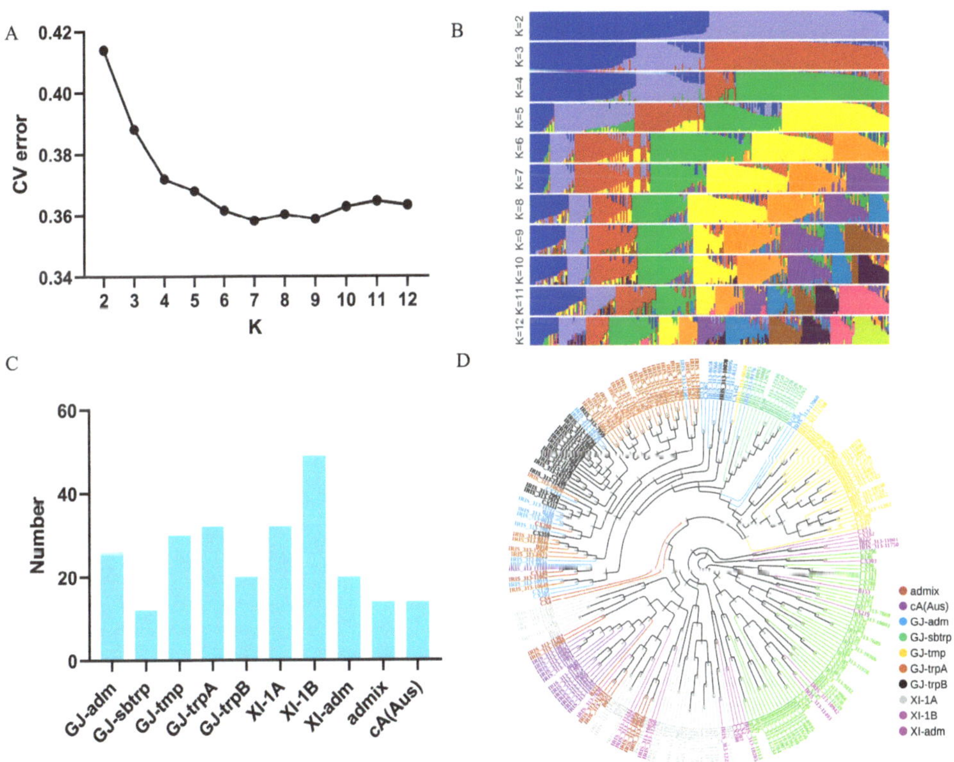

Figure 1. Population structure of the 249 rice accessions. (**A**) Plot of the CV error with k ranging from 2 to 12. The k means the subgroups of the populations. (**B**) Population structure based on different ks. Each color represents a different subgroup. (**C**) The number of rice accessions in different subgroups based on K = 7. GJ-sbtrp = Geng/japonica subtropical subpopulation, GJ-trp = Geng/japonica tropical subpopulation, GJ-tmp = Geng/japonica temperate subpopulation, GJ-adm = Geng/japonica admixed types between two or more GJ subpopulations, XI-1 = Xian/indica subpopulation 1, XI-adm = Xian/indica admixed types between two or more XI subpopulations, cA (Aus) = centrum-Aus population, admix = admixed between any two or more of the XI, GJ, or cA (Aus) populations. (**D**) Phylogenetic tree of the 249 rice accessions; samples are colored by their assignment to k = 7 subgroups from Admixture.

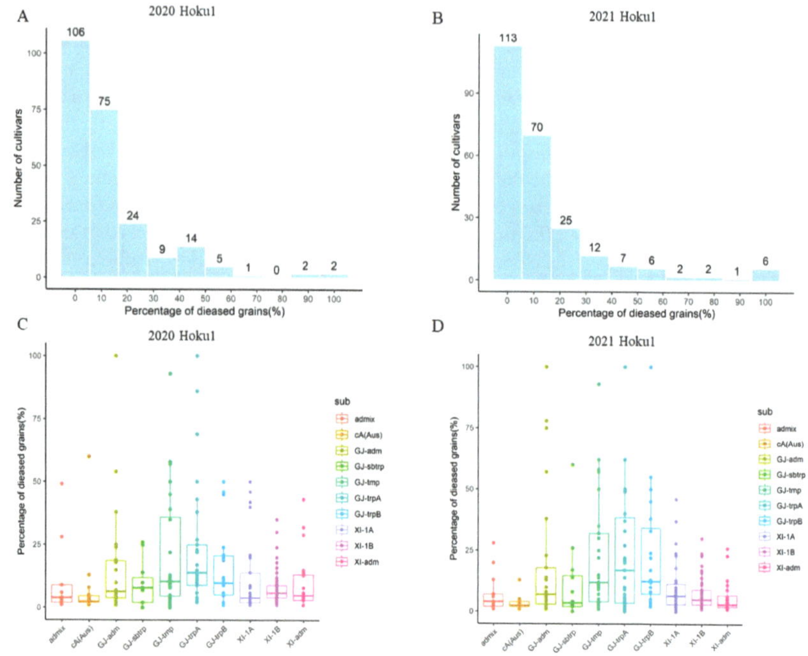

Figure 2. The panicle blast resistance of the 249 rice accessions. (**A**,**B**) The frequency distribution of panicle blast phenotypes for 249 rice varieties inoculated with the Hoku1 strain in the years 2020 and 2021, respectively. (**C**,**D**) Box plots for panicle blast resistance, based on the K = 7 subgroups. The horizontal line in the center of each box denotes the median. The upper and lower limits of each box represent quartiles.

2.2. Identification of Blast-Resistant Loci in Whole Genome

A genome-wide association study (GWAS) was conducted to identify blast-resistant loci (BRL) using a mixed linear model (MLM) based on 3,883,371 high-quality SNPs (MAF > 0.05, missing rate < 50%). The kinship and population structure were used as covariates to control for false positives; the QQ plot still exhibits inflation, perhaps because disease resistance is a result of polygenicity. In 2020, a total of 2873 SNPs ($-\log10p > 4$) showed a significant association with the blast resistance trait, and 3118 SNPs were significantly associated in 2021. Among them, 727 SNPs were detected in both years. Based on the significant associated SNPs, 37 and 38 BRLs were identified in the two respective years. In 2020, the identified 37 loci were distributed on all 12 chromosomes. In 2021, no loci were identified on chromosomes 8 and 9 (Figure 3A,D). Out of these loci, 10 BRLs had been previously reported, and 7 BRLs were consistently identified in both years. Specifically, BRL3, BRL5, BRL8, and BRL11 were co-localized with known blast resistance genes *Pi63*, *Pi56*, *pb1*, and *Pita*, respectively [3,27–29] (Table 1). BRL1, BRL14, BRL15, BRL19, and BRL20 have been reported by Li et al. (2019) [30], and BRL17 has been reported by Liu et al. (2020) [25]. BRL2 and BRL13, BRL4 and BRL17, BRL6 and BRL18, BRL7 and BRL19, BRL9 and BRL21, BRL10 and BRL22, as well as BRL12 and BRL23, were identified repeatedly in 2020 and 2021 (Table 1).

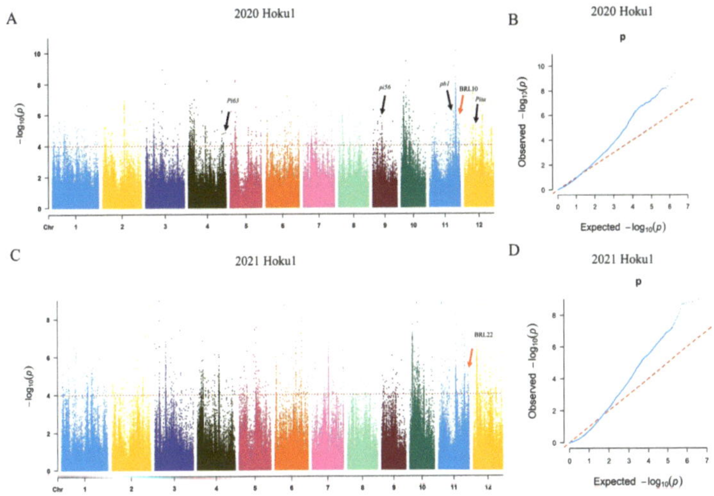

Figure 3. GWAS of the 249 rice accessions for identification of the rice blast-resistant candidate genes. (**A**,**C**) Manhattan plot of the SNPs on 12 rice chromosomes. The *x*-axis represents the genomic position of the SNPs along each chromosome; the *y*-axis indicates the transformed *p*-value, −log10*p*. The black arrows indicate the cloned R genes, while the red arrows indicate the repeat loci BRL10 and BRL22 newly identified in this study. (**B**,**D**) Quantile–Quantile plots of expected and observed *p*-value, −log10*p*.

Table 1. Part of the loci identified in 249 rice varieties in 2020 and 2021 by inoculating with the Hoku1.

Locus	Chr	Position	Top SNP	*p*-Value	Years	Locus Reference
BRL1	3	15255983-15434525	118389	4.86×10^{-5}	2020	[30]
BRL2	4	4088528-4505882	4088552	3.4708×10^{-6}	2020	
BRL3	4	32972226-33188952	33188952	1.626×10^{-5}	2020	*Pi63*
BRL4	6	1454195-1604484	1490561	2.461×10^{-6}	2020	
BRL5	9	8674158-9304868	9069693	1.812×10^{-5}	2020	*pi56*
BRL6	10	4221890-5647528	4798719	1.5076×10^{-8}	2020	
BRL7	10	7667073-8691546	7837737	6.7119×10^{-9}	2020	
BRL8	11	22734774-23104761	22905810	9.0682×10^{-7}	2020	*pb1*
BRL9	11	24235297-25570330	24926857	7.8047×10^{-11}	2020	
BRL10	11	28018096-28187112	28103725	4.9804×10^{-6}	2020	
BRL11	12	10653088-10797158	10734066	9.7844×10^{-6}	2020	*Pita*
BRL12	12	15203722-15771634	15673466	6.6541×10^{-7}	2020	
BRL13	4	4112569-4517761	4444275	1.648×10^{-6}	2021	
BRL14	4	5459995-5745230	5590542	1.4732×10^{-6}	2021	[30]
BRL15	5	7403759-7735177	7735177	5.0795×10^{-7}	2021	[30]
BRL16	5	16077078-16438269	16438239	3.0808×10^{-8}	2021	[25]
BRL17	6	1216276-1514569	1490561	2.0864×10^{-6}	2021	
BRL18	10	4203994-5367946	4749066	5.9358×10^{-7}	2021	
BRL19	10	6200492-9329504	6525986	1.571×10^{-9}	2021	[30]
BRL20	11	8017796-8518194	8409165	1.5501×10^{-6}	2021	[30]
BRL21	11	24202727-24319531	24319531	1.3067×10^{-7}	2021	
BRL22	11	28136278-28187112	28186726	1.203×10^{-5}	2021	
BRL23	12	15535945-16059900	15673466	1.0786×10^{-7}	2021	

2.3. Analysis of the Candidate Genes in BRL10 and BRL22

To further ascertain the resistance-related genes, we analyzed the loci consistently identified in both years. BRL10 and BRL22, located at the end of the long arm of chromosome 11, emerged as repeat loci in both years and were specifically focused on in this study. The combined positions of BRL10 and BRL22, spanning from 28018096 to 28187112, were selected for subsequent pairwise linkage disequilibrium (LD) analysis. An LD block with D′ > 0.85, extending from position 28136278 to 28187112, was chosen as the candidate region (Figure 4A). According to the Rice Genome Annotation Project (http://rice.uga.edu/, accessed on 1 June 2020), this genomic region contains four candidate genes, including LOC_Os11g46870, LOC_Os11g46880, LOC_Os11g46900, and LOC_Os11g46890 (Figure 4A, Table 2). LOC_Os11g46870, LOC_Os11g46880, and LOC_Os11g46900 encode kinase proteins, while LOC_Os11g46890 encodes a protein with an unknown function (Figure 4A, Table 2). Within this region, seven associated SNPs were detected, including six SNPs in the intergenic region and one nonsynonymous SNP in LOC_Os11g46890, causing a Glycine (Gly) to Serine (Ser) substitution at position 183 (Table 3). Because few SNPs are located in the genes, it is challenging to determine disease-resistant casual genes through haplotype analysis. To further investigate the potential resistance genes in this locus, we examined the inducing expression patterns of four candidate genes challenged with the rice blast pathogen. The expression of LOC_Os11g46890 was not detected. Among the remaining three genes, LOC_Os11g46880 showed a significant induced expression pattern in response to the rice blast pathogen but not the other two genes (Figure 4B–D). The gradual increase in the expression level of this gene indicated a positive response to rice blast fungus. Based on this observation, we consider LOC_Os11g46880 as the potential disease-resistant-related gene in this locus.

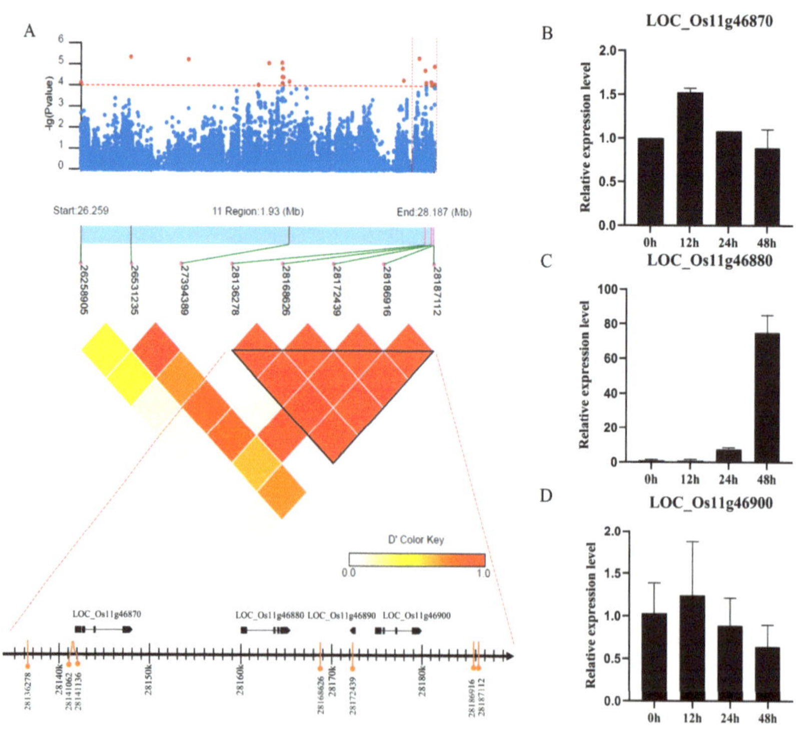

Figure 4. Identification of the rice blast-resistant candidate genes. (**A**) Local Manhattan plot (**top**) and LD heatmap (**bottom**) on BRL10 (**left**). Candidate genes and significant SNPs in the selected LD block (**right**). (**B–D**) qRT-PCR analysis of the candidate genes during *M. oryzae* infection. The x-axis indicates the time points post-inoculation. The y-axis indicates the relative transcription level of candidate genes. Error bars, mean ± SD (n = 3).

Table 2. The annotation of the candidate genes.

Gene	Annotation	Position	Length
LOC_Os11g46880	protein kinase domain containing	Chr11:28159898-28165301	711
LOC_Os11g46870	protein kinase	Chr11:28142191-28148337	680
LOC_Os11g46900	wall-associated receptor kinase	Chr11:28174980-28179995	707
LOC_Os11g46890	expressed protein	Chr11:28172388-28172985	199

Table 3. Types of variations observed in significant SNPs within the candidate region.

SNP	Ref	Alt	Mutant	Gene	Change
Chr11-28136278	G	A	Intergenic	LOC_Os11g46860-LOC_Os11g46870	-
Chr11-28141062	C	T	Intergenic	LOC_Os11g46860-LOC_Os11g46870	-
Chr11-28141136	G	A	Intergenic	LOC_Os11g46860-LOC_Os11g46870	-
Chr11-28168626	G	A	Intergenic	LOC_Os11g46880-LOC_Os11g46890	-
Chr11-28172439	C	T	Nonsynonymous	LOC_Os11g46890	Gly183Ser
Chr11-28186916	G	T	Intergenic	LOC_Os11g46900-LOC_Os11g46910	-
Chr11-28187112	C	A	Intergenic	LOC_Os11g46900-LOC_Os11g46910	-

2.4. Pb4 Positively Regulates Resistance to Rice Blast

LOC_Os11g46880 is predicted to contain an extracellular N-terminal signal peptide, GUB domain, EGF domain, EGF-Ca^{2+} domain, and intracellular Ser/Thr protein kinase domain, indicative of a typical WAK protein. There are 125 and 25 WAKs that have been annotated in rice and *Arabidopsis thaliana*, respectively. In rice, the 125 WAKs can be classified into 67 WAK-RLKs containing both extracellular EGF-like domains and the cytoplasmic kinase domain, 28 WAK-RLCKs containing only the cytoplasmic kinase domain, 13 WAK-RLPs containing only the extracellular EGF-like domain, 12 WAK-short genes, which share 40% identity with a longer WAK protein but lack a domain, and 5 WAK-pseudogenes with stop codons or frameshifts in the coding region. The kinase domain of the 67 WAK-RLKs and 27 WAK-RLCKs in rice, along with the 24 *Arabidopsis thaliana* WAKs and LOC_Os11g46880, were used to conduct an ML evolutionary tree. The evolutionary tree revealed that most WAKs were species-specific, and genes in the WAK-RLK or WAK-RLCK were more formed with a clade, indicating a significant correlation between the kinase domain and the extracellular domain. Notably, LOC_Os11g46880 was most similar to the WAK-RLK gene *OsWAK122* in a rice WAK-RLK clade (Figure 5A). To further investigate the role of LOC_Os11g46880 in the infection process of rice blast fungus, we generated overexpression transgenic plants of LOC_Os11g46880. Transgenic lines 22, 23, and 25, which had a higher expression, exhibited significantly enhanced resistance to rice blast compared to the wild type (Figure 5B–D). It was indicated that LOC_Os11g46880 positively regulated resistance to rice panicle blast, and we named it Pb4, following the panicle resistance gene Pb3. Given that Pb4 is a WAK-RLK and most RLKs are located in the cell membrane, we conducted subcellular localization studies of *N. benthamiana* and confirmed its subcellular location in the cell membrane (Figure 6C). Previous reports showed that OsWAK's expression is under the control of CEBiP, and the extracellular domain of CEBiP can interact with the extracellular domain of CERK [31,32]. In yeast two-hybrid assays, we found that the extracellular domain of Pb4 also interacted with the extracellular domain of CEBiP. However, the full-length Pb4 did not interact with the full-length CEBiP in yeast, which might be due to the presence of a transmembrane domain (Figure 6A,B). Given that CEBiP acts as a receptor for chitin in rice, we hypothesize that the expression of *Pb4* may also be induced by chitin. By treating rice seedlings with chitin, the expression of *Pb4* was indeed induced (Figure 5E). Several WAKs have been reported to act as receptors for OGs. We then treated rice with OGs and found that the expression of *Pb4* was significantly induced (Figure 5F). Taken together, as a WAK-RLK, *Pb4* positively regulates rice resistance against rice panicle blast. The interaction between the extracellular domain of PB4 and CEBiP, as well as the induced expression by chitin and polygalacturonic acid, suggest that *Pb4* may participate in the immune response of rice to chitin and oligogalacturonides.

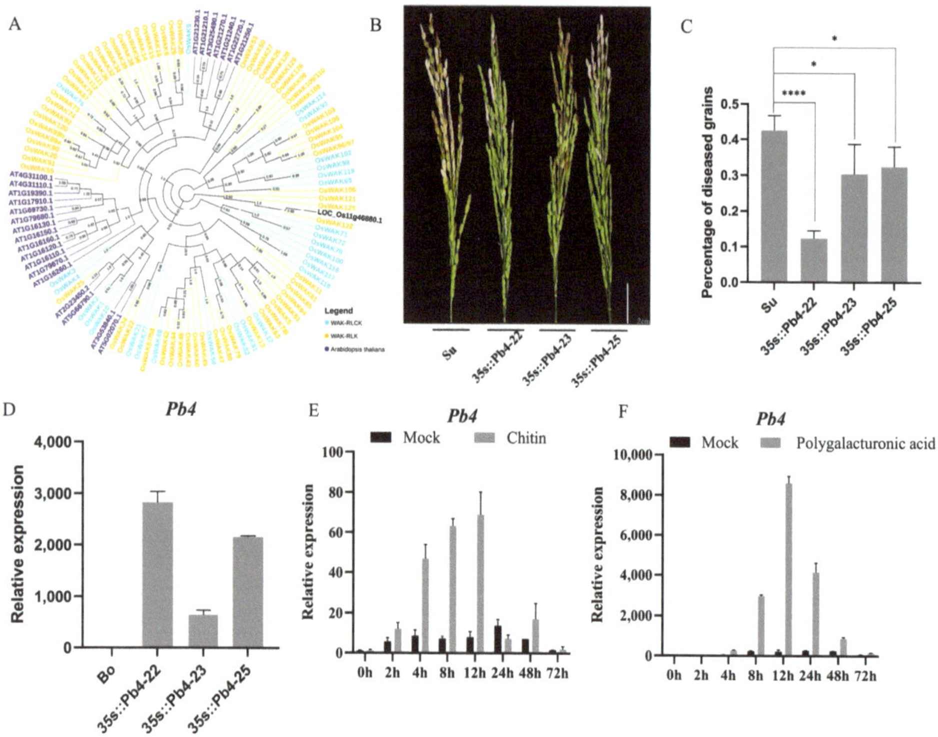

Figure 5. *Pb4* positively regulates resistance to rice blast. (**A**) Evolutionary tree of Pb4, along with 67 WAK-RLKs, 27 WAK-RLCKs (1 WAK-RLCK was not predicted a kinase domain) in rice, and 24 WAKs (1 WAK was not predicted a kinase domain) with kinase domain in *Arabidopsis*. The purple branches represent *Arabidopsis* WAKs, the yellow branches represent rice WAK-RLKs, the blue branches represent rice WAK-RLCKs, and the black branch corresponds to LOC_Os11g46880, which is Pb4. (**B**,**C**) Increased blast resistance of *Pb4*. Panicles were inoculated with *M. oryzae* at the booting stage. The diseased grains were counted at 10 dpi. Scale bar, 2 cm. * $p < 0.05$, **** $p < 0.0001$. (**D**) Expression levels of *Pb4* in overexpression plants. Error bars, mean ± SD ($n = 3$). (**E**,**F**) Two-week-old rice seedlings were treated with chitin (10 μg/mL) or polygalacturonic acid (100 μg/mL). Error bars indicate mean ± standard deviation ($n = 3$). Two-week-old rice was treated with chitin or polygalacturonic acid. Error bars, mean ± SD ($n = 3$).

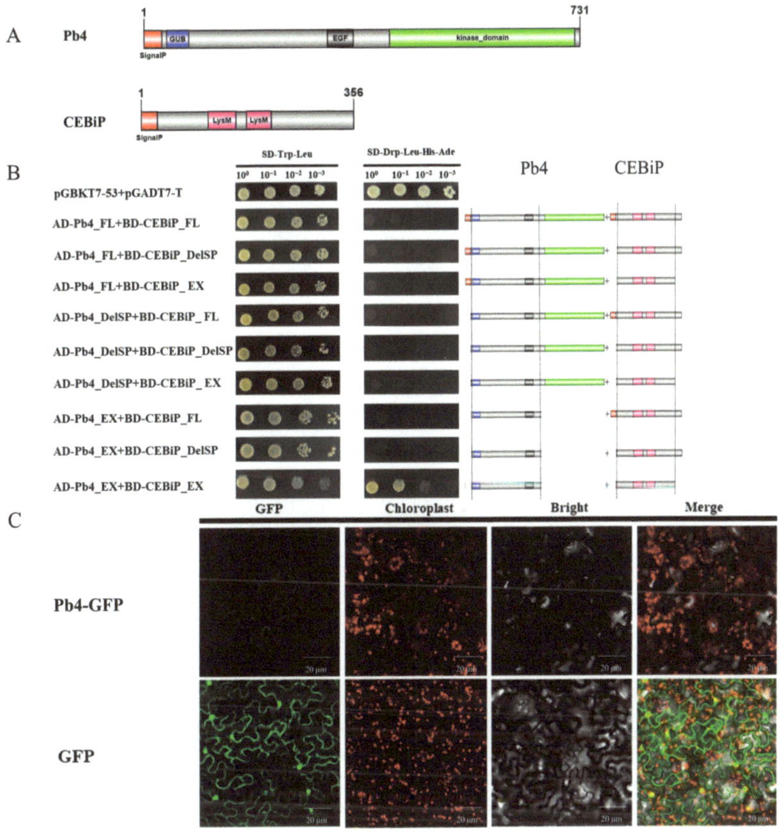

Figure 6. Pb4 interacted with CEBiP and was located in the cell membrane. (**A**) Protein structure diagram of Pb4 and CEBiP. (**B**) Interaction between the extracellular domains of Pb4 and CEBiP in yeast. The right shows the protein structure diagrams of truncated Pb4 and CEBiP. FL: full length; DelSP: Delete Signal Peptide; EX: extracellular (excluding SP); CY: cytoplasmic. (**C**) Subcellular localization of Pb4-GFP on the cell membrane in tobacco cells.

3. Discussion

Rice blast disease, caused by the fungus *Magnaporthe oryzae*, is one of the most devastating diseases that significantly impact rice production globally. Among various disease symptoms, the seedling blast and panicle blast are the most common; however, the economic and production losses caused by panicle blast are more significant [33]. Though more than 100 genes have been identified for seedling blast, very few genes conferring resistance to panicle blast were cloned [33]. Developing and utilizing resistant rice varieties is the most efficient and cost-effective approach to control blast. However, currently, only a few of the cloned R genes can be applied in breeding due to a lack of broad-spectrum and durable resistance. Among the cloned R genes, *Pi1, Pi5, Pi33, Pi54, Piz, Piz-t, Pi2, Pi9, Pi40,* and *Pigm* confer broad-spectrum resistance to leaf blast [34–41]. *Pb1, Pb2, Pb3, Pi25, Pi64, Pi68(t),* and *Pigm* loci confer resistance to panicle blast, and only *Pigm* provides broad-spectrum resistance to panicle blast [3–8,42].

Compared to traditional linkage mapping, GWAS is time-saving and labor-saving and can be used to discover novel disease-associated-resistant genes with higher resolution [23]. In this research, we identified 68 blast resistance loci through GWAS assays using 249 rice varieties selected from the 3K rice population. Through the LD block and expression analysis of the candidate genes after inoculating with rice blast fungus, we identified LOC_Os11g46880 as a resistance-related gene for further research and named it *Pb4*. *Pb4* encodes a typical WAK-RLK protein with an N-terminal signal peptide, extracellular GUB domain, EGF domain, EGF-Ca^{2+} domain, and an intracellular Ser/Thr protein kinase domain. The overexpression transgenic plants of *Pb4* showed enhanced resistance to panicle blast with a reduced diseased grains rate compared to the wild type.

As typical RLK proteins, WAKs are a superfamily with 125 annotated genes in rice and 25 genes in *Arabidopsis thaliana*. According to the conserved domains, the 125 WAKs in rice have been classified into five subfamilies, including 67 WAK-RLKs, 28 WAK-RLCKs, 13 WAK-RLPs, 12 WAK-short genes, and 5 WAK-pseudogenes. Previous reports suggest that WAKs are involved in the recognition of carbohydrate-based DAMPs or PAMPs as PRRs [43] and are connected with the cell wall through pectin binding sites [44]. In *Arabidopsis thaliana*, both AtWAK1 and AtWAK2 can bind to pectin through their extracellular non-EGF domains in vitro [45,46]. However, so far, only AtWAK1 has been confirmed as a receptor of oligogalacturonides (OGs) using domain-swapping methods in vivo, which activates the intracellular kinase domain and triggers downstream defense responses [46]. On the other hand, AtWAK2 is required for the activation of numerous genes, including MAPK6 phosphorylation, in the protoplasts upon pectin induction [45]. In this study, the expression of *Pb4* was significantly induced by polygalacturonic acid, suggesting it may be involved in OGs recognition as a PRR protein.

Several studies have directly demonstrated the roles of WAKs in the resistance responses against bacterial and fungal pathogens. In *Arabidopsis*, overexpression of *AtWAK1* enhances resistance to *Botrytis cinerea* [47], while *AtWAKL22* confers dominant resistance to Fusarium wilt disease [48]. In rice, the recently cloned disease resistance gene *Xa4*, encoding a WAK protein, can promote cellulose synthesis, inhibit cell wall loosening, strengthen the plant cell wall, and provide durable resistance against rice bacterial blight [49]. Subcellular localization and apoplastic fractionation experiments in *N. benthamiana* revealed that OsWAK1 is associated with the cell wall, and its expression is induced by the rice blast fungus, salicylic acid (SA), methyl jasmonate (MeJA), and mechanical injury. *OsWAK14*, *OsWAK91*, and *OsWAK92* mediate rice blast resistance, triggering ROS burst and the expression of defense-related genes. As a large gene family in rice, WAKs may form a heteromeric complex to conduct function. Yeast two-hybrid experiments have shown that OsWAK14, OsWAK91, and OsWAK92 can indeed form both homomeric and heteromeric complexes [50,51]. In cotton, GhWAK7A can interact with GhLYK5 and GhCERK1 in the uninfected state. However, during fungal pathogen infection, GhLYK5 can recognize chitin, which is released from fungal cell walls, and interact with GhCERK1 to promote the phosphorylation of GhWAK7A. Phosphorylated GhLYK5 may facilitate and maintain the formation of the GhLYK5-GhCERK1 complex, thus further activating cytoplasmic signaling and eliciting defense responses [16]. We also found that the expression of *Pb4* is inducible by chitin, and the extracellular domain of Pb4 can interact with the extracellular domain of CEBiP in yeast two-hybrid assays. As an RLP protein without an intracellular kinase domain, CEBiP often interacts with an RLK protein, such as CERK1, to transmit signals from the extracellular to intracellular, leading to the activation of downstream signaling pathways. The interaction between Pb4 and CEBiP may indicate they form complexes to recognize external signals and activate intracellular defense signaling.

In summary, we identified a novel non-NLR panicle resistance gene through GWAS, which belongs to the WAK gene family. As an RLK protein, Pb4 may act as a receptor for the OGs derived from the cell wall and form a co-receptor with CEBiP to recognize the chitin released from the *M. oryzae* cell wall to activate PTI. It might also confer broad-spectrum resistance to pathogens through strengthening the plant's cell wall. The specific mechanisms underlying intracellular signal transmission, as well as the potential recognition of other PAMPs or DAMPs by WAKs and the resistance spectrum of Pb4, require further investigation. In summary, this study identified a novel panicle blast-resistant gene for breeding and provides a theoretical basis for understanding the role of WAKs in mediating rice resistance against rice blast disease.

4. Materials and Methods

4.1. Plant and Fungal Materials

A total of 249 cultivars from the 3K rice accessions with a relatively uniform growth period were selected to grow in Nanjing, Jiangsu province, which were provided by the Shanghai Institute of Plant Physiology and Ecology. The rice blast strain Hoku1 used in this study was provided by the Institute of Crop Science, Chinese Academy of Agricultural Sciences. The blast fungi were grown on corn–rice–straw agar plates at 28 °C for seven days and then transferred in black light (20 W) in an incubator at 28 °C for sporulation and culture for 7–10 days to promote spores production. For panicle blast inoculation, rice plants in the booting stage were inoculated with the conidial suspension by using an injection method as described previously [52]. Three plants of each line and three panicles per plant were inoculated with a conidial suspension (1×10^5 conidia/mL). The disease severity was visually assessed, and the percentage of diseased grains was determined two weeks after inoculation according to the method previously described [53].

4.2. Population Structure and GWAS Analysis

The population structure of the 249 cultivars was analyzed using ADMIXTURE (version 1.3.0) on the core SNPs [54]. The k (number of groups) was set from 2 to 12, and k = 7 was chosen due to its cross-validation error being minimum. The group membership for each sample was defined by applying the threshold of ≥0.65 to this matrix. If the sum of the components for the subpopulations within the major groups XI and GJ was ≥0.65, the samples were classified as XI-adm or GJ-adm, respectively, and the remaining samples were deemed admixed (admix). The construction of phylogenetic trees was performed using FastTree based on the maximum likelihood method using the core SNPs [55].

GWAS analysis of panicle blast resistance was conducted using the mixed linear model in Tassel 5.0 and the filtered SNPs (MAF > 0.05, missing rate < 50%) [54]. The p-value < 10^{-4} was chosen as the threshold to define significant SNPs, and a region with more than eight consecutive SNPs (the distance between SNPs was less than 200 kb) was named as a locus. The kinship and population structure (k = 7) were used as covariates. Manhattan and QQ plots were plotted using the R package "cmplot" [56]. A linkage disequilibrium (LD) heatmap was generated with LDBlockShow [57].

4.3. RNA Extraction and Real-Time PCR

For the expression pattern analysis of *Pb4*, samples were collected at different time points after the rice blast pathogen inoculation or treated with chitin at a concentration of 10 μg/mL and polygalacturonic acid at a concentration of 100 μg/mL, immediately frozen in liquid nitrogen, and stored at −80 °C. Total RNA was extracted using the ATGPure® Cell/Tissue RNA Extraction Kit (Code: R201). The first-strand cDNA synthesis was then conducted using HiScript® II RT SuperMix (Code: R223-01). For the qRT-PCR, the AceQ® qPCR SYBR Green Master Mix (Product Code: Q111-02) was employed, and the entire qRT-PCR process was carried out on an LC480 II qPCR system (Roche fluorescence quantitative LightCycler480, San Francisco, CA, USA). The expression level of *Actin* was used as an internal control.

4.4. Construction of Transgenic Rice Plants

Full-length cDNA of *Pb4* was inserted into the pCAMBIA1300s vector to generate the overexpression transgenic vector driven by the 35s promoter. The verified vector plasmid was then transformed into susceptible rice plants, Suyunuo, using the Agrobacterium-mediated method. The primers YZ-Pb4-F/R and q-Pb4-F/R were used to validate positive transgenic plants and analyze the expression levels of *Pb4* (Supplemental Table S1).

4.5. Subcellular Localization

Full-length cDNA of *Pb4*, fused with GFP, was constructed into the PCAMBIA1300s-GFP vector driven by the 35s promoter to generate the subcellular localization vector. The Agrobacterium carrying the Pb4-GFP and P19 plasmid were respectively prepared in the infiltration medium (10 mM $MgCl_2$, 10 mM MES, 100 μM Acetosyringone, pH = 5.6) to an OD_{600} = 1.0. They were mixed in equal amounts and injected into *N. benthamiana* leaves. After 48 h, the samples were observed using a Leica laser scanning confocal microscope.

4.6. Y2H Assay

Y2H assays were conducted using the Gold Yeast Two-Hybrid System (Clontech) protocol. Different lengths of *Pb4* and *CEBiP* cDNA fragments were cloned into the pGADT7 and pGBKT7 vectors, respectively. pGBKT7 and pGADT7 vectors were co-transformed into the Y2H Gold yeast strain. The transformed yeast cells were grown on selection plates (SD/-Leu/-Trp/-His/-Ade) to detect the interactions.

Supplementary Materials: The supporting information can be downloaded at: https://www.mdpi.com/article/10.3390/ijms25020830/s1.

Author Contributions: Experimental design, Y.F., L.M. and Y.B.; field phenotyping, Y.F., L.M., X.P. and P.T.; data analysis, Y.F.; gene cloning and experimental validation, Y.F., W.W., K.L., Z.X. and C.L.; percentage diseased grains investigation, Y.F., L.M., X.P. and P.T.; writing—original draft preparation, Y.F. and Z.W.; writing—review and editing, Y.B., J.W. and H.Z. All authors have read and agreed to the published version of the manuscript.

Funding: This research was supported by grants from the National Key Project for Jiangsu Agriculture Science and Technology Innovation Fund (CX(2019)1008), the National Key Research and Development Program of China (2020YFE0202900), the Natural Science Foundation of China (32072033, 31871602, 31171516), the Natural Science Foundation of Jiangsu Province (BK20211212), the Hainan Provincial Natural Science Foundation of China (323MS070), the Guidance Foundation and the Sanya Institute of Nanjing Agricultural University (NAUSY-MS06), the Fundamental Research Funds for the Central Universities (XUEKEN2023021) and the Seed Industry Revitalization Project of Jiangsu Province (JBGS2021040).

Institutional Review Board Statement: Not applicable.

Informed Consent Statement: Not applicable.

Data Availability Statement: Data are contained within the article and Supplementary Materials.

Acknowledgments: We thank Yongfeng Liu for providing the rice blast isolates.

Conflicts of Interest: The authors declare no conflicts of interest.

References

1. Dean, R.; Van Kan, J.A.; Pretorius, Z.A.; Hammond-Kosack, K.E.; Di Pietro, A.; Spanu, P.D.; Rudd, J.J.; Dickman, M.; Kahmann, R.; Ellis, J.; et al. The Top 10 fungal pathogens in molecular plant pathology. *Mol. Plant Pathol.* **2012**, *13*, 414–430. [CrossRef] [PubMed]
2. Li, W.; Deng, Y.; Ning, Y.; He, Z.; Wang, G.L. Exploiting Broad-Spectrum Disease Resistance in Crops: From Molecular Dissection to Breeding. *Annu. Rev. Plant Biol.* **2020**, *71*, 575–603. [CrossRef] [PubMed]
3. Hayashi, N.; Inoue, H.; Kato, T.; Funao, T.; Shirota, M.; Shimizu, T.; Kanamori, H.; Yamane, H.; Hayano-Saito, Y.; Matsumoto, T.; et al. Durable panicle blast-resistance gene Pb1 encodes an atypical CC-NBS-LRR protein and was generated by acquiring a promoter through local genome duplication. *Plant J.* **2010**, *64*, 498–510. [CrossRef] [PubMed]
4. Yu, Y.; Ma, L.; Wang, X.; Zhao, Z.; Wang, W.; Fan, Y.; Liu, K.; Jiang, T.; Xiong, Z.; Song, Q.; et al. Genome-Wide Association Study Identifies a Rice Panicle Blast Resistance Gene, *Pb2*, Encoding NLR Protein. *Int. J. Mol. Sci.* **2022**, *23*, 5668. [CrossRef] [PubMed]
5. Ma, L.; Yu, Y.; Li, C.; Wang, P.; Liu, K.; Ma, W.; Wang, W.; Fan, Y.; Xiong, Z.; Jiang, T.; et al. Genome-Wide Association Study Identifies a Rice Panicle Blast Resistance Gene *Pb3* Encoding NLR Protein. *Int. J. Mol. Sci.* **2022**, *23*, 14032. [CrossRef]
6. Chen, J.; Shi, Y.; Liu, W.; Chai, R.; Fu, Y.; Zhuang, J.; Wu, J. A *Pid3* allele from rice cultivar Gumei2 confers resistance to *Magnaporthe oryzae*. *J. Genet. Genom.* **2011**, *38*, 209–216. [CrossRef]
7. Ma, J.; Lei, C.; Xu, X.; Hao, K.; Wang, J.; Cheng, Z.; Ma, X.; Ma, J.; Zhou, K.; Zhang, X.; et al. *Pi64*, Encoding a Novel CC-NBS-LRR Protein, Confers Resistance to Leaf and Neck Blast in Rice. *Mol. Plant Microbe Interact.* **2015**, *28*, 558–568. [CrossRef]

8. Devi, S.; Singh, K.; Umakanth, B.; Vishalakshi, B.; Rao, K.V.S.; Suneel, B.; Sharma, S.K.; Kadambari, G.K.M.; Prasad, M.S.; Senguttvel, P.; et al. Identification and Characterization of a Large Effect QTL from *Oryza glumaepatula* Revealed Pi68(t) as Putative Candidate Gene for Rice Blast Resistance. *Rice* **2020**, *13*, 17. [CrossRef]
9. Zhou, X.; Liao, H.; Chern, M.; Yin, J.; Chen, Y.; Wang, J.; Zhu, X.; Chen, Z.; Yuan, C.; Zhao, W.; et al. Loss of function of a rice TPR-domain RNA-binding protein confers broad-spectrum disease resistance. *Proc. Natl. Acad. Sci. USA* **2018**, *115*, 3174–3179. [CrossRef]
10. Gao, M.; He, Y.; Yin, X.; Zhong, X.; Yan, B.; Wu, Y.; Chen, J.; Li, X.; Zhai, K.; Huang, Y.; et al. Ca^{2+} sensor-mediated ROS scavenging suppresses rice immunity and is exploited by a fungal effector. *Cell* **2021**, *184*, 5391–5404.e17. [CrossRef]
11. Fukuoka, S.; Saka, N.; Koga, H.; Ono, K.; Shimizu, T.; Ebana, K.; Hayashi, N.; Takahashi, A.; Hirochika, H.; Okuno, K.; et al. Loss of function of a proline-containing protein confers durable disease resistance in rice. *Science* **2009**, *325*, 998–1001. [CrossRef] [PubMed]
12. Chen, X.; Shang, J.; Chen, D.; Lei, C.; Zou, Y.; Zhai, W.; Liu, G.; Xu, J.; Ling, Z.; Cao, G.; et al. AB-lectin receptor kinase gene conferring rice blast resistance. *Plant J.* **2006**, *46*, 794–804. [CrossRef] [PubMed]
13. Sun, X.; Cao, Y.; Yang, Z.; Xu, C.; Li, X.; Wang, S.; Zhang, Q. *Xa26*, a gene conferring resistance to *Xanthomonas oryzae* pv. *oryzae* in rice, encodes an LRR receptor kinase-like protein. *Plant J.* **2004**, *37*, 517–527. [CrossRef]
14. Hu, K.; Cao, J.; Zhang, J.; Xia, F.; Ke, Y.; Zhang, H.; Xie, W.; Liu, H.; Cui, Y.; Cao, Y.; et al. Improvement of multiple agronomic traits by a disease resistance gene via cell wall reinforcement. *Nat. Plants* **2017**, *3*, 17009. [CrossRef] [PubMed]
15. Kanneganti, V.; Gupta, A.K. Wall associated kinases from plants—An overview. *Physiol. Mol. Biol. Plants* **2008**, *14*, 109–118. [CrossRef] [PubMed]
16. Verica, J.A.; He, Z.H. The cell wall-associated kinase (WAK) and WAK-like kinase gene family. *Plant Physiol.* **2002**, *129*, 455–459. [CrossRef] [PubMed]
17. Park, A.R.; Cho, S.K.; Yun, U.J.; Jin, M.Y.; Lee, S.H.; Sachetto-Martins, G.; Park, O.K. Interaction of the *Arabidopsis* receptor protein kinase Wak1 with a glycine-rich protein, AtGRP-3. *J. Biol. Chem.* **2001**, *276*, 26688–26693. [CrossRef]
18. Decreux, A.; Thomas, A.; Spies, B.; Brasseur, R.; Van Cutsem, P.; Messiaen, J. In vitro characterization of the homogalacturonan-binding domain of the wall-associated kinase WAK1 using site-directed mutagenesis. *Phytochemistry* **2006**, *67*, 1068–1079. [CrossRef]
19. Cabrera, J.C.; Boland, A.; Messiaen, J.; Cambier, P.; Van Cutsem, P. Egg box conformation of oligogalacturonides: The time-dependent stabilization of the elicitor-active conformation increases its biological activity. *Glycobiology* **2008**, *18*, 473–482. [CrossRef]
20. Kohorn, B.D.; Johansen, S.; Shishido, A.; Todorova, T.; Martinez, R.; Defeo, E.; Obregon, P. Pectin activation of MAP kinase and gene expression is WAK2 dependent. *Plant J.* **2009**, *60*, 974–982. [CrossRef]
21. Zhai, K.; Deng, Y.; Liang, D.; Tang, J.; Liu, J.; Yan, B.; Yin, X.; Lin, H.; Chen, F.; Yang, D.; et al. RRM Transcription Factors Interact with NLRs and Regulate Broad-Spectrum Blast Resistance in Rice. *Mol. Cell* **2019**, *74*, 996–1009.e7. [CrossRef] [PubMed]
22. Harkenrider, M.; Sharma, R.; De Vleesschauwer, D.; Tsao, L.; Zhang, X.; Chern, M.; Canlas, P.; Zuo, S.; Ronald, P.C. Overexpression of Rice Wall-Associated Kinase 25 (*OsWAK25*) Alters Resistance to Bacterial and Fungal Pathogens. *PLoS ONE* **2016**, *11*, e0147310. [CrossRef] [PubMed]
23. Demirjian, C.; Vailleau, F.; Berthome, R.; Roux, F. Genome-wide association studies in plant pathosystems: Success or failure? *Trends Plant Sci.* **2023**, *28*, 471–485. [CrossRef] [PubMed]
24. Wang, W.; Mauleon, R.; Hu, Z.; Chebotarov, D.; Tai, S.; Wu, Z.; Li, M.; Zheng, T.; Fuentes, R.R.; Zhang, F.; et al. Genomic variation in 3010 diverse accessions of Asian cultivated rice. *Nature* **2018**, *557*, 43–49. [CrossRef] [PubMed]
25. Liu, M.H.; Kang, H.; Xu, Y.; Peng, Y.; Wang, D.; Gao, L.; Wang, X.; Ning, Y.; Wu, J.; Liu, W.; et al. Genome-wide association study identifies an NLR gene that confers partial resistance to *Magnaporthe oryzae* in rice. *Plant Biotechnol. J.* **2020**, *18*, 1376–1383. [CrossRef]
26. Xu, Y.; Bai, L.; Liu, M.; Liu, Y.; Peng, S.; Hu, P.; Wang, D.; Liu, Q.; Yan, S.; Gao, L.; et al. Identification of two novel rice S genes through combination of association and transcription analyses with gene-editing technology. *Plant Biotechnol. J.* **2023**, *21*, 1628–1641. [CrossRef]
27. Bryan, G.T.; Wu, K.S.; Farrall, L.; Jia, Y.; Hershey, H.P.; McAdams, S.A.; Faulk, K.N.; Donaldson, G.K.; Tarchini, R.; Valent, B. A single amino acid difference distinguishes resistant and susceptible alleles of the rice blast resistance gene *Pi-ta*. *Plant Cell* **2000**, *12*, 2033–2046. [CrossRef]
28. Liu, Y.; Liu, B.; Zhu, X.; Yang, J.; Bordeos, A.; Wang, G.; Leach, J.E.; Leung, H. Fine-mapping and molecular marker development for *Pi56(t)*, a NBS-LRR gene conferring broad-spectrum resistance to *Magnaporthe oryzae* in rice. *Theor. Appl. Genet.* **2013**, *126*, 985–998. [CrossRef]
29. Xu, X.; Hayashi, N.; Wang, C.T.; Fukuoka, S.; Kawasaki, S.; Takatsuji, H.; Jiang, C.J. Rice blast resistance gene *Pikahei-1(t)*, a member of a resistance gene cluster on chromosome 4, encodes a nucleotide-binding site and leucine-rich repeat protein. *Mol. Breed.* **2014**, *34*, 691–700. [CrossRef]
30. Li, C.; Wang, D.; Peng, S.; Chen, Y.; Su, P.; Chen, J.; Zheng, L.; Tan, X.; Liu, J.; Xiao, Y.; et al. Genome-wide association mapping of resistance against rice blast strains in South China and identification of a new *Pik* allele. *Rice* **2019**, *12*, 47. [CrossRef]
31. Delteil, A.; Gobbato, E.; Cayrol, B.; Estevan, J.; Michel-Romiti, C.; Dievart, A.; Kroj, T.; Morel, J.B. Several wall-associated kinases participate positively and negatively in basal defense against rice blast fungus. *BMC Plant Biol.* **2016**, *16*, 17. [CrossRef]

32. Kouzai, Y.; Mochizuki, S.; Nakajima, K.; Desaki, Y.; Hayafune, M.; Miyazaki, H.; Yokotani, N.; Ozawa, K.; Minami, E.; Kaku, H.; et al. Targeted Gene Disruption of Reveals Its Indispensable Role in Chitin Perception and Involvement in the Peptidoglycan Response and Immunity in Rice. *Mol. Plant Microbe Interact.* **2014**, *27*, 975–982. [CrossRef]
33. Kalia, S.; Rathour, R. Current status on mapping of genes for resistance to leaf- and neck-blast disease in rice. *3 Biotech.* **2019**, *9*, 209. [CrossRef] [PubMed]
34. Xiao, N.; Wu, Y.Y.; Li, A.L. Strategy for Use of Rice Blast Resistance Genes in Rice Molecular Breeding. *Rice Sci.* **2020**, *27*, 263–277.
35. Chen, H.L.; Chen, B.T.; Zhang, D.P.; Xie, Y.F.; Zhang, Q. Pathotypes of Pyricularia grisea in Rice Fields of Central and Southern China. *Plant Dis.* **2001**, *85*, 843–850. [CrossRef] [PubMed]
36. Berruyer, R.; Adreit, H.; Milazzo, J.; Gaillard, S.; Berger, A.; Dioh, W.; Lebrun, M.H.; Tharreau, D. Identification and fine mapping of *Pi33*, the rice resistance gene corresponding to the *Magnaporthe grisea* avirulence gene *ACE1*. *Theor. Appl. Genet.* **2003**, *107*, 1139–1147. [CrossRef] [PubMed]
37. Jeon, J.S.; Chen, D.; Yi, G.H.; Wang, G.L.; Ronald, P.C. Genetic and physical mapping of *Pi5(t)*, a locus associated with broad-spectrum resistance to rice blast. *Mol. Genet. Genom.* **2003**, *269*, 280–289. [CrossRef] [PubMed]
38. Qu, S.; Liu, G.; Zhou, B.; Bellizzi, M.; Zeng, L.; Dai, L.; Han, B.; Wang, G.L. The broad-spectrum blast resistance gene *Pi9* encodes a nucleotide-binding site-leucine-rich repeat protein and is a member of a multigene family in rice. *Genetics* **2006**, *172*, 1901–1914. [CrossRef]
39. Zhou, B.; Qu, S.; Liu, G.; Dolan, M.; Sakai, H.; Lu, G.; Bellizzi, M.; Wang, G.L. The eight amino-acid differences within three leucine-rich repeats between *Pi2* and *Piz-t* resistance proteins determine the resistance specificity to *Magnaporthe grisea*. *Mol. Plant Microbe Interact.* **2006**, *19*, 1216–1228. [CrossRef]
40. Jeung, J.U.; Kim, B.R.; Cho, Y.C.; Han, S.S.; Moon, H.P.; Lee, Y.T.; Jena, K.K. A novel gene, *Pi40(t)*, linked to the DNA markers derived from NBS-LRR motifs confers broad spectrum of blast resistance in rice. *Theor. Appl. Genet.* **2007**, *115*, 1163–1177. [CrossRef]
41. Thakur, S.; Singh, P.K.; Das, A.; Rathour, R.; Variar, M.; Prashanthi, S.K.; Singh, A.K.; Singh, U.D.; Chand, D.; Singh, N.K.; et al. Extensive sequence variation in rice blast resistance gene *Pi54* makes it broad spectrum in nature. *Front. Plant Sci.* **2015**, *6*, 345. [CrossRef] [PubMed]
42. Deng, Y.; Zhai, K.; Xie, Z.; Yang, D.; Zhu, X.; Liu, J.; Wang, X.; Qin, P.; Yang, Y.; Zhang, G.; et al. Epigenetic regulation of antagonistic receptors confers rice blast resistance with yield balance. *Science* **2017**, *355*, 962–965. [CrossRef] [PubMed]
43. Rebaque, D.; Del Hierro, I.; Lopez, G.; Bacete, L.; Vilaplana, F.; Dallabernardina, P.; Pfrengle, F.; Jorda, L.; Sanchez-Vallet, A.; Perez, R.; et al. Cell wall-derived mixed-linked beta-1,3/1,4-glucans trigger immune responses and disease resistance in plants. *Plant J.* **2021**, *106*, 601–615. [CrossRef] [PubMed]
44. Aziz, A.; Gauthier, A.; Bezier, A.; Poinssot, B.; Joubert, J.M.; Pugin, A.; Heyraud, A.; Baillieul, F. Elicitor and resistance-inducing activities of beta-1,4 cellodextrins in grapevine, comparison with beta-1,3 glucans and alpha-1,4 oligogalacturonides. *J. Exp. Bot.* **2007**, *58*, 1463–1472. [CrossRef] [PubMed]
45. Casasoli, M.; Spadoni, S.; Lilley, K.S.; Cervone, F.; De Lorenzo, G.; Mattei, B. Identification by 2-D DIGE of apoplastic proteins regulated by oligogalacturonides in *Arabidopsis thaliana*. *Proteomics* **2008**, *8*, 1042–1054. [CrossRef] [PubMed]
46. Souza, C.A.; Li, S.; Lin, A.Z.; Boutrot, F.; Grossmann, G.; Zipfel, C.; Somerville, S.C. Cellulose-Derived Oligomers Act as Damage-Associated Molecular Patterns and Trigger Defense-Like Responses. *Plant Physiol.* **2017**, *173*, 2383–2398. [CrossRef] [PubMed]
47. Denoux, C.; Galletti, R.; Mammarella, N.; Gopalan, S.; Werck, D.; De Lorenzo, G.; Ferrari, S.; Ausubel, F.M.; Dewdney, J. Activation of defense response pathways by OGs and Flg22 elicitors in Arabidopsis seedlings. *Mol. Plant* **2008**, *1*, 423–445. [CrossRef]
48. Davis, K.R.; Darvill, A.G.; Albersheim, P.; Dell, A. Host-Pathogen Interactions: XXIX. Oligogalacturonides Released from Sodium Polypectate by Endopolygalacturonic Acid Lyase Are Elicitors of Phytoalexins in Soybean. *Plant Physiol.* **1986**, *80*, 568–577. [CrossRef]
49. Simpson, S.D.; Ashford, D.A.; Harvey, D.J.; Bowles, D.J. Short chain oligogalacturonides induce ethylene production and expression of the gene encoding aminocyclopropane 1-carboxylic acid oxidase in tomato plants. *Glycobiology* **1998**, *8*, 579–583. [CrossRef]
50. Galletti, R.; Denoux, C.; Gambetta, S.; Dewdney, J.; Ausubel, F.M.; De Lorenzo, G.; Ferrari, S. The AtrbohD-mediated oxidative burst elicited by oligogalacturonides in *Arabidopsis* is dispensable for the activation of defense responses effective against Botrytis cinerea. *Plant Physiol.* **2008**, *148*, 1695–1706. [CrossRef]
51. Galletti, R.; Ferrari, S.; De Lorenzo, G. Arabidopsis MPK3 and MPK6 play different roles in basal and oligogalacturonide- or flagellin-induced resistance against *Botrytis cinerea*. *Plant Physiol.* **2011**, *157*, 804–814. [CrossRef]
52. Fang, N.; Wei, X.; Shen, L.; Yu, Y.; Li, M.; Yin, C.; He, W.; Guan, C.; Chen, H.; Zhang, H.; et al. Fine mapping of a panicle blast resistance gene *Pb-bd1* in Japonica landrace Bodao and its application in rice breeding. *Rice* **2019**, *12*, 18. [CrossRef]
53. Alexander, D.H.; Novembre, J.; Lange, K. Fast model-based estimation of ancestry in unrelated individuals. *Genome Res.* **2009**, *19*, 1655–1664. [CrossRef]
54. Price, M.N.; Dehal, P.S.; Arkin, A.P. FastTree: Computing large minimum evolution trees with profiles instead of a distance matrix. *Mol. Biol. Evol.* **2009**, *26*, 1641–1650. [CrossRef]
55. Bradbury, P.J.; Zhang, Z.; Kroon, D.E.; Casstevens, T.M.; Ramdoss, Y.; Buckler, E.S. TASSEL: Software for association mapping of complex traits in diverse samples. *Bioinformatics* **2007**, *23*, 2633–2635. [CrossRef]

56. Yin, L.; Zhang, H.; Tang, Z.; Xu, J.; Yin, D.; Zhang, Z.; Yuan, X.; Zhu, M.; Zhao, S.; Li, X.; et al. rMVP: A Memory-efficient, Visualization-enhanced, and Parallel-accelerated Tool for Genome-wide Association Study. *Genom. Proteom. Bioinform.* **2021**, *19*, 619–628. [CrossRef]
57. Dong, S.S.; He, W.M.; Ji, J.J.; Zhang, C.; Guo, Y.; Yang, T.L. LDBlockShow: A fast and convenient tool for visualizing linkage disequilibrium and haplotype blocks based on variant call format files. *Brief. Bioinform.* **2021**, *22*, bbaa227. [CrossRef]

International Journal of
Molecular Sciences

MDPI

Article

The RING-Type Domain-Containing Protein GNL44 Is Essential for Grain Size and Quality in Rice (*Oryza sativa* L.)

Lei He [1,2], Tao Chen [1,2], Wenhua Liang [1,2], Chunfang Zhao [1,2], Ling Zhao [1,2], Shu Yao [1,2], Lihui Zhou [1,2], Zhen Zhu [1,2], Qingyong Zhao [1,2], Kai Lu [1,2], Cailin Wang [1,2], Li Zhu [3,*] and Yadong Zhang [1,2,*]

1 Institute of Food Crops, Key Laboratory of Jiangsu Province for Agrobiology, East China Branch of National Center of Technology Innovation for Saline-Alkali Tolerant Rice, Jiangsu Academy of Agricultural Science, Nanjing 210014, China; clwang@jaas.ac.cn (C.W.)

2 Zhongshan Biological Breeding Laboratory, Nanjing 210014, China

3 State Key Laboratory of Rice Biology, China National Rice Research Institute, Hangzhou 311400, China

* Correspondence: zhuli05@caas.cn (L.Z.); zhangyd@jaas.ac.cn (Y.Z.)

Abstract: Grain size in rice (*Oryza sativa* L.) shapes yield and quality, but the underlying molecular mechanism is not fully understood. We functionally characterized *GRAIN NUMBER AND LARGE GRAIN SIZE 44* (*GNL44*), encoding a RING-type protein that localizes to the cytoplasm. The *gnl44* mutant has fewer but enlarged grains compared to the wild type. *GNL44* is mainly expressed in panicles and developing grains. Grain chalkiness was higher in the *gnl44* mutant than in the wild type, short-chain amylopectin content was lower, middle-chain amylopectin content was higher, and appearance quality was worse. The amylose content and gel consistency of *gnl44* were lower, and protein content was higher compared to the wild type. Rapid Visco Analyzer results showed that the texture of cooked *gnl44* rice changed, and that the taste value of *gnl44* was lower, making the eating and cooking quality of *gnl44* worse than that of the wild type. We used *gnl44*, *qgl3*, and *gs3* monogenic and two-gene near-isogenic lines to study the effects of different combinations of genes affecting grain size on rice quality-related traits. Our results revealed additive effects for these three genes on grain quality. These findings enrich the genetic resources available for rice breeders.

Keywords: appearance quality; eating and cooking quality; grain yield; grain width; rice

Citation: He, L.; Chen, T.; Liang, W.; Zhao, C.; Zhao, L.; Yao, S.; Zhou, L.; Zhu, Z.; Zhao, Q.; Lu, K.; et al. The RING-Type Domain-Containing Protein GNL44 Is Essential for Grain Size and Quality in Rice (*Oryza sativa* L.). *Int. J. Mol. Sci.* **2024**, *25*, 589. https://doi.org/10.3390/ijms25010589

Academic Editor: Jiří Fajkus

Received: 22 November 2023
Revised: 15 December 2023
Accepted: 23 December 2023
Published: 2 January 2024

1. Introduction

Rice (*Oryza sativa* L.) is an important food crop that provides the daily dietary calories for much of the human population; as such, its yield is crucial to ensuring food security. Grain weight, grain number per panicle, and panicle number per plant are the most important agronomic components of grain yield. Numerous genes regulating grain number have been identified in rice over the years [1–5]. These genes function in various periods of spike development. Nevertheless, knowledge of the genetic and molecular mechanisms that establish grain number in rice is still limited.

Grain weight is determined by the combination of grain length, width, and thickness. Many quantitative trait loci (QTLs) and genes related to grain size have been functionally characterized in rice. Grain size-related genes that have been cloned to date affect one of five pathways: the ubiquitin proteasome pathway, phytohormone signaling or biosynthesis, transcriptional regulation, mitogen-activated protein kinase (MAPK) signaling, or G protein signaling [6]. For instance, *GRAIN WIDTH 2* (*GW2*) encodes a RING-type protein that negatively regulates grain width, as plants carrying a loss-of-function *gw2* allele produce wider grains [7]; *DECREASED GRAIN SIZE1* (*DGS1*) encodes a C_3HC_4-type RING protein that regulates rice grain size and participates in the ubiquitin proteasome pathway [8]. *PROTEIN PHOSPHATASE WITH KELCH-LIKE REPEATS 1* (*OsPPKL1*), also named *GRAIN LENGTH 3.1* (*GL3.1*), encodes a protein phosphatase containing two Kelch domains that negatively regulate rice grain length through a modulation of brassinosteroid

(BR) signaling [9]. *GW5* and *GRAIN SIZE 5* (*GS5*), *OsGSK5/TGW3/GL3.3*, and *OsGSK2* are also involved in BR signaling [10–13]. Similarly, *THOUSAND GRAIN WEIGHT 6* (*TGW6*) and *BIG GRAIN 1* (*BG1*) regulate grain size through auxin signaling [14,15]. *Grain Length, Width and Weight 7* (*GLW7*) encodes the plant-specific transcription factor OsSPL13, which regulates grain size by controlling cell proliferation in the spikelet hull [16]. The MAPK kinase kinase 10–MAPK kinase 4–MAPK6 (OsMKKK10–OsMKK4–OsMAPK6) regulatory module plays an important role in regulating grain size [4,17,18]. GRAIN SIZE AND NUMBER 1 (GSN1), also named MAPK PHOSPHATASE 1 (OsMKP1), negatively regulates the OsMKKK10–OsMKK4–OsMPK6 cascade to coordinate the balance between grain size and grain number. The rice genome harbors one gene each for *Gα* and *Gβ* and five *Gγ* genes. Both Gα and Gβ proteins can positively regulate grain size [19]. Gγs can be classified into three distinct types according to their C-terminal structures. Group I (RGG1) and Group II (RGG2) participate in the regulation of abiotic stress [20,21]. DENSE AND ERECT PANICLE 1 (DEP1), GGC2, and GRAIN SIZE3 (GS3) belong to Group III and can regulate grain size and rice yield. Moreover, the manipulation of *DEP1*, *GGC2*, or *GS3* transcript levels can affect grain size [22–24].

In addition to yield, grain size can also affect the quality of rice. In general, large grains are associated with high yield but poor quality, especially the appearance quality of milled rice (grains with the outer hull removed), which includes chalkiness [25,26]. Slender rice is popular with consumers for its transparent appearance and lack of chalkiness. The genotype at *GW7* can improve rice quality by changing the pattern of cell division, making rice grains more slender and less chalky [18]. Although several grain size genes have been cloned in rice, the genetic and molecular mechanisms that determine grain size remain largely unknown [27]. Functional identification of other grain size genes or QTLs is important to further our understanding of the molecular mechanisms regulating grain-related traits and to help meet the demand for high-yielding and superior-quality cultivars.

Although numerous QTLs/genes related to grain size have been functionally characterized, there have been few studies about the genetic interactions of these grain size-related genes [28]. Gao et al. (2015) used the near-isogenic lines (NILs) NIL *GS3*, NIL-*qgl3*, and NIL-*GS3/qgl3* to study the interaction effects of *GS3* and *qGL3*, finding that *GS3* and *qGL3* have an additive effect on regulating rice grain length [29]. He et al. (2023) used NILs that differed by a single gene or two genes to study the genetic effects of the grain size-related genes *gw2*, *gs3*, and *qgl3* on grain size, finding that combinations of these three genes had additive effects on grain size [30]. These studies focused on the effects of grain size-related genes on grain size and yield. However, less is known about the effects of grain size-related genes on rice quality.

Here, we report that *GNL44*, encoding a RING-type E3 ubiquitin ligase, regulates grain size and quality in rice. We analyzed the effects of *gnl44*, *qgl3*, and *gs3* on rice quality-related traits using monogenic and two-gene NILs. Our results revealed additive effects for these three genes on grain quality. The findings may facilitate our understanding of the mechanisms underlying the determination of grain size and quality. Our results also provide the necessary theoretical basis and genetic resources for breeding high-yield and superior-quality rice cultivars.

2. Results

2.1. The *gnl44* Mutant Produces Fewer but Larger Grains

The *gnl44* mutant was isolated from EMS-mutagenized M_2 populations of the WT *japonica* cultivar Wuyunjing. *gnl44* mutant plants were slightly shorter than the WT (Figure 1A,E). Panicle length, the number of primary branches, and the number of secondary branches were also lower in *gnl44* compared to the WT (Figure 1B,F–H). The grain number per panicle of *gnl44* was about 43.95% lower than that of the WT (Figure 1I). In addition, the *gnl44* mutant produced large grains (Figure 1C,D). The length of *gnl44* grains increased by 9.8% (Figure 1J) and their width increased by 21.2% compared to the WT

(Figure 1K). The thickness of the *gnl44* grains increased by 15.3% compared to the WT (Figure 1L). The *gnl44* grains were also significantly heavier than the WT grains (Figure 1M). However, the grain yield per plant in *gnl44* was decreased by about 21.8% compared to the WT (Figure 1N). These results indicate that *gnl44* negatively regulates grain number and grain size in rice.

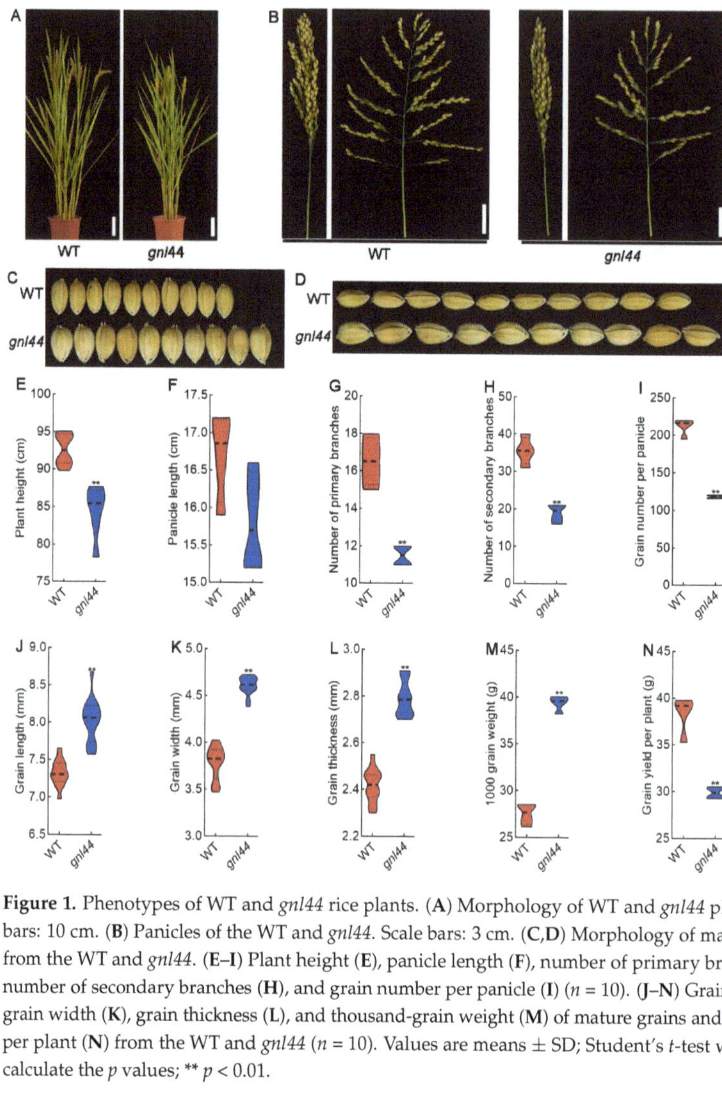

Figure 1. Phenotypes of WT and *gnl44* rice plants. (**A**) Morphology of WT and *gnl44* plants. Scale bars: 10 cm. (**B**) Panicles of the WT and *gnl44*. Scale bars: 3 cm. (**C**,**D**) Morphology of mature grains from the WT and *gnl44*. (**E–I**) Plant height (**E**), panicle length (**F**), number of primary branches (**G**), number of secondary branches (**H**), and grain number per panicle (**I**) (n = 10). (**J–N**) Grain length (**J**), grain width (**K**), grain thickness (**L**), and thousand-grain weight (**M**) of mature grains and grain yield per plant (**N**) from the WT and *gnl44* (n = 10). Values are means ± SD; Student's t-test was used to calculate the p values; ** $p < 0.01$.

2.2. GNL44 Affects Cell Division and Cell Expansion

Given that *gnl44* had longer and wider grains surrounded by the spikelet hull compared to the WT (Figure 2A), we produced paraffin sections from developing grains to analyze the parenchyma cell layer in *gnl44* and WT. The length of the outer parenchyma cell layer and the number of cells were significantly increased in *gnl44* compared to the WT (Figure 2B–E). In addition, we measured the length and width of cells in the outer and inner glumes of WT and *gnl44*, revealing that the outer epidermal cells in *the gnl44* lemmas were longer and wider than those of the WT. Similarly, the inner epidermal cells of *gnl44* were longer and wider than those of the WT (Figure 2F–K). Thus, GNL44 regulates grain size by limiting cell expansion in spikelet hulls.

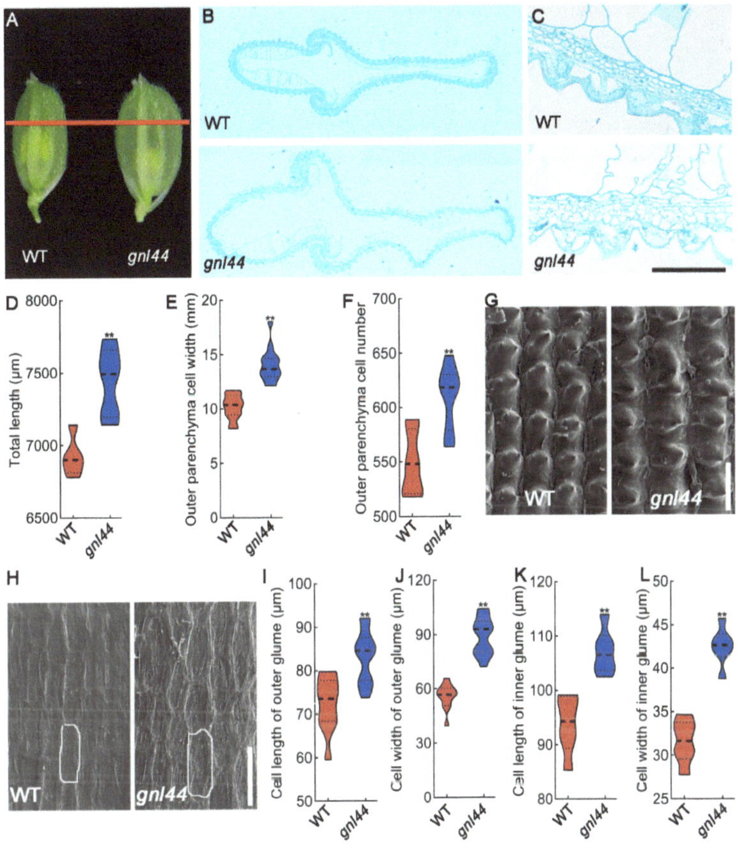

Figure 2. Histological analysis of spikelet hulls in the WT and *gnl44*. (**A**) Young spikelet hulls of the WT and *gnl44*. The red line indicates the position of the cross-section shown in (**B**). (**B**) Cross-sections of spikelet hulls (10×). (**C**) Magnified view of the cross-section in ((**B**), 40×); scale bar: 50 μm. (**D–F**) Total length (**D**), cell width (**E**), and cell number (**F**) in the outer parenchyma layer (n = 20). (**G,H**) Scanning electron microscopy analysis of the outer ((**G**), 120×) and inner ((**H**), 150×) surfaces of glumes. Scale bars: 100 μm. (**I–L**) Cell length (**I,K**) and width (**J,L**) in outer (**I,J**) and inner (**K,L**) glumes (n = 50). Values are means ± SD; Student's t-test was used to calculate the p values; ** $p < 0.01$.

2.3. Cloning of GNL44

To determine the genetic basis of the *gnl44* phenotype, we crossed the *gnl44* mutant with the WT rice cultivar WYJ. All F_1 plants showed a wild-type phenotype (normal grains), and the segregation (normal grains: large grains) ratio in the F_2 progeny was 3:1 (Table 1). These results suggest that the *gnl44* mutation is recessive and affects a single nuclear gene.

Table 1. Genetic analysis of *gnl44*.

Crossing Combination	Phenotype of F_1 Plants	Number of F_2 Plants		χ^2 (3:1 = 3.84)
		Normal Grains	Large Grains	
gnl44 × Wuyunjing	Normal grains	487	119	2.5782
Wuyunjing × *gnl44*	Normal grains	503	122	2.7002

We cloned *GNL44* using the MutMap approach. We sequenced genomic DNA from a pool of F_2 plants with the *gnl44* phenotype from the *gnl44* × WYJ cross. We quantified the frequency of single-nucleotide polymorphisms (SNPs) in the sequencing results, as described previously [2]. We detected 2979 SNPs and 227 small insertions and deletions (InDels) between WYJ and the pooled F_2 plants with the *gnl44* phenotype. Using these SNPs, we calculated the SNP index in the pooled F_2 plants across the entire genome (Figure 3A). The mapping interval of *GNL44* on chromosome 2 is 520 kb. Of all SNPs in this interval, only one located in the coding region of a predicted gene had an SNP index of 1, with clear and tight linkage with neighboring SNPs. This SNP in *gnl44* is a mutation from A to G located in the 7th exon of *Os02g0244100*, causing a premature termination of translation (Figure 3B–D). These results indicate that *Os02g0244100* is the candidate gene for *GNL44*.

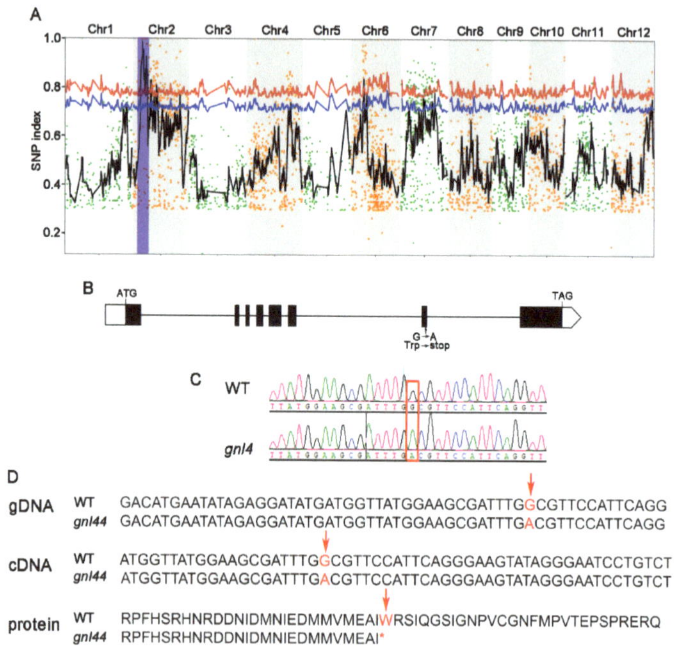

Figure 3. Positional cloning of *GNL44*. (**A**) SNP index and corresponding two-sided confidence intervals: 95% (blue line) and 99% (red lines). Different color dots are the original values of the SNP index of different chromosomes, and the black curve are the fitted values of SNP index. The regions indicated in blue shading correspond to the candidate genomic interval. (**B**) *GNL44* gene structure. Black boxes: exons; white boxes: untranslated regions; black lines: introns. ATG and TGA represent the start codon and the stop codon, respectively. (**C**) Sanger sequencing electropherograms for the WT and *gnl44*. The mutation site is highlighted by the red box. (**D**) Mutation in *GNL44* genomic DNA, the *GNL44* cDNA, and in the predicted GNL44 or gnl44 protein in the WT and *gnl44*, arrow indicated the mutation site, * indicated the stop codon.

2.4. Complementation Test

To test whether the mutation in *Os02g0244100* was responsible for the observed *gnl44* phenotypes, we carried out a genetic complementation test. We introduced a WT genomic fragment containing the entire coding region of *GNL44* together with 1662 bp of upstream and 787 bp of downstream sequence into the *gnl44* mutant by Agrobacterium-mediated transformation, yielding 5 transgenic lines (GNL44COM/*gnl44*).

We quantified similar values for grain number per panicle, grain length, grain width, and thousand-grain weight in the WT and the complementation lines (Figure 4A–G). These results indicate that *GNL44* is *Os02g0244100*, also named *GW2*. *GNL44* is annotated by The Rice Annotation Project (rapdb.dna.affrc.go.jp, accessed on 6 May 2021) as encoding a RING-type E3 ubiquitin ligase.

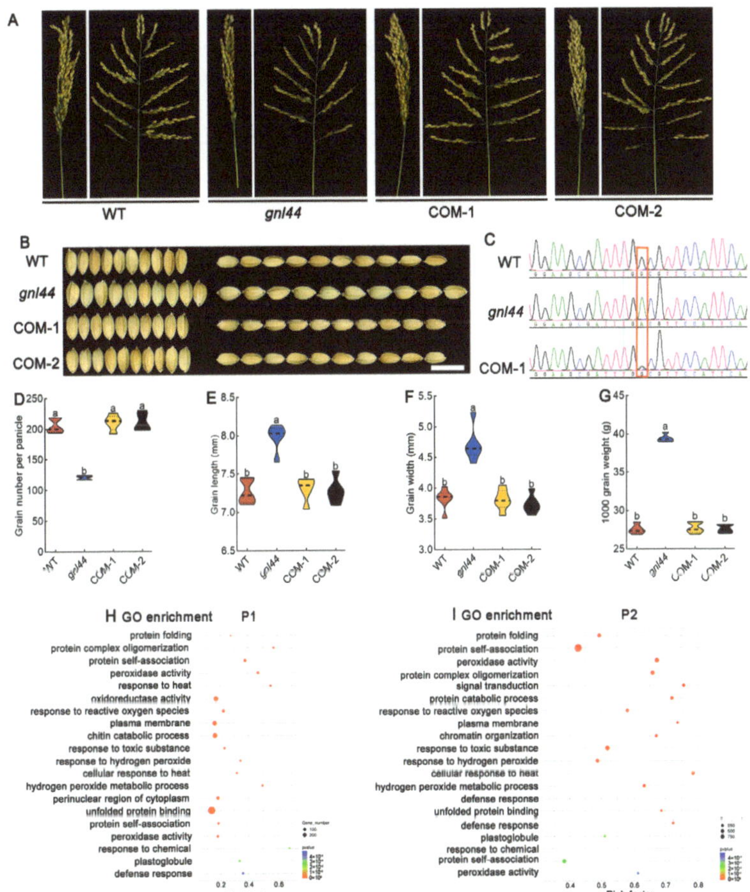

Figure 4. Complementation of the *gnl44* mutant. (**A**) Photographs of panicle from the WT, *gnl44*, and two independent *GNL44* complementation lines (COM-1, COM-2). (**B**) Morphology of mature grains from the WT, the *gnl44* mutant, and the two complementation lines. Scale bar: 1 cm. (**C**) Sanger sequencing electropherograms for the WT, *gnl44*, and COM-1. The mutation site is highlighted by the red box. (**D**–**G**) Grain number per panicle (**D**), grain length (**E**), grain width (**F**), and thousand-grain weight (**G**) of mature grains of A ($n = 10$). Values are means ± SD. Different letters indicate significant differences as determined by analysis of variance (ANOVA) followed by an LSD test for the comparison of means ($p < 0.05$). (**H**,**I**) Gene Ontology (GO) enrichment analysis of differentially expressed genes (DEGs) between WT and *gnl44* at different stages of panicle development.

To study the molecular mechanism underlying the role of *GNL44* in grain-type formation, we performed transcriptome sequencing of panicles at different stages of development. We prepared and sequenced RNA samples from WT and *gnl44* panicles at stages P1 (3 cm) and P2 (10 cm). The results indicated that the target genes of *GNL44* are mainly involved in protein metabolism pathways such as protein folding, protein self-association, and protein complex oligomerization (Figure 4H,I). Peroxidase activity and oxidoreductase activity were also significantly enriched (Figure 4H,I). These results suggest that the *GNL44* mutation has a significant impact on important biological processes such as protein-related metabolic pathways and redox pathways, thereby affecting the growth and development of glumes during the early stage of panicle development.

2.5. Expression of GNL44 and Subcellular Localization of GNL44

We measured *GNL44* expression in different tissues collected from WT plants and observed that *GNL44* was preferentially expressed in panicles and developing seeds (Figure 5A). To assess the subcellular localization of GNL44, we transfected rice protoplasts with a construct encoding a GNL44–green fluorescent protein (GFP) fusion or with *35S:GFP* as control (Figure 5B). We detected green fluorescence for GNL44-GFP in the cytoplasm, indicating that GNL44 is a cytosolic protein.

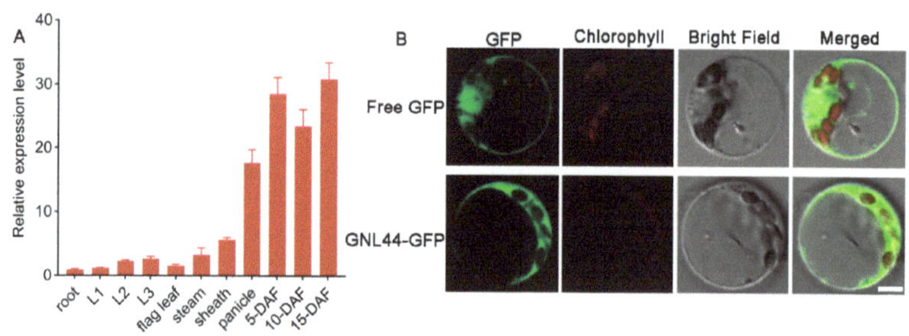

Figure 5. *GNL44* expression pattern in the WT and subcellular localization of GNL44-GFP. (**A**) Relative *GNL44* transcript levels in various organs of WT plants. L1–L3: first–third leaf of rice seedlings; DAF: days after fertilization. The expression in roots was set to 1. (**B**) Subcellular localization of GNL44-GFP in transfected rice protoplasts. Free GFP from the p35S:GFP construct served as control. GFP: GFP fluorescence; chlorophyll: chlorophyll autofluorescence. Scale bar: 10 μm.

2.6. GNL44 Affects Grain Appearance Quality of Milled Rice

To explore the effect of the *GNL44* genotype on rice quality, we characterized the appearance of rice from WT and *gnl44*. The width and length of brown grains from *gnl44* were higher than those from the WT (Figure 6A–D). The transparency of the milled rice from *gnl44* was lower than that from WT. Some milled rice from *gnl44* showed large areas with a white core in endosperm (Figure 6E,F). In agreement, chalky grain rate and chalkiness degree in the *gnl44* mutant were significantly higher than in WT (Figure 6G,H). To gain a better picture of grain ultrastructure, we observed the structure of starch granules from WT and *gnl44* grains by scanning electron microscopy. The starch granules of WT were closely arranged and regularly organized in contrast to the starch granules of *gnl44*, which were irregular and loosely arranged (Figure 6E,F). These results suggest that the appearance quality of milled rice is reduced in the *gnl44* mutant.

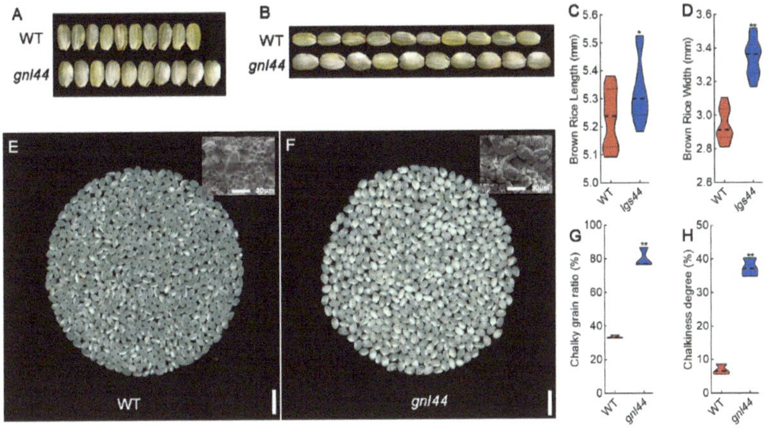

Figure 6. Grain appearance quality of WT and *gnl44* milled rice. (**A**,**B**) Photographs showing the length (**A**) and width (**B**) of brown rice from the WT and *gnl44*. Scale bars: 10 mm. (**C**,**D**) Length (**C**) and width (**D**) of brown rice from the WT and *gnl44*. Values are means ± SD ($n = 3$). (**E**,**F**) Increased chalkiness in milled rice of *gnl44* compared to the WT. Scale bars: 10 mm. Scanning electron microscopy of cross-sections of milled rice from the WT and *gnl44* are shown in upper right corners (scale bars, 30 μm). (**G**,**H**) Percentage of milled rice with chalkiness (**G**) and chalkiness degree of milled rice from the WT and *gnl44* (**H**). Student's *t*-test was used to calculate the *p* values; ** $p < 0.01$, * $p < 0.05$.

2.7. Eating and Cooking Quality of the gnl44 Mutant Is Worse Than That of the WT

To clarify the effects of *GNL44* on eating and cooking quality, we performed a rice quality assay. Rice flour produced from *gnl44* grains had lower amylose content, higher protein content, and lower gel consistency compared to the WT (Figure 7A–C).

The length distribution of amylopectin chains is important for determining the physicochemical properties of starch as well as the texture characteristics of cooked rice. Amylopectin can be divided into four types based on chain length: fa chains (degree of polymerization [DP] 6–12), fb1 chains (DP 13–24), fb2 chains (DP 25–36), and fb3 chains (DP > 37) [31,32]. The hardness of cooked rice is negatively correlated with the proportion of shorter fa chains present in branched starch but positively correlated with the proportion of fb2 chains. A higher proportion of long-chain branched starch increases the hardness and reduces the stickiness of cooked rice [33,34].

We measured the length distribution of amylopectin chains in rice flour produced from WT and *gnl44* grains. WT flour had higher contents of DP 6–12 amylopectin chains than *gnl44* flour, except for DP 8 chains. By contrast, WT flour had fewer DP 13–24 chains than *gnl44* flour did, except for DP 15. The contents of DP 25–36 chains were slightly lower in the WT than in *gnl44*, except for DP 36, while aggregates of DP >37 chains accumulated to similar levels in the WT and *gnl44* (Figure 7D). These results suggest that *gnl44* mainly affects the biosynthesis of fa and fb1 chains and that the grain quality of milled rice is reduced in the *gnl44* mutant.

Amylose content, protein content, gel consistency, and length distribution of amylopectin chains are all closely related to eating and cooking quality (ECQ), which prompted us to understand the effect of *gnl44* on the texture of cooked rice by carrying out an RVA experiment. The values for peak viscosity and breakdown viscosity of *gnl44* were lower than those of the WT. However, cold viscosity, final viscosity, and setback viscosity all increased in *gnl44* relative to the WT (Figure 7E, Table S2). Furthermore, the taste value of *gnl44* was lower than that of the WT (Figure 7F). These results suggest that the ECQ of *gnl44* is decreased compared to that of the WT.

The above results indicate that the mutation in *GNL44* in rice leads to poor appearance quality and lower ECQ. To determine which genes might mediate this effect, we explored the downstream target genes of GNL44 by RNA-seq analysis. We collected seeds from WT and *gnl44* plants at 5 and 15 days after flowering (DAF) for transcriptome analysis. Protein metabolism pathways such as protein folding, protein catabolic process, and protein complex oligomerization as well as redox pathways such as peroxidase activity and response to reactive oxygen species (ROS) were enriched among the differentially expressed genes (DEGs) (Figure 7G,H). These results indicate that *GNL44* affects rice appearance quality and ECQ via protein-related metabolic pathways and redox-related biological processes.

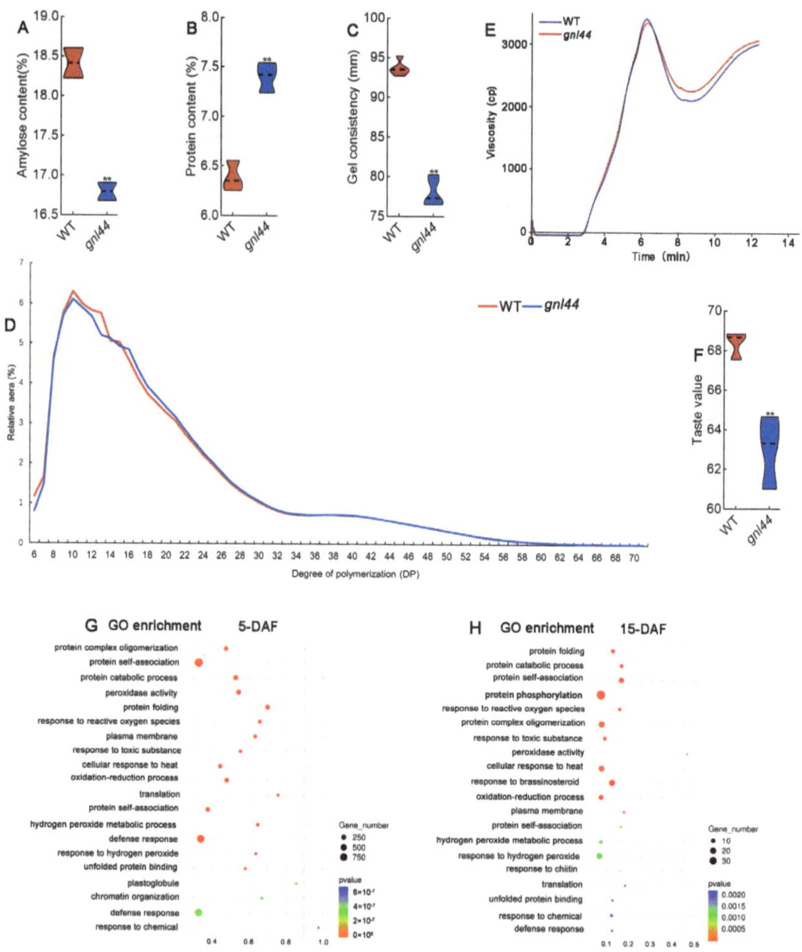

Figure 7. ECQ of the WT and *gnl44*. (**A–C**) Amylose content, protein content, and gel consistency from WT and *gnl44* flour. (**D**) Chain length distribution of amylopectin in the WT and *gnl44*. (**E**) RVA spectra of WT and *gnl44* flour. (**E**) Rapid Visco Analyzer profile characteristics of WT and *gnl44*. (**F**) Taste value of WT and *gnl44* grains. Values are means ± SD ($n = 3$ in (**A–C**,**F**,**G**)). Student's t-test was used to calculate the p values; ** $p < 0.01$. (**G**,**H**) Gene Ontology (GO) enrichment analysis of differentially expressed genes (DEGs) between WT and *gnl44* at different stages after flowering.

2.8. Rice Quality of NIL-*gnl44*, NIL-*gs3*, and NIL-*qgl3*

High yield and superior quality are the primary goals of rice breeders and can be affected by grain size [35]. We previously reported that the large-grain rice cultivar TD70 (thousand-grain weight of 68.6 g) carries superior alleles at the loci *qGL3* and *GS3*, which control grain size, while Kasalath (thousand-grain weight of 19.1 g) has normal alleles at these loci. TD70 is therefore an ideal material for studying the effects of genetic interactions between different grain size genes on milled rice quality [36].

We wished to explore the effect of the interaction between *GNL44* and *qGL3* and *GS3* on rice quality by constructing monogenic NILs of *qgl3* and *gs3* with Kasalath as the receptor and TD70 as the donor. We also constructed NIL-*gnl44* with Kasalath as the receptor and the *gnl44* mutant as the donor.

The width of the milled grain from NIL-*gnl44* and the length of the milled grain from NIL-*qgl3* were significantly higher compared to that of Kasalath (Figure 8A–C). The chalky grain ratio and chalkiness degree of NIL-*gnl44* were higher than those of Kasalath, and those of NIL-*qgl3* were significantly lower than those of Kasalath (Figure 8D,E). A rice quality assay showed that the amylose contents of NIL-*gnl44* and NIL-*gs3* were higher than that of Kasalath. The protein contents of NIL-*gnl44* and NIL-*qgl3* were also higher than that of Kasalath, while that of NIL-*gs3* was lower than that of Kasalath (Figure 8F,G). The Gel consistency of the three NILs was lower than that of Kasalath (Figure 8H). Moreover, the RVA experiments showed that the peak viscosity of NIL-*gnl44* and NIL-*gs3* were higher, and that of NIL-*qgl3* was lower, compared to that of Kasalath. The cold viscosity of NIL-*gs3* was elevated compared to that of Kasalath but was similar to that of Kasalath in NIL-*gnl44* and NIL-*gs3*. The breakdown viscosities of NIL-*gs3* and NIL-*qgl3* were lower than that of Kasalath. The final viscosity and setback viscosity were higher in all three monogenic NILs relative to Kasalath. The consistency viscosity of NIL-*gnl44* was elevated, similar in NIL-*gs3*, and slightly increased in NIL-*qgl3* compared to that of Kasalath (Figure 8I,J). These results suggest that *GNL44* and *qGL3* mainly affect the chalkiness and gel consistency and that *GS3* mainly affects protein content.

2.9. Additive Effects of *gnl44*, *gs3*, and *qgl3* on Rice Quality

We also constructed two-gene NILs to evaluate the effects of different combinations of grain size genes on rice quality. The length and width of milled grain from these two-gene NILs increased in general compared to those of Kasalath (Figure 9A–C). The chalky grain ratio and chalkiness degree of the two-gene NILs NIL-*gnl44*/*gs3* and NIL-*gnl44*/*qgl3* in which *gnl44* was pyramided were higher than those of the corresponding monogenic NILs NIL-*gs3* and NIL-*qgl3*. However, the chalky grain ratio and chalkiness degree of the two-gene NILs NIL-*gnl44*/*qgl3* and NIL-*gs3*/*qgl3* in which *qgl3* was pyramided were lower than those of the corresponding monogenic NILs NIL-*gnl44* and NIL-*gs3* (Figure 9D,E). The amylose content of all three two-gene NILs was decreased compared to that of Kasalath (Figure 9F). The protein content of the two-gene NILs NIL-*gnl44*/*gs3* and NIL-*gs3*/*qgl3* was higher than that of the corresponding monogenic NILs NIL-*gnl44* and NIL-*qgl3* (Figure 9G). The gel consistency of NIL-*gnl44*/*gs3* and NIL-*gs3*/*qgl3* was lower than that of the corresponding monogenic NILs NIL-*gnl44* and NIL-*qgl3*. The gel consistency of NIL-*gnl44*/*qgl3* was the highest among the seven lines (three two-gene NILs, three monogenic NILs, and Kasalath) (Figure 9H).

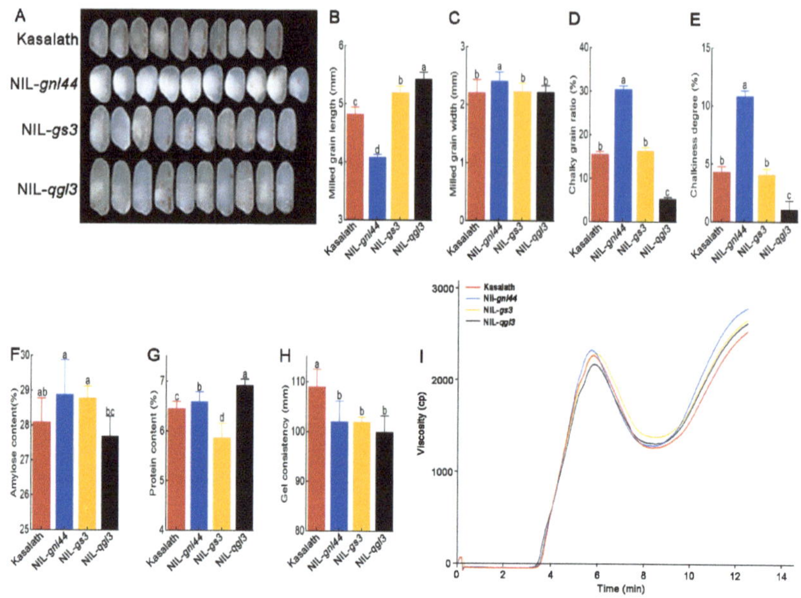

J Rapid Visco Analyzer profile characteristics of Kasalath and monogenic NILs

lines	Peak viscosity (cP)	Cold viscosity (cP)	Breakdown viscosity (cP)	Final viscosity (cP)	Setback viscosity (cP)	Consistency viscosity (cP)	Peak Time (min)	Pasting temperature (°C)
Kasalath	2274±3.79	1272±9.01	1002±12.72	2543±7.22	269±11.00	1271±3.21	5.8±0.067	79.75±0.380
NIL-*gnl44*	2327.5±2.50**	1291±0.00	1036.5±2.50	2796±0.00**	468.5±2.50**	1505±0.00**	5.77±0.033	79.45±0.400
NIL-*gs3*	2306.5±10.50	1394.5±4.50**	912±6.00**	2688±16.00*	381.5±5.50**	1293.5±11.50	5.9±0.033	79.85±0.050
NIL-*qgl3*	2210±36.00	1289±25.00	921±61.00**	2642±11.00*	432±25.00**	1353±36.00*	5.8±0.067	78.275±0.025

Values are Mean±SD (n = 3). Student's t-test was used to generate P values; * $P < 0.05$ and ** $P < 0.01$.

Figure 8. Grain quality of monogenic NILs. (**A**) Photographs showing milled rice from Kasalath, NIL-*gnl44*, NIL-*qgl3*, and NIL-*gs3*. (**B–H**) Milled grain length (**B**), milled grain width (**C**), chalky grain ratio (**D**), chalkiness degree (**E**), amylose content (**F**), protein content (**G**), and gel consistency (**H**) of Kasalath, NIL-*gnl44*, NIL-*qgl3*, and NIL-*gs3* grains or rice flour. Values are means ± SD (n = 10 in (**B**,**C**); n = 5 in (**D**,**E**); n = 3 in (**F**,**G**); n = 4 in (**H**)). (**I**) RVA spectra of Kasalath and monogenic NILs. (**J**) RVA profile characteristics of Kasalath and monogenic NILs. Different letters indicate significant differences as determined by ANOVA followed by an LSD test for the comparison of means ($p < 0.05$).

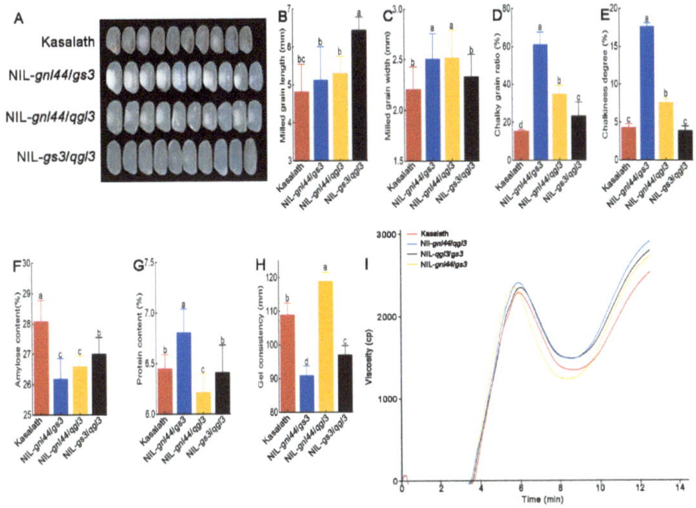

J Rapid Visco Analyzer profile characteristics of Kasalath and two-gene NILs

lines	Peak viscosity (cP)	Cold viscosity (cP)	Breakdown viscosity (cP)	Final viscosity (cP)	Setback viscosity (cP)	Consistency viscosity (cP)	Peak Time (min)	Pasting temperature (°C)
Kasalath	2288±2.84	1343±9.51	945±10.66	2530±8.62	269±9.32	1282±8.14	5.87±0.044	80.7±0.25
NIL-*gnl44/gs3*	2352.5±1.50**	1285.5±49.50*	1067±51.00**	2707.5±22.50*	355±21.00**	1422±72.00**	5.73±0.067	79.5±0.350
NIL-*gnl44/qgl3*	2374±32.00**	1538±57.00**	836±89.00*	2847.5±62.50**	473.5±30.50**	1309.5±119.50	5.97±0.100	79.55±0.400
NIL-*gs3/qgl3*	2345±0.00**	1414.5±69.50	930.5±69.50	2819.5±26.50**	474.5±26.50**	1405±96.00**	5.87±0.067	79.53±0.430

Values are Mean±SD (n = 3). Student's t-test was used to generate P values; * $P < 0.05$ and ** $P < 0.01$.

Figure 9. Grain quality of two-gene NILs. (**A**) Photographs showing milled rice from Kasalath, NIL-*lgs44/qgl3*, NIL-*lgs44/gs2*, and NIL-*qgl3/gs3*. (**B–H**) Milled grain length (**B**), milled grain width (**C**), chalky grain ratio (**D**), chalkiness degree (**E**), amylose content (**F**), protein content (**G**), and gel consistency (**H**) of Kasalath, NIL-*lgs44/qgl3*, NIL-*lgs44/gs2*, and NIL-*qgl3/gs3* grains or flour. Values are means ± SD (n = 10 in (**B**,**C**); n = 5 in (**D**,**E**); n = 3 in (**F**,**G**); n = 4 in (**H**)). (**I**) RVA spectra of Kasalath and two-gene NILs. (**J**) RVA profile characteristics of Kasalath and two-gene NILs. Different letters indicate significant differences as determined by ANOVA followed by an LSD test for the comparison of means ($p < 0.05$).

RVA results showed that the peak viscosity was elevated in all two-gene NILs relative to their corresponding monogenic NILs and Kasalath. The cold viscosity of NIL-*gnl44/qgl3* was higher, of NIL-*gs3/qgl3* was similar, and of NIL-*gnl44/gs3* was slightly decreased relative to that of Kasalath (Figure 9I,J). The breakdown viscosity increased in NIL-*gnl44/gs3* and slightly decreased in NIL-*gnl44/gs3* compared to Kasalath. The final viscosity and setback viscosity increased for all three two-gene NILs relative to Kasalath. The consistency viscosity of NIL-*gnl44/gs3* and NIL-*gs3/qgl3* also increased (Figure 9I,J). Taken together, these findings suggest that *GNL44*, *GS3*, and *qGL3* differentially affect rice quality.

We performed a two-way analysis of variance (ANOVA) to dissect the genetic interactions between *gnl44*, *qgl3*, and *gs3* (Tables 2 and S3). We determined that the combination of *gnl44* and *qgl3* leads to a genetic interaction that influences chalky grain ratio, protein content, gel consistency, and amylose content. *gnl44* and *gs3* had significant genetic interactions for chalky grain ratio, chalkiness degree, protein content, and amylose content. Conversely, *gs3* and *qgl3* exhibited genetic interactions for milled grain length, chalky grain ratio, and chalkiness degree (Tables 2 and S3).

Table 2. Two-way analysis of variance for genetic interactions among *gnl44*, *qgl3*, and *gs3*.

Trait	Variation	*df*	SS	MS	F	*p*
Milled grain length	*gs3*	1	4.66	4.66	24.33	<0.0001
	qgl3	1	8.96	8.96	46.8	<0.0001
	gs3 × *qgl3*	1	1.13	1.13	5.92	0.0201
	error	36	6.89	0.19		
Chalky grain ratio	*gnl44*	1	4428	4428	417.08	<0.0001
	gs3	1	1245	1245	117.29	<0.0001
	gnl44 × *gs3*	1	1118	1118	105.27	<0.0001
	error	16	169.9	10.62		
	gnl44	1	2488.2	2488.2	510.55	<0.0001
	qgl3	1	38.4	38.4	7.87	0.0127
	gnl44 × *qgl3*	1	281.25	281.25	57.71	<0.0001
	error	16	78.0	4.87		
	gs3	1	446.0	446.0	34.61	<0.0001
	qgl3	1	13.7	13.7	1.06	0.318
	gs3 × *qgl3*	1	371.1	371.1	28.79	<0.0001
	error	16	206.2	12.9		
Chalkiness degree	*gnl44*	1	499.6	499.6	2435.97	<0.0001
	gs3	1	54.65	54.65	266.46	<0.0001
	gnl44 × *gs3*	1	60.4	60.41	294.56	<0.0001
	error	16	3.3	0.20		
	gs3	1	8.3	8.3	25.74	0.0001
	qgl3	1	15.8	15.8	49.16	<0.0001
	gs3 × *qgl3*	1	10.6	10.6	32.98	<0.0001
	error	16	5.2	0.32		
Protein content	*gnl44*	1	0.89	0.89	18.34	0.0027
	gs3	1	0.10	0.10	2.09	0.1864
	gnl44 × *gs3*	1	0.49	0.49	10.11	0.013
	error	8	0.39	0.048		
	gnl44	1	0.235	0.235	9.18	0.0163
	qgl3	1	0.006	0.006	0.22	0.6517
	gnl44 × *qgl3*	1	0.529	0.529	20.64	0.0019
	error	8	0.205	0.026		
Gel consistency	*gnl44*	1	121.0	121.0	10.62	0.0068
	qgl3	1	81.0	81.0	7.11	0.0205
	gnl44 × *qgl3*	1	728.7	728.7	63.99	<0.0001
	error	12	136.7	11.39		
Amylose content	*gnl44*	1	2.47	2.47	5	0.0557
	gs3	1	3.56	3.56	7.23	0.0276
	gnl44 × *gs3*	1	8.74	8.74	17.72	0.003
	error	8	3.945	0.49		
	gnl44	1	0.12	0.12	0.25	0.6284
	qgl3	1	5.87	5.87	12.59	0.0075
	gnl44 × *qgl3*	1	2.99	2.99	6.42	0.0351
	error	8	3.73	0.47		

Df: degree of freedom; SS: sum of squares; MS: mean of square; F: F value of ANOVA.

3. Discussion

3.1. *The gnl44 Mutant Has Fewer but Larger Grains*

In this study, we identified the mutant *gnl44* with a larger grain size from an EMS mutant library of the *japonica* rice cultivar Wuyunjing. Compared to WT, the *gnl44* mutant had a 20.2% increase in grain width and a 9.8% increase in grain length (Figure 1E,F). However, due to a concomitant drop in grain number per panicle, the final grain yield per plant decreased significantly compared to WT (Figure 1N). Using the MutMap approach for gene mapping, together with complementation experiments, we established that a G-to-A mutation in exon 7 of *Os02g0244100* led to the early termination of translation and was responsible for the mutant phenotype (Figures 3 and 4). Subcellular localization revealed that GNL44 localized to the cytoplasm (Figure 5B). The results of tissue model expression showed that *gnl44* was highly expressed in the panicle and developing endosperm (Figure 5A). The Rice Annotation Project (rapdb.dna.affrc.go.jp, accessed on 6 May 2021) notes that *GNL44* encodes the RING-type E3 ubiquitin ligase GW2; the *gnl44* mutant therefore defines a new allele of *GW2*.

Lu et al. (2013) analyzed the *GW2* sequence across 127 rice varieties and reported that most SNPs/InDels were located in the intronic and promoter regions. Moreover, the variation in the promoter region did not affect *GW2* expression [37]. Zhang et al. (2015) and Dixit et al. (2013) analyzed *GW2* haplotypes in different rice resources and obtained similar results [36,38]. In this study, using EMS mutagenesis, we obtained a new mutant allele of *GW2* named *gnl44*. The *gnl44* mutant exhibits a severe reduction in grain number but also a larger grain phenotype, making *gnl44* a rare new allele of *GW2*. Song et al. (2007) cloned rice *GW2* using the cultivars WY3 and FAZ1. The authors determined that grain length, grain width, grain thickness, and thousand-grain weight were higher in NIL-*gw2* compared to FAZ1. However, grain number per panicle decreased and grain yield per plant increased in NIL-*gw2* [7]. Huang et al. (2022) independently cloned *GW2* using the cultivar HZ and observed that NIL-*gw2* had larger grains and a higher thousand-grain weight compared to HZ. However, there was no significant change in the grain number per panicle of NIL-*gw2*, and the yield per plant increased [39]. In the current study, we isolated the new *GW2* EMS mutant *gnl44*. Although *gnl44* grains were wider, longer, and heavier compared to WT grains, the number of grains per panicle decreased by 43.95% in *gnl44* compared to WT, resulting in a 21.8% decrease in grain yield per plant. Among the lines examined in these three studies, the phenotype of the NIL generated by Song et al. resulted from premature termination due to mutations in exon 4 of *GW2*, that of *gnl44* isolated in the current study resulted from premature termination due to mutations in exon 7, and that of the NIL isolated by Huang et al. resulted from a single base substitution in this gene. These findings suggest that exons 7 and 8 of *GW2* might be related to the establishment of grain number per panicle in rice.

GO enrichment analysis of the transcriptome data from young panicles at different developmental stages indicated that protein folding and other protein-related pathways were significantly enriched among the DEGs (Figure 4H,I). This finding is consistent with the observation that GNL44, an E3 ubiquitin ligase, is involved in protein degradation. Hao et al. (2021) found that GNL44/GW2 ubiquitinates the glutaredoxin protein OsGRX8 and targets it for degradation to regulate grain size. The modified expression of *OsGRX8* affects plant responses to various abiotic stresses [40]. Indeed, we found that many abiotic stress-related pathways were enriched in the transcriptome results (Figures 4H,I and 7G,H). These results suggest that GNL44/GW2 may be involved in abiotic stress responses.

3.2. *gnl44 Affects Appearance Quality and ECQ*

GW2 is a major gene affecting grain size, and previous research has mostly focused on its regulation of grain size [7,39,41,42]. Few studies have explored the regulation of rice quality by *GW2* in depth. In this study, we investigated the appearance and cooking qualities of the *gnl44* mutant. Compared to WT, chalky grain ratio and chalkiness percentage of the *gnl44* mutant increased (Figure 6G,H). The starch in *gnl44* mutant grains was loosely arranged (Figure 6E,F). Moreover, the observed distribution of amylopectin chain length suggests that *GNL44* mainly affected the starch biosynthesis of fa and fb1 chains (Figure 7D). The appearance quality of milled *gnl44* grain was worse than that of WT. In addition, the amylose and protein contents of *gnl44* grains was significantly decreased compared to WT (Figure 7A). The results of RVA showed that the peak viscosity and breakdown viscosity of *gnl44* were also decreased (Figure 7D,E). The taste value of the *gnl44* mutant was lower than that of WT (Figure 7F). These results indicate that *GNL44* mutation leads to poor appearance and taste quality in milled rice.

Song et al. (2007) [7] previously showed that amylose content, protein content, and gel consistency did not differ between NIL-*gw2* and FAZ1. However, the chalky grain ratio of NIL-*gw2* was increased compared to FAZ1. Based on these results, Song et al. [7] postulated that the *GW2* allele from cultivar WY3 had little influence on appearance and would not reduce cooking quality. Huang et al. (2022) [39] showed that amylose content, chalkiness degree, and chalky grain ratio similarly did not differ between NIL-*gw2* and HZ. However, the gel consistency of NIL-*gw2* was higher than that of HZ, while its protein content was lower than that of HZ. This observation was taken to indicate that the jf42-*GW2* allele did not affect the appearance quality of milled rice.

Wx is the major QTL affecting the appearance quality, eating quality, and cooking quality of rice [43,44]. To date, at least eight *Wx* alleles have been functionally characterized, including Wx^{lv}, Wx^{a}, Wx^{in}, Wx^{b}, $Wx^{op/hp}$, Wx^{mq}, Wx^{mp}, and *wx* [44–48]. In this study, the *Wx* genotype of WYJ, *gnl44*, and TD70 was Wx^{b}, and the *Wx* genotype of Kasalath, the monogenic NILs, and the two-gene NILs was Wx^{lv}. We determined that in the Wx^{b} background, the AC content decreased in the presence of the *GNL44/GW2* mutation (Figure 7A), while the same mutation increased the AC content in the Wx^{lv} background (Figure 8F). These results indicate that although *GNL44/GW2* can affect AC, *Wx* is still the major gene determining AC.

Rice grain length and grain width are important factors affecting appearance quality. We investigated the grain length-to-width ratio in all three *GW2* studies. The grain length-to-width ratio was reported to be about 3.3 (Song et al., 2007 [7]), 2.91 (Huang et al., 2022 [39]), or 1.98 (this work). Therefore, the grain length-to-width ratio of *gnl44* in this study was significantly lower than that of *gw2* alleles reported in the other two studies, offering a possible explanation for the deterioration in appearance quality and cooking quality of *gnl44*.

Achary et al. [49] used clustered regularly interspaced short palindromic repeat (CRISPR)/CRISPR-associated nuclease 9 (Cas9)-mediated gene editing to target the fourth exon of *GW2*. The endosperm of the *GW2*-KO mutant seeds showed a thicker aleurone layer, with higher protein content in the seeds. The accumulation of essential dietary minerals (Fe, Zn, K, P, Ca) in the rice endosperm of the *OsGW2-KO* mutant also increased [49]. This result indicates that *GW2* not only affects the appearance and cooking quality but also the mineral content of rice.

The excessive accumulation of ROS in rice endosperm can trigger programmed cell death (PCD), thereby inhibiting the accumulation of storage substances and leading to chalkiness. The imbalance of redox homeostasis caused by oxidative stress is the main cause of grain chalkiness [50,51]. In this study, analysis of transcriptome data revealed the significant enrichment of genes involved in the oxidative stress response and protein metabolism-related pathways in *gnl44* grains (Figure 7G,H). These results suggest that *GNL44* may regulate grain appearance and ECQ by affecting the metabolism of rice quality-related proteins and redox homeostasis.

3.3. Potential Value of *gnl44* for Rice Quality Improvement

In rice breeding, it is necessary to select the appropriate grain size gene to achieve a desired improvement in yield and quality [35]. Haplotype analysis indicates that the functional variation in *GW2* has largely not been exploited by breeders [37,38,52]. Therefore, evaluating the effects of different grain size genes, especially *gw2* and others, is of great significance for rice breeding in general and for rice quality in particular.

In this study, we isolated a rare new allele of *gw2*—*gnl44*. To assess the potential breeding value of *gnl44*, we used monogenic NILs and two-gene NILs carrying *gnl44*, *gs3*, or *qgl3* alleles to study their additive effects on traits related to rice quality. The amylose content of all three two-gene NILs decreased compared to that of Kasalath (Figure 9F). The amylose and protein contents of NIL-*gnl44/qgl3* were lower than those of Kasalath, but its gel consistency was higher than that of Kasalath (Figure 9F–H). Milled rice is reported to have a low amylose content, low protein content, and high gel consistency with good cooking quality [53]. We hypothesize that pyramiding of *gnl44* and *qgl3* can be a useful approach to improving the cooking quality of rice. However, the grain yield per plant of NIL-*gnl44*/*qgl3* was lower than that of Kasalath (Table S4), making it unsuitable for breeding rice varieties with high yield and good quality. The chalkiness degree, chalky grain ratio, and gel consistency of NIL-*gs2*/*qgl3* were lower than those of Kasalath, while its amylose and protein contents were higher than those of Kasalath (Figure 9D–H). Moreover, the grain yield per plant of NIL-*gs2*, *qgl3* was comparable to that of Kasalath (Table S4). Therefore, we believe that these two genes can be applied to the breeding of new rice varieties with high yield and good quality. The chalkiness grain rate and chalkiness degree of NIL-*gnl44*/*gs3* were much higher than those of Kasalath, and the highest among the three two-gene NILs were characterized here (Figure 9D,E). The appearance quality of NIL-*gnl44*, *gs3* was poor. Compared to Kasalath, NIL-*gnl44*/*gs3* had a lower amylose content, lower gel consistency, and higher protein content (Figure 9F–H). The grain yield per plant of NIL-*gnl44* and *gs3* was also lower than that of Kasalath (Table S4). Therefore, *gnl44* and *gs3* are not suitable for breeding new rice varieties with high yield and good quality.

4. Materials and Methods

4.1. Plant Materials and Growth Conditions

The *gnl44* mutant was obtained by mutagenizing the *japonica* rice (*Oryza sativa*) cultivar Wuyunjing (WYJ) with 0.8% ethyl methane sulfonate (EMS). We selected the *gnl44* mutant from the M2 generation. After 5 years of cultivation, its phenotype is stable and was used in this study. All plants in this study were cultivated at the Jiangsu Academy of Agricultural Sciences located in Nanjing, Jiangsu province (118°5′ E, 32°0′ N). The following agronomic traits were determined for field-grown plants at maturity: plant height, panicle length, grain number per panicle, primary and secondary branches per panicle, grain width, grain length, grain thickness, grain yield per plant, and thousand-grain weight [7].

4.2. Histological Analysis

Young spikelet hulls of wild-type (WT) and *gnl44* plants were fixed in FAA (50% ethanol, 5% glacial acetic acid, and 5% formaldehyde, all *v*/*v*) for 48 h. The samples were treated as described by Li et al. [54]. The stained sections were observed with a microscope (Olympus BX51, Tokyo, Japan). For glume cell observation, glumes of WT and *gnl44* were fixed in 2.5% (*w*/*v*) glutaraldehyde (2.5% [*w*/*v*] glutaraldehyde, 19.5% 2 M NaH_2PO_4, and 30.5% 2 M Na_2HPO_4) for 48 h and observed by scanning electron microscopy (Zeiss, EVO-LS10, Oberkochen, Germany).

4.3. Positional Cloning of GNL44

An F_2 population was produced by crossing *gnl44* with the WT cultivar. WYJ was used to clone *GNL44*. The whole genomes of a mixed pool of 50 plants with the *gnl44* phenotype and wild-type WYJ were sequenced using a NextSeq 500 system (Illumina, San Diego, CA, USA). The MutMap was used to isolate *GNL44* as previously described [55]. The reference rice genome IRGSP1.0 was used in this study.

For complementation of the *gnl44* mutant, a 8330 bp genomic fragment of the candidate *GNL44* gene, which contains the entire genomic coding region of *GNL44*; 1662 bp of upstream sequence; and 787 bp of downstream sequence were amplified with the GNL44-COMF/R primers and sequenced (Table S1). The amplified fragment was cloned into the *BamHI* site of the pCAMBIA1300 binary vector. The resulting pCAMBIA1300-GNL44 plasmid was introduced into Agrobacterium (*Agrobacterium tumefaciens*) strain EHA105. *gnl44* calli were produced by the induction of mature embryos according to the descriptions of Toki (1997) and Toki (2006) [56,57]. Then, *gnl44* calli were transformed by an Agrobacterium-mediated transformation method.

4.4. Subcellular Localization of GNL44

The full-length *GNL44* coding sequence without the stop codon was amplified and cloned in-frame and upstream of the *GFP* coding sequence into the *Sal* I site of the p*35S::GFP* vector. The inserted fragment was amplified by PCR using the specific primer pairs shown in Table S1. The resulting construct was transfected into protoplasts isolated from 20-day-old 93-11 rice seedlings, which were incubated overnight in the dark as described previously [58]. GFP fluorescence was visualized with a confocal laser scanning microscope (Zeiss LSM 710).

4.5. RT-qPCR and RNA-seq Analysis

Total RNA was extracted from rice tissues using a Total RNA Miniprep kit (Axygene, Hangzhou, China) for RT-qPCR. First-strand cDNA was synthesized from total RNA using a ReverTra Ace qPCR-RT kit (Toyobo, Osaka, Japan), according to the manufacturer's instructions. qPCR was performed using SYBR premix Ex Taq II (Takara, Kusatsu, Japan) on an Applied Biosystems 7900HT instrument. The primers used for qPCR are listed in Table S1. Rice *UBQ5* was used as an internal control. Data were analyzed following the relative quantification method [59]. Values are means $\pm$ SD of three biological replicates. Student's *t*-test was used for statistical analysis.

For RNA-seq analysis, RNA samples were prepared from 3 cm (P1) and 10 cm (P2) panicles and the seeds of WT and *gnl44* plants at 5 and 10 days after flowering (DAF). RNA-seq analysis and Gene Ontology (GO) enrichment analysis were performed as described previously [60].

4.6. Determination of Taste Value and Rapid Visco Analyzer (RVA) Profiles

An RCTA-11A Taste Analyzer (Satake, Hiroshima, Japan) was used to analyze the taste value of milled rice according to the description of Chen et al. [61]. A 30 g aliquot of washed milled rice was transferred into a 50 mL aluminum box containing 40 mL distilled water and cooked in an electric rice cooker. After 20 min at equilibrium, the taste value of cooked rice was evaluated.

The RVA profiles of rice flour were investigated with a Rapid Visco Analyzer (Techmaster, Newport Scientific; Warriewood, Sydney, Australia). A 2.0 g aliquot of rice flour and 25 mL distilled water were mixed well in the RVA aluminum canister. The sample was heated according to the method described by Li et al. [62]. Finally, peak viscosity, cold viscosity, final viscosity, breakdown viscosity, setback viscosity, and pasting temperature were measured.

4.7. Measurement of Physicochemical Characteristics

The amylose content of rice flour was determined as described by Shi et al. [63]. The protein content was estimated from the nitrogen content using the Kjeldahl method with a conversion factor of 5.95 [64]. The gel consistency was measured following the method of Tan et al. [65].

4.8. Amylopectin Branch Chain Length Distribution Analysis

A high-performance anion-exchange chromatography system (ICS-6000, Thermo Fisher Scientific, Waltham, MA, USA) coupled with a Dionex™ CarboPac™ PA10 anion-exchange column was used following the method of Li et al. [62] with modifications to analyze the length distribution of amylopectin chains.

Starch samples (5 mg) were suspended in 1 mL double-distilled water and boiled for 30 min. Gelatinized polyglucan (500 μL) was mixed with 10 μL 1 m sodium acetate buffer, 2 μL sodium azide (2% solution, *w*/*v*), and 5 μL isoamylase (1000 U/μL 15284, Sigma Aldrich Corporation, St. Louis, MO, USA) and stored in a thermotank at 40 °C for 24 h. Samples were transferred to a boiling water bath for 10 min to terminate the reaction. After ultra-filtration, 500 μL of the solution was taken for analysis of the distribution of amylopectin branch chain length. The branching degree of starch was measured using a Bruker BioSpin GmbH nuclear magnetic resonance spectrometer (Bruker, Rheinstetten, Germany) following the method of Zou et al. [66].

4.9. Construction of NILs

All NILs in this study used Kasalath as the recipient material. For NIL-*gnl44*, the *gnl44* mutant was used as a donor; for NIL-*gs3* and NIL-*qgl3*, TD70 was used as donor. After five generations of repeated backcrossing, the BC_5F_1 population was obtained, from which the BC_5F_3 population was obtained after two generations of selfing.

From the BC_5F_3 generation, we developed near-isogenic lines for *qgl3*(NIL-*qgl3*) and *gs3*((NIL-*gs3*)), which carry an ∼103.5- and ∼118.4-kb segment of TD70 containing the *qgl3* and *gs3* loci in the Kasalath genetic background, respectively, using molecular marker-assisted selection (MAS). NIL-gnl44 was constructed with a very small *gnl44* chromosomal region (~123.6 kb) containing the *gnl44* locus in the Kasalath genetic background using MAS. An analysis of the genes in these segments revealed that they do not contain known grain size- or quality-related genes.

For the two-gene NILs, NIL-*gnl44*, NIL-*gs3*, and NIL-*qgl3* were crossed to obtain F_1 seeds. In 2018, F_1 plants were grown and F_2 seeds were harvested in Nanjing, Jiangsu Province, China. F_2 plants were planted in Hainan Province, China, in 2018, and plants homozygous for *gnl44* and *gs3* (NIL-*gnl44*/*gs3*) or *gnl44* and *qgl3* (NIL-*gnl44*/*qgl3*) were selected from the F_2 family using MAS.

5. Conclusions

In this study, we functionally characterized the effects of *GNL44* on grain size, appearance quality, and ECQ in rice. The *gnl44* mutant had fewer but larger grains than the WT. The appearance quality and taste value of the *gnl44* mutant were reduced. The phenotype of *gnl44* is caused by a mutation in the *Os02g0244100* gene. We also explored the effects of *gnl44*, *gs3*, and *qgl3* on rice quality using monogenic NILs and two-gene NILs. Analysis of the monogenic NILs showed that *gnl44* and *qgl3* mainly affected chalkiness and gel consistency, and *gs3* mainly affected protein content. Analysis of the grain quality and ANOVA of the two-gene NILs showed that *gnl44* and *gs3* had significant genetic interactions for chalky grain ratio, chalkiness degree, protein content, and amylose content. The genetic interaction of *gnl44* and *qgl3* influences chalky grain ratio, protein content, gel consistency, and amylose content. The genetic interaction of *gs3* and *qgl3* affects milled grain length, chalky grain ratio, and chalkiness degree. Our results lay the foundation for breeding rice cultivars with high yields and superior quality.

Supplementary Materials: The following supporting information can be downloaded at: https://www.mdpi.com/article/10.3390/ijms25010589/s1.

Author Contributions: Investigation, T.C., W.L., S.Y. and C.W.; Methodology, C.Z., L.Z. (Ling Zhao), L.Z. (Lihui Zhou), Z.Z., Q.Z., K.L. and L.Z. (Li Zhu); Supervision, Y.Z.; Writing—original draft, L.H. All authors have read and agreed to the published version of the manuscript.

Funding: This work was supported by The National Natural Science Foundation of China (Grant No. 31901485), the Natural Science Foundation of Jiangsu Province (BK20211137), Revitalization of Jiangsu Seed Industry in China (JBGS [2021] 001), China Agriculture Research System (Grant No. CARS-01), and Zhongshan Biological Breeding Laboratory (BM2022008-003), The Key Agricultural research Project (NK2022010302, NK202205060105) and Zhongshan Biological Breeding Laboratory (BM2022008-003).

Institutional Review Board Statement: Not applicable.

Informed Consent Statement: Not applicable.

Data Availability Statement: The data presented in this study are available upon request from the corresponding author.

Conflicts of Interest: The authors declare no conflicts of interest.

References

1. Guo, T.; Chen, K.; Dong, N.-Q.; Shi, C.-L.; Ye, W.-W.; Gao, J.-P.; Shan, J.-X.; Lin, H.-X. GRAIN SIZE AND NUMBER1 Negatively Regulates the OsMKKK10-OsMKK4-OsMPK6 Cascade to Coordinate the Trade-off between Grain Number per Panicle and Grain Size in Rice. *Plant Cell* **2018**, *30*, 871–888. [CrossRef] [PubMed]
2. Huang, L.; Hua, K.; Xu, R.; Zeng, D.; Wang, R.; Dong, G.; Zhang, G.; Lu, X.; Fang, N.; Wang, D.; et al. The LARGE2-APO1/APO2 regulatory module controls panicle size and grain number in rice. *Plant Cell* **2021**, *33*, 1212–1228. [CrossRef] [PubMed]
3. Liu, X.; Zhou, S.; Wang, W.; Ye, Y.; Zhao, Y.; Xu, Q.; Zhou, C.; Tan, F.; Cheng, S.; Zhou, D.-X. Regulation of Histone Methylation and Reprogramming of Gene Expression in the Rice Inflorescence Meristem. *Plant Cell* **2015**, *27*, 1428–1444. [CrossRef] [PubMed]
4. Xu, R.; Duan, P.; Yu, H.; Zhou, Z.; Zhang, B.; Wang, R.; Li, J.; Zhang, G.; Zhuang, S.; Lyu, J.; et al. Control of Grain Size and Weight by the OsMKKK10-OsMKK4-OsMAPK6 Signaling Pathway in Rice. *Mol. Plant* **2018**, *11*, 860–873. [CrossRef] [PubMed]
5. Ikeda-Kawakatsu, K.; Maekawa, M.; Izawa, T.; Itoh, J.-I.; Nagato, Y. ABERRANT PANICLE ORGANIZATION 2/RFL, the rice ortholog of Arabidopsis LEAFY, suppresses the transition from inflorescence meristem to floral meristem through interaction with APO1. *Plant J.* **2012**, *69*, 168–180. [CrossRef] [PubMed]
6. Li, N.; Xu, R.; Li, Y. Molecular Networks of Seed Size Control in Plants. *Annu. Rev. Plant Biol.* **2019**, *70*, 435–463. [CrossRef] [PubMed]
7. Song, X.-J.; Huang, W.; Shi, M.; Zhu, M.-Z.; Lin, H.-X. A QTL for rice grain width and weight encodes a previously unknown RING-type E3 ubiquitin ligase. *Nat. Genet.* **2007**, *39*, 623–630. [CrossRef]
8. Zhu, X.; Zhang, S.; Chen, Y.; Mou, C.; Huang, Y.; Liu, X.; Ji, J.; Yu, J.; Hao, Q.; Yang, C.; et al. Decreased grain size1, a C3HC4-type RING protein, influences grain size in rice (*Oryza sativa* L.). *Plant Mol. Biol.* **2021**, *105*, 405–417. [CrossRef]
9. Qi, P.; Lin, Y.-S.; Song, X.-J.; Shen, J.-B.; Huang, W.; Shan, J.-X.; Zhu, M.-Z.; Jiang, L.; Gao, J.-P.; Lin, H.-X. The novel quantitative trait locus *GL3.1* controls rice grain size and yield by regulating Cyclin-T1;3. *Cell Res.* **2012**, *22*, 1666–1680. [CrossRef]
10. Liu, J.; Chen, J.; Zheng, X.; Wu, F.; Lin, Q.; Heng, Y.; Tian, P.; Cheng, Z.; Yu, X.; Zhou, K.; et al. *GW5* acts in the brassinosteroid signalling pathway to regulate grain width and weight in rice. *Nat. Plants* **2017**, *3*, 17043. [CrossRef]
11. Xia, D.; Zhou, H.; Liu, R.; Dan, W.; Li, P.; Wu, B.; Chen, J.; Wang, L.; Gao, G.; Zhang, Q.; et al. GL3.3, a Novel QTL Encoding a GSK3/SHAGGY-like Kinase, Epistatically Interacts with GS3 to Produce Extra-long Grains in Rice. *Mol. Plant* **2018**, *11*, 754–756. [CrossRef] [PubMed]
12. Ying, J.-Z.; Ma, M.; Bai, C.; Huang, X.-H.; Liu, J.-L.; Fan, Y.-Y.; Song, X.-J. TGW3, a Major QTL that Negatively Modulates Grain Length and Weight in Rice. *Mol. Plant* **2018**, *11*, 750–753. [CrossRef] [PubMed]
13. Hu, Z.; Lu, S.-J.; Wang, M.-J.; He, H.; Sun, L.; Wang, H.; Liu, X.-H.; Jiang, L.; Sun, J.-L.; Xin, X.; et al. A Novel QTL qTGW3 Encodes the GSK3/SHAGGY-Like Kinase OsGSK5/OsSK41 that Interacts with OsARF4 to Negatively Regulate Grain Size and Weight in Rice. *Mol. Plant* **2018**, *11*, 736–749. [CrossRef] [PubMed]
14. Ishimaru, K.; Hirotsu, N.; Madoka, Y.; Murakami, N.; Hara, N.; Onodera, H.; Kashiwagi, T.; Ujiie, K.; Shimizu, B.-i.; Onishi, A.; et al. Loss of function of the IAA-glucose hydrolase gene *TGW6* enhances rice grain weight and increases yield. *Nat. Genet.* **2013**, *45*, 707–711. [CrossRef] [PubMed]
15. Liu, L.; Tong, H.; Xiao, Y.; Che, R.; Xu, F.; Hu, B.; Liang, C.; Chu, J.; Li, J.; Chu, C. Activation of *Big Grain1* significantly improves grain size by regulating auxin transport in rice. *Proc. Natl. Acad. Sci. USA* **2015**, *112*, 11102–11107. [CrossRef] [PubMed]
16. Si, L.; Chen, J.; Huang, X.; Gong, H.; Luo, J.; Hou, Q.; Zhou, T.; Lu, T.; Zhu, J.; Shangguan, Y.; et al. *OsSPL13* controls grain size in cultivated rice. *Nat. Genet.* **2016**, *48*, 447–456. [CrossRef] [PubMed]
17. Duan, P.; Rao, Y.; Zeng, D.; Yang, Y.; Xu, R.; Zhang, B.; Dong, G.; Qian, Q.; Li, Y. *SMALL GRAIN 1*, which encodes a mitogen-activated protein kinase kinase 4, influences grain size in rice. *Plant J.* **2014**, *77*, 547–557. [CrossRef] [PubMed]
18. Liu, S.; Hua, L.; Dong, S.; Chen, H.; Zhu, X.; Jiang, J.; Zhang, F.; Li, Y.; Fang, X.; Chen, F. OsMAPK6, a mitogen-activated protein kinase, influences rice grain size and biomass production. *Plant J.* **2015**, *84*, 672–681. [CrossRef]
19. Utsunomiya, Y.; Samejima, C.; Takayanagi, Y.; Izawa, Y.; Yoshida, T.; Sawada, Y.; Fujisawa, Y.; Kato, H.; Iwasaki, Y. Suppression of the rice heterotrimeric G protein β-subunit gene, *RGB1*, causes dwarfism and browning of internodes and lamina joint regions. *Plant J.* **2011**, *67*, 907–916. [CrossRef]
20. Swain, D.M.; Sahoo, R.K.; Srivastava, V.K.; Tripathy, B.C.; Tuteja, R.; Tuteja, N. Function of heterotrimeric G-protein γ subunit RGG1 in providing salinity stress tolerance in rice by elevating detoxification of ROS. *Planta* **2017**, *245*, 367–383. [CrossRef]
21. Yadav, D.K.; Islam, S.M.S.; Tuteja, N. Rice heterotrimeric G-protein gamma subunits (RGG1 and RGG2) are differentially regulated under abiotic stress. *Plant Signal. Behav.* **2012**, *7*, 733–740. [CrossRef] [PubMed]
22. Mao, H.; Sun, S.; Yao, J.; Wang, C.; Yu, S.; Xu, C.; Li, X.; Zhang, Q. Linking differential domain functions of the GS3 protein to natural variation of grain size in rice. *Proc. Natl. Acad. Sci. USA* **2010**, *107*, 19579–19584. [CrossRef] [PubMed]
23. Fan, C.; Xing, Y.; Mao, H.; Lu, T.; Han, B.; Xu, C.; Li, X.; Zhang, Q. *GS3*, a major QTL for grain length and weight and minor QTL for grain width and thickness in rice, encodes a putative transmembrane protein. *Theor. Appl. Genet.* **2006**, *112*, 1164–1171. [CrossRef] [PubMed]
24. Sun, S.; Wang, L.; Mao, H.; Shao, L.; Li, X.; Xiao, J.; Ouyang, Y.; Zhang, Q. A G-protein pathway determines grain size in rice. *Nat. Commun.* **2018**, *9*, 851. [CrossRef] [PubMed]
25. Harberd, N.P. Shaping Taste: The Molecular Discovery of Rice Genes Improving Grain Size, Shape and Quality. *J. Genet. Genom.* **2015**, *42*, 597–599. [CrossRef]

26. Zhao, D.-S.; Li, Q.-F.; Zhang, C.-Q.; Zhang, C.; Yang, Q.-Q.; Pan, L.-X.; Ren, X.-Y.; Lu, J.; Gu, M.-H.; Liu, Q.-Q. GS9 acts as a transcriptional activator to regulate rice grain shape and appearance quality. *Nat. Commun.* **2018**, *9*, 1240. [CrossRef] [PubMed]
27. Li, N.; Li, Y. Signaling pathways of seed size control in plants. *Curr. Opin. Plant Biol.* **2016**, *33*, 23–32. [CrossRef]
28. Zuo, J.; Li, J. Molecular Genetic Dissection of Quantitative Trait Loci Regulating Rice Grain Size. *Annu. Rev. Genet.* **2014**, *48*, 99–118. [CrossRef]
29. Gao, X.; Zhang, X.; Lan, H.; Huang, J.; Wang, J.; Zhang, H. The additive effects of GS3 and qGL3 on rice grain length regulation revealed by genetic and transcriptome comparisons. *BMC Plant Biol.* **2015**, *15*, 156. [CrossRef]
30. Lei, H.; Wenhua, L.; Jiang, H.; Chunfang, Z.; Shu, Y.; Tao, C.; Zhen, Z.; Qingyong, Z.; Kai, L.; Ling, Z.; et al. Additive Effects of QTLs/Genes on Rice Grain Size Traits Revealed by Genetic Comparisons. *Rice Sci.* **2023**, *30*, 171–175. [CrossRef]
31. Hanashiro, I.; Abe, J.-i.; Hizukuri, S. A periodic distribution of the chain length of amylopectin as revealed by high-performance anion-exchange chromatography. *Carbohydr. Res.* **1996**, *283*, 151–159. [CrossRef]
32. Zhu, F. Relationships between amylopectin internal molecular structure and physicochemical properties of starch. *Trends Food Sci. Technol.* **2018**, *78*, 234–242. [CrossRef]
33. Patindol, J.; Gu, X.; Wang, Y.-J. Chemometric analysis of cooked rice texture in relation to starch fine structure and leaching characteristics. *Starch-Stärke* **2010**, *62*, 188–197. [CrossRef]
34. Li, C.; Luo, J.-X.; Zhang, C.-Q.; Yu, W.-W. Causal relations among starch chain-length distributions, short-term retrogradation and cooked rice texture. *Food Hydrocoll.* **2020**, *108*, 106064. [CrossRef]
35. Ren, D.; Ding, C.; Qian, Q. Molecular bases of rice grain size and quality for optimized productivity. *Sci. Bull.* **2023**, *68*, 314–350. [CrossRef] [PubMed]
36. Zhang, Y.-D.; Zhang, Y.-H.; Dong, S.-L.; Chen, T.; Zhao, Q.-Y.; Zhu, Z.; Zhou, L.-H.; Yao, S.; Zhao, L.; Yu, X.; et al. QTL Mapping for Grain Size Traits Based on Extra-Large Grain Rice Line TD70. *Rice Sci.* **2013**, *20*, 400–406. [CrossRef]
37. Lu, L.; Shao, D.; Qiu, X.; Sun, L.; Yan, W.; Zhou, X.; Yang, L.; He, Y.; Yu, S.; Xing, Y. Natural variation and artificial selection in four genes determine grain shape in rice. *New Phytol.* **2013**, *200*, 1269–1280. [CrossRef] [PubMed]
38. Dixit, N.; Dokku, P.; Amitha Mithra, S.V.; Parida, S.K.; Singh, A.K.; Singh, N.K.; Mohapatra, T. Haplotype structure in grain weight gene GW2 and its association with grain characteristics in rice. *Euphytica* **2013**, *192*, 55–61. [CrossRef]
39. Huang, J.; Chen, Z.; Lin, J.; Guan, B.; Chen, J.; Zhang, Z.; Chen, F.; Jiang, L.; Zheng, J.; Wang, T.; et al. gw2.1, a new allele of GW2, improves grain weight and grain yield in rice. *Plant Sci.* **2022**, *325*, 111495. [CrossRef]
40. Sharma, R.; Priya, P.; Jain, M. Modified expression of an auxin-responsive rice CC-type glutaredoxin gene affects multiple abiotic stress responses. *Planta* **2013**, *238*, 871–884. [CrossRef]
41. Choi, B.S.; Kim, Y.J.; Markkandan, K.; Koo, Y.J.; Song, J.T.; Seo, H.S. GW2 Functions as an E3 Ubiquitin Ligase for Rice Expansin-Like 1. *Int. J. Mol. Sci.* **2018**, *19*, 1904. [CrossRef] [PubMed]
42. Hao, J.; Wang, D.; Wu, Y.; Huang, K.; Duan, P.; Li, N.; Xu, R.; Zeng, D.; Dong, G.; Zhang, B.; et al. The GW2-WG1-OsbZIP47 pathway controls grain size and weight in rice. *Mol. Plant* **2021**, *14*, 1266–1280. [CrossRef] [PubMed]
43. Zhou, H.; Xia, D.; Zhao, D.; Li, Y.; Li, P.; Wu, B.; Gao, G.; Zhang, Q.; Wang, G.; Xiao, J.; et al. The origin of Wxla provides new insights into the improvement of grain quality in rice. *J. Integr. Plant Biol.* **2021**, *63*, 878–888. [CrossRef] [PubMed]
44. Zhang, C.; Yang, Y.; Chen, S.; Liu, X.; Zhu, J.; Zhou, L.; Lu, Y.; Li, Q.; Fan, X.; Tang, S.; et al. A rare Waxy allele coordinately improves rice eating and cooking quality and grain transparency. *J. Integr. Plant Biol.* **2021**, *63*, 889–901. [CrossRef] [PubMed]
45. Cai, X.-L.; Wang, Z.-Y.; Xing, Y.-Y.; Zhang, J.-L.; Hong, M.-M. Aberrant splicing of intron 1 leads to the heterogeneous 5′ UTR and decreased expression of waxy gene in rice cultivars of intermediate amylose content. *Plant J.* **1998**, *14*, 459–465. [CrossRef] [PubMed]
46. Liu, L.; Ma, X.; Liu, S.; Zhu, C.; Jiang, L.; Wang, Y.; Shen, Y.; Ren, Y.; Dong, H.; Chen, L.; et al. Identification and characterization of a novel Waxy allele from a Yunnan rice landrace. *Plant Mol. Biol.* **2009**, *71*, 609–626. [CrossRef] [PubMed]
47. Yang, J.; Wang, J.; Fan, F.-J.; Zhu, J.-Y.; Chen, T.; Wang, C.-L.; Zheng, T.-Q.; Zhang, J.; Zhong, W.-G.; Xu, J.-L. Development of AS-PCR marker based on a key mutation confirmed by resequencing of Wx-mp in Milky Princess and its application in japonica soft rice (*Oryza sativa* L.) breeding. *Plant Breed.* **2013**, *132*, 595–603. [CrossRef]
48. Zhang, C.; Zhu, J.; Chen, S.; Fan, X.; Li, Q.; Lu, Y.; Wang, M.; Yu, H.; Yi, C.; Tang, S.; et al. Wxlv, the Ancestral Allele of Rice Waxy Gene. *Mol. Plant* **2019**, *12*, 1157–1166. [CrossRef]
49. Achary, V.M.M.; Reddy, M.K. CRISPR-Cas9 mediated mutation in GRAIN WIDTH and WEIGHT2 (GW2) locus improves aleurone layer and grain nutritional quality in rice. *Sci. Rep.* **2021**, *11*, 21941. [CrossRef]
50. He, W.; Li, W.; Luo, X.; Tang, Y.; Wang, L.; Yu, F.; Lin, Q. Rice FERONIA-LIKE RECEPTOR 3 and 14 affect grain quality by regulating redox homeostasis during endosperm development. *J. Exp. Bot.* **2023**, *74*, 3003–3018. [CrossRef]
51. Wu, B.; Yun, P.; Zhou, H.; Xia, D.; Gu, Y.; Li, P.; Yao, J.; Zhou, Z.; Chen, J.; Liu, R.; et al. Natural variation in WHITE-CORE RATE 1 regulates redox homeostasis in rice endosperm to affect grain quality. *Plant Cell* **2022**, *34*, 1912–1932. [CrossRef] [PubMed]
52. Zhang, Y.; Zeng, J.; Liang, Y.; Zhao, C.; Chen, T.; Zhao, Q.; Zhu, Z.; Zhou, L.; Yao, S.; Zhao, L.; et al. Functional Marker Development and Effect Analysis of Grain Size Gene GW2 in Extreme Grain Size Germplasm in Rice. *Rice Sci.* **2015**, *22*, 65–70.
53. Li, P.; Chen, Y.-H.; Lu, J.; Zhang, C.-Q.; Liu, Q.-Q.; Li, Q.-F. Genes and Their Molecular Functions Determining Seed Structure, Components, and Quality of Rice. *Rice* **2022**, *15*, 18. [CrossRef] [PubMed]

54. Li, Y.-F.; Zeng, X.-Q.; Li, Y.; Wang, L.; Zhuang, H.; Wang, Y.; Tang, J.; Wang, H.-L.; Xiong, M.; Yang, F.-Y.; et al. MULTI-FLORET SPIKELET 2, a MYB Transcription Factor, Determines Spikelet Meristem Fate and Floral Organ Identity in Rice. *Plant Physiol.* **2020**, *184*, 988–1003. [CrossRef]
55. Abe, A.; Kosugi, S.; Yoshida, K.; Natsume, S.; Takagi, H.; Kanzaki, H.; Matsumura, H.; Yoshida, K.; Mitsuoka, C.; Tamiru, M.; et al. Genome sequencing reveals agronomically important loci in rice using MutMap. *Nat. Biotechnol.* **2012**, *30*, 174–178. [CrossRef]
56. Toki, S.; Hara, N.; Ono, K.; Onodera, H.; Tagiri, A.; Oka, S.; Tanaka, H. Early infection of scutellum tissue with Agrobacterium allows high-speed transformation of rice. *Plant J.* **2006**, *47*, 969–976. [CrossRef]
57. Toki, S. Rapid and efficient Agrobacterium-mediated transformation in rice. *Plant Mol. Biol. Rep.* **1997**, *15*, 16–21. [CrossRef]
58. Zhang, Y.; Su, J.; Duan, S.; Ao, Y.; Dai, J.; Liu, J.; Wang, P.; Li, Y.; Liu, B.; Feng, D.; et al. A highly efficient rice green tissue protoplast system for transient gene expression and studying light/chloroplast-related processes. *Plant Methods* **2011**, *7*, 30. [CrossRef]
59. Zhao, J.; Long, T.; Wang, Y.; Tong, X.; Tang, J.; Li, J.; Wang, H.; Tang, L.; Li, Z.; Shu, Y.; et al. RMS2 Encoding a GDSL Lipase Mediates Lipid Homeostasis in Anthers to Determine Rice Male Fertility1 [OPEN]. *Plant Physiol.* **2020**, *182*, 2047–2064. [CrossRef]
60. Dong, Z.; Chen, Y. Transcriptomics: Advances and approaches. *Sci. China Life Sci.* **2013**, *56*, 960–967. [CrossRef]
61. Li, B.; Du, X.; Fei, Y.; Wang, F.; Xu, Y.; Li, X.; Li, W.; Chen, Z.; Fan, F.; Wang, J.; et al. Efficient Breeding of Early-Maturing Rice Cultivar by Editing PHYC via CRISPR/Cas9. *Rice* **2021**, *14*, 86. [CrossRef] [PubMed]
62. Li, H.; Wang, R.; Zhang, Q.; Li, G.; Shan, Y.; Ding, S. Morphological, structural, and physicochemical properties of starch isolated from different lily cultivars grown in China. *Int. J. Food Prop.* **2019**, *22*, 737–757. [CrossRef]
63. Shi, K.; Gu, X.; Lu, W.; Lu, D. Effects of weak-light stress during grain filling on the physicochemical properties of normal maize starch. *Carbohydr. Polym.* **2018**, *202*, 47–55. [CrossRef] [PubMed]
64. Tao, K.; Yu, W.W.; Prakash, S.; Gilbert, R.G. High-amylose rice: Starch molecular structural features controlling cooked rice texture and preference. *Carbohydr. Polym.* **2019**, *219*, 251–260. [CrossRef]
65. Tan, Y.F.; Li, J.X.; Yu, S.B.; Xing, Y.Z.; Xu, C.G.; Zhang, Q. The three important traits for cooking and eating quality of rice grains are controlled by a single locus in an elite rice hybrid, Shanyou 63. *Theor. Appl. Genet.* **1999**, *99*, 642–648. [CrossRef]
66. Zou, J.; Xu, M.J.; Wen, L.R.; Yang, B. Structure and physicochemical properties of native starch and resistant starch in Chinese yam (*Dioscorea opposita* Thunb.). *Carbohydr. Polym.* **2020**, *237*, 116188. [CrossRef]

International Journal of *Molecular Sciences*

MDPI

Article

Conserved Residues Lys64 and Glu78 at the Subunit Surface of Tau Glutathione Transferase in Rice Affect Structure and Enzymatic Properties

Xue Yang [1], Zhe Zhang [1], Lei Wu [1], Meiying Yang [1], Siyuan Li [1,*] and Jie Gao [2,*]

[1] College of Life Sciences, Jilin Agricultural University, Changchun 130118, China; xyang@jlau.edu.cn (X.Y.); 20200957@mails.jlau.edu.cn (Z.Z.); lwu@jlau.edu.cn (L.W.); yangmeiying@jlau.edu.cn (M.Y.)

[2] CAS Key Laboratory of Tropical Forest Ecology, Xishuangbanna Tropical Botanical Garden, Chinese Academy of Sciences, Menglun 666303, China

* Correspondence: siyuanl@jlau.edu.cn (S.L.); gaojie@xtbg.org.cn (J.G.)

Abstract: Glutathione transferases (GSTs) are a superfamily of dimeric proteins associated with the detoxification of various reactive electrophiles and responsive to a multitude of stressors. We individually substituted Lys64 and Glu78 with Ala using site-directed mutagenesis to understand the role of subunit interactions in the structure and enzymatic properties of a rice GST (OsGSTU17). The wild-type OsGSTU17 lost the conserved hydrogen bond between subunits in tau class GSTs due to conserved Tyr92 replaced with Phe92, but still exhibited high substrate activities, and thermal stability remained in its dimeric structure. The significant decrease in thermal stability and obvious changes in the structure of mutant K64A implied that conserved Lys64 might play an essential role in the structural stability of tau class GSTs. The mutant E78A, supposed to be deprived of hydrogen and salt bonds between subunits, appeared in the soluble form of dimers, even though its tertiary structure altered and stability declined dramatically. These results suggest that the hydrogen and ionic bonds provided by conserved residues are not as important for OsGSTU17 dimerization and enzymatic properties. These results further supplement our understanding of the relationship between the structure and function of GSTs and provide a theoretical basis for improving crop resistance through targeted modification of GSTs.

Keywords: glutathione transferases; hydrogen bond; ionic bond; site-directed mutagenesis; structural changes; activity changes

Citation: Yang, X.; Zhang, Z.; Wu, L.; Yang, M.; Li, S.; Gao, J. Conserved Residues Lys64 and Glu78 at the Subunit Surface of Tau Glutathione Transferase in Rice Affect Structure and Enzymatic Properties. *Int. J. Mol. Sci.* **2024**, *25*, 398. https://doi.org/10.3390/ijms25010398

Academic Editors: Yongmei Bao and Shimin Zuo

Received: 27 November 2023
Revised: 24 December 2023
Accepted: 25 December 2023
Published: 28 December 2023

1. Introduction

Glutathione transferases (GSTs, EC 2.5.1.8), mainly in the form of dimers, are found in all classes of eukaryotic and prokaryotic organisms and catalyze a broad range of reactions involved in the addition of glutathione to substrate compounds. Typically, these are involved in nucleophilic substitution using the sulfhydryl group of reduced glutathione (GSH) to attack the electrophilic center of toxic substances, which results in the formation of hydrophilic metabolites that are readily excreted. Additionally, GSTs play a significant role in primary and secondary metabolism, stress tolerance, and cell signaling; they are also considered vigilant stress monitors capable of rapidly responding to biotic and abiotic stimuli in the surroundings [1–3]. Based on immunological characterizations, sequence similarity, and biochemical properties, plant GSTs are divided into 14 classes, among which four classes were previously considered to be plant-specific [3]. However, since highly homologous proteins with phi class GSTs were identified in fungi, bacteria, and protozoa, only tau, lambda, and DHAR class GSTs were recognized as plant-specific [4,5].

Despite complex structural variations between distinct classes, GSTs are mainly dimeric, and each subunit contains two domains: a thioredoxin-like (βαβαββα motif) N-terminal domain and an α-helix C-terminal domain connected by a linker region

(Figure 1) [6]. The highly conserved N-terminal domain encompasses conserved polar amino acid residues that exist across all cytosolic GSTs and constitute the GSH binding site (G-site), liable for triggering the GSH binding and subsequent activation of its thiol group [7,8]. The less conserved C-terminal domain forms a pocket-like hydrophobic substrate binding site (H-site), responsible for binding a variety of hydrophobic substrates, which makes it more diverse [9]. The steps of GST detoxification reactions consist of binding the reduced GSH and hydrophobic toxic substances at its G-site and H-site, respectively, to activate the thiol group of GSH, allowing for the nucleophilic attack upon the toxic substances [10]. Although each subunit has active sites of its own, the function of the enzymes can be achieved only if they form a dimer, except for classes DHAR and Lambda, which function as monomeric proteins [11]. Subunit interactions responsible for maintaining dimerization are essential for tau class GST functions.

The quaternary structure of GSTs is formed by interconnections between complementary regions in opposite subunits via hydrophobic interactions, hydrogen bonds, and salt linkages. So far, extensive investigations have been focused on the formation of quaternary structures of animal GSTs, such as the structures in pi, mu, delta, and epsilon class GSTs, in which either a lock-and-key motif located on a subunit interface or a unique subunit interface motif was identified for specific subunit recognition and dimer stabilization [12–15]. However, little is known about the underlying mechanism for maintaining the quaternary structures in plant GSTs. The only knowledge is that a significant number of conserved residues located on the interfaces of subunits that form hydrogen, ionic bonds, or hydrophobic interactions might be involved in the dimerization of the tau class GSTs [16,17]. Previous structural studies of tau class GSTs have shown that, usually, a hydrogen bond between the carboxyl group in the main chain of proline and the hydroxyl group in the side chain of tyrosine from the other subunit and salt bridges between the carboxylate group in the side chain of glutamate and guanidinium groups of two arginine in the second subunit are required to maintain the dimers [16–19]. Wang et al. found that the detriments of such interactions could lead to a complete loss of enzymatic stability and/or activities [20]. However, high enzymatic thermal stability and activities exhibited by a tau class GST OsGSTU17 in rice, which naturally lacks the hydrogen bond interaction between subunits by proline and tyrosine, were observed [21]. Therefore, the role of hydrogen bonds, salt bridges, and conserved residues on the functional tau class GSTs needs to be further elucidated. OsGSTU17 has a positive drought resistance effect in rice [22], and we took it as the research target, of which a conserved key tyrosine residue of tau class GSTs is replaced by a phenylalanine residue naturally, to understand the mechanism of the maintenance of dimers in the tau class GSTs through site-directed mutagenesis, with the aim of improving rice's ability to resist abiotic stress through gene editing technology in the future.

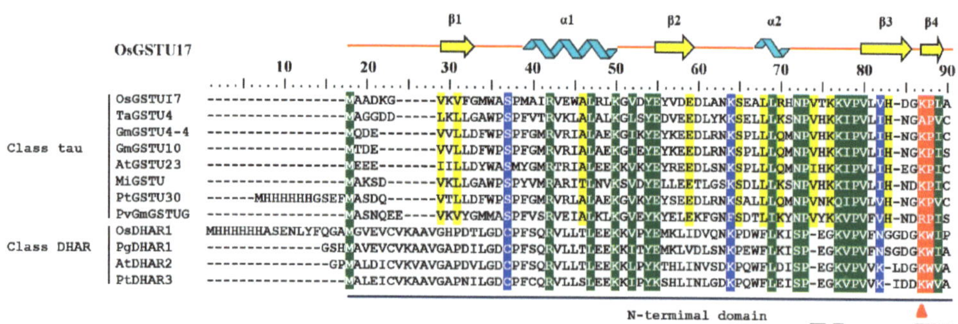

Figure 1. *Cont.*

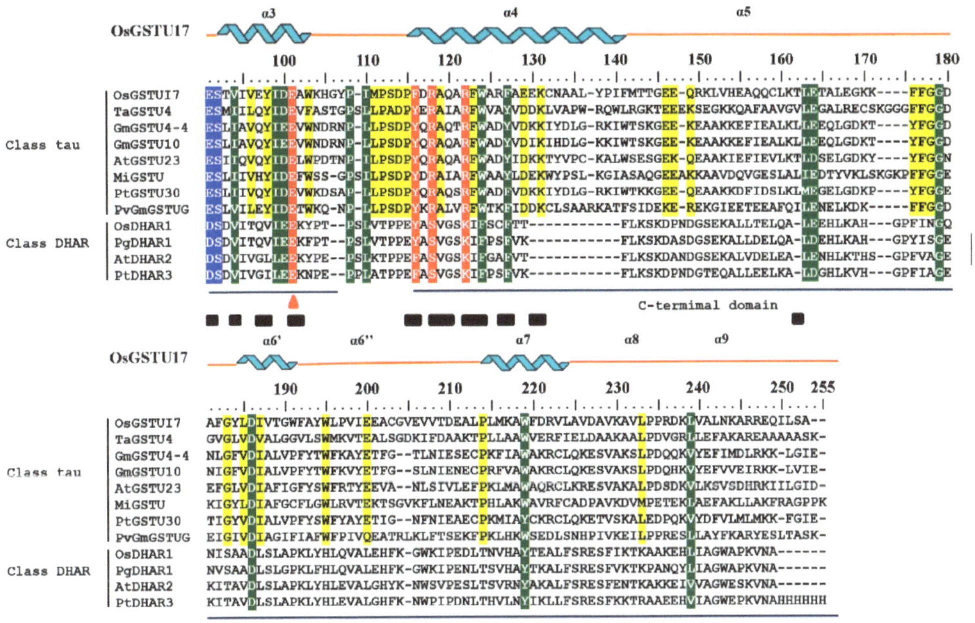

Figure 1. Sequence alignment of tau and DHAR class GSTs and the secondary structural element predictions of OsGSTU17. α-helices and β-strands were represented as helices and arrows, respectively. The conserved residues predicted to form hydrogen bonds and salt bridges between subunits were shaded in red. The conserved residues of the G-site were shaded in blue. The mutated sites were marked with ▲ in red. The residues buried at the interfaces of dimers were marked with ■ in black [16]. The conserved residues in all GSTs were shaded in green, whereas those among tau class GSTs were shaded in yellow. The protein sequences used in the alignment were OsGSTU17 (*Oryza sativa*, AF402804), TaGSTU4 (*Triticum tauschii*, 1GWC), GmGSTU4-4 (*Glycine max*, 2VO4), GmGSTU10 (*Glycine max*, 4CHS), AtGSTU23 (*Arabidopsis thaliana*, 6EP6), MiGSTU (*Mangifera indica*, 5G5F), PtGSTU30 (*Pinus tabulaeformis*, 5J4U), PvGmGSTUG (*Phaseolus vulgaris*, 6GHF), OsDHAR1 (*Oryza sativa*, 5D9T), PgDHAR1 (*Pennisetum glaucum*, 5EVO), AtDHAR2 (*Arabidopsis thaliana*, 5LOL), and PtDHAR3 (*Populus trichocarpa*, 2N5F).

2. Results

2.1. Prediction of Amino Acid Residues Involved in Subunit Interactions

A significant number of conserved residues residing on the interfaces of tau class GST subunits were identified according to the sequence alignment with proteins having crystal structures in the protein data bank (PDB) (Figure 1). Three-dimensional structural alignment between the predicted tertiary structure of OsGSTU17 and structure of TaGSTU4 (PDB:1GWC) (Figure 2) revealed that the conserved amino acid residues, Pro65, Glu78, Phe92 (replacing conserved Tyr93 in TaGSTU4), Arg94, and Arg98 (alignment positions of 88, 101, 116, 118, and 122, respectively, in Figure 1) at the subunit interface of OsGSTU17 were superimposed on the residues at corresponding positions of TaGSTU4. Since these conserved amino acid residues mentioned above are responsible for formation hydrogen and salt bonds in TaGSTU4 to maintain interactions between subunits [16,17], it is speculated that these corresponding residues in OsGSTU7 might be involved in the dimerization of OsGSTU7. Interestingly, despite the natural absence of the hydrogen bond in OsGSTU17, this enzyme exhibited high stability and substrate activities [21], contrary to Wang's observation [20], which showed that the replacement of the key residue Tyr93 with Phe led to an almost abolishment of its detoxification capability. Whether the dramatic decline in the enzymatic activities of mutant Y93F of TaGSTU4 is due to the natural replacement of a polar, positively charged Lys64 adjacent to Pro65 with a non-polar, hydrophobic Ala64 residue in TaGSTU4 needs to be clarified. Here, we explored site mutagenesis to substitute conserved residues Lys64 and Glu78 (alignment positions of 87 and 101, respectively, in Figure 1) in OsGSTU17 with Ala separately to understand the role of these amino acids in the interactions between subunits in OsGSTU17.

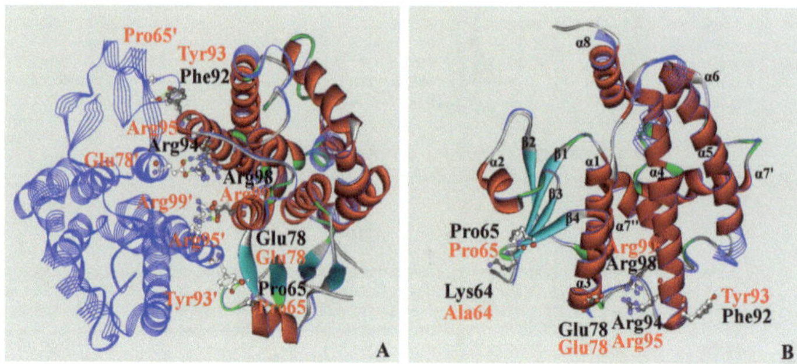

Figure 2. Comparison of the structure of OsGSTU17 and TaGSTU4 (PDB: 1GWC). (**A**) Overlapping of the monomeric structure of OsGSTU17 with the dimeric structure of TaGSTU4; (**B**) Comparison of spatial conformations of conserved amino acids on the surface of subunits of OsGSTU17 and TaGSTU4 proteins. The structure of OsGSTU17 was displayed in flat ribbons with α-helix shown in red and β-sheet shown in cyan. The structure of TaGSTU4 was displayed in blue line ribbons. Amino acid residues involved in hydrogen bonds and ionic bonds were displayed in balls and sticks. The carbon atoms of TaGSTU4 and OsGSTU17 were represented by white and gray balls, respectively. Nitrogen and oxygen atoms were represented by blue and red balls, respectively. Hydrogen bonds and ionic bonds were indicated by green dashes. Functional residues involved with subunit interactions in OsGSTU17 and TaGSTU4 were marked in black and red, respectively.

2.2. *Expression and Purification of Mutant Proteins*

The two mutants, K64A and E78A, of OsGSTU17 were expressed in *Escherichia coli* BL21 (DE3), then purified by affinity chromatography. SDS-PAGE showed that the mutants were expressed in soluble forms (Figure 3), and both denatured proteins appeared as single bands on the gel, with a molecular weight of about 25.0 kDa, which was consistent with the expected theoretical molecular weight of the OsGSTU17 mutant monomer (Figure 3).

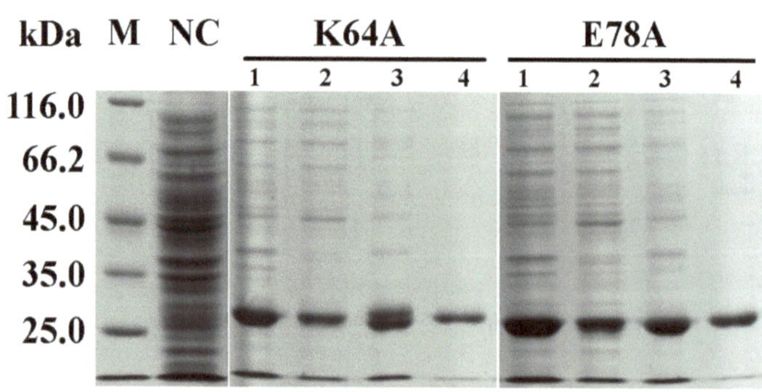

Figure 3. Analysis of the expression and purification of mutant proteins by SDS-PAGE. Lane M: Protein marker; Lane NC: Negative control (bacteria containing modified pET30 plasmids); Lane 1: Bacteria containing recombinant plasmids after being induced by IPTG; Lane 2: Supernatant of cell lysate by ultrasound; Lane 3: Precipitate of cell lysate by ultrasound; and Lane 4: Purified recombinant mutant proteins.

2.3. Substrate Activities of Mutant Proteins

Seven typical substrates were selected to examine the detoxification and glutathione peroxidase (GPOX) activities of the mutants and wild-type OsGSTU17 (Table 1). 7-chloro-4-nitrobenzo-2-oxa-1,3-diazole (NBD-Cl), 1-chloro-2,4-dinitrobenzene (CDNB), 4-nitrobenzyl chloride (NBC), 1,2-dichloro-4-nitrobenzene (DCNB), ethacrynic acid (ECA), and 4-nitrophenyl acetate (4-NPA) were used to detect the detoxification activities of proteins, while cumene hydroperoxide (Cum-OOH) was the usual substrate for glutathione peroxidase. Generally, compared with the wild-type OsGSTU17, the mutant proteins showed significant changes in substrate activities yet preserved their substrate specificity profiling as the same as that of the wild-type. Specifically, mutant K64A and E78A showed significant decreases in substrate activities towards NBC, retaining 3.8% and 10.1% of those of the wild-type, respectively. Towards NBD-Cl, the enzymatic activity of mutant E78A decreased by approximately 50%, and a slight decrease was observed in mutant K64A. In contrast, both mutant proteins exhibited increased activities towards CDNB, yet they were not statistically significant. The activity of mutant K64A towards Cum-OOH increased about 2.3-fold; however, the substrate activity of E78A towards Cum-OOH remained unchanged with that of the wild-type.

Table 1. Specific activities of K64A and E78A mutants and wild-type.

Substrates	Specific Activity (μmol min^{-1} mg^{-1})		
	WT [21]	K64A	E78A
NBD-Cl	0.203 ± 0.006	0.180 ± 0.001	0.087 ± 0.001
CDNB	0.113 ± 0.019	0.158 ± 0.003	0.138 ± 0.007
NBC	1.153 ± 0.046	0.044 ± 0.013	0.117 ± 0.005
Cum-OOH	0.013 ± 0.002	0.030 ± 0.001	0.014 ± 0.001
DCNB	ND	ND	ND
4-NPA	ND	ND	ND
ECA	ND	ND	ND

Note: The values shown are means ± S.D., calculated from three replicates. The values for the wild-type were obtained from our previous study [21]. NBD-Cl represents 7-chloro-4-nitrobenzo-2-oxa-1,3-diazole; CDNB represents 1-chloro-2,4-dinitrobenzene; NBC represents 4-nitrobenzyl chloride; Cum-OOH represents cumene hydroperoxide; DCNB represents 1,2-dichloro-4-nitrobenzene; 4-NPA represents 4-nitrophenyl acetate; and ECA represents ethacrynic acid.

2.4. Kinetic Parameters of Mutant Proteins

The impacts of the substitution of Lys64 with Ala in OsGSTU17 on the enzymatic kinetics were explored (Table 2). Compared to the wild-type OsGSTU17, the affinities of the mutant K64A to GSH and NBD-Cl decreased significantly, with the K_m values increasing by 4- and 2-fold and the catalytic efficiencies (k_{cat}/K_m) decreasing by 80% and 46%, respectively. However, the replacement of the residue did not significantly affect the enzymatic V_{max} and k_{cat} to the substrates, indicating that the substitution of Lys64 with Ala in OsGSTU17 resulted in the disruption of the enzymatic substrate affinity, which might be the primary reason for the decline of its catalytic efficiency. The activity of mutant E78A towards the substrates was too low for the kinetic studies.

Table 2. Kinetic analysis of mutant K64A and wild-type.

	GSH				NBD-Cl			
	K_m (mM)	V_{max} (μM min^{-1} mg^{-1})	k_{cat} (S^{-1})	k_{cat}/K_m (mM^{-1}S^{-1})	K_m (mM)	V_{max} (μM min^{-1} mg^{-1})	k_{cat} (S^{-1})	k_{cat}/K_m (mM^{-1}S^{-1})
WT [21]	0.058 ± 0.006	0.225 ± 0.011	0.264	4.552	0.324 ± 0.016	0.219 ± 0.006	0.257	0.793
K64A	0.233 ± 0.001	0.226 ± 0.001	0.171	0.924	0.507 ± 0.025	0.287 ± 0.010	0.217	0.428

Note: The values shown are means ± S.D., calculated from three replicates. The values for the wild-type were obtained from our previous study [21].

2.5. Structural and Thermodynamic Analysis of Mutant Proteins

Structural changes in OsGSTU17 induced by amino acid replacements were analyzed by circular dichroism (CD) spectroscopy, ANS binding assay, native PAGE (polyacrylamide gel electrophoresis), and thermal stability assessment.

CD spectra have been frequently adopted for the rapid determination of protein secondary structure and folding properties to examine the effects of mutations on protein conformation or stability [23]. Differences in far-UV CD absorbance (190–260 nm) reflect distinct variations in the secondary structures [24]. No significant change in secondary structures between mutants and wild-type OsGSTU17 was detected by far-UV CD spectroscopy analysis in this study (Figure 4). The percentages of α-helix, β-sheet, and random coils roughly calculated from the far-UV CD spectra showed that minor differences existed between the mutants and wild-type and were mainly located in the α-helix and antiparallel β-sheet (Table S1). The near-UV CD spectra (250–320 nm) reflect the properties of disulfide bonds and aromatic residues and inform the spatial conformation of aromatic residues, which were used to advise on protein tertiary structures [25–27]. Aromatic residues phenylalanine, tyrosine, and tryptophan exhibit distinct absorbance peaks in the wavelength regions of 250–270 nm, 270–290 nm, and 280–300 nm, respectively. The counts of residues and their conformational and environmental alterations would affect the CD signals, which would facilitate hinting at the subtle changes in protein tertiary structure caused by residue substitutions. According to the results of near-UV CD spectra (Figure 5), the mutation of the conserved amino acid residue Lys64 caused a significant change in the spatial conformation of tyrosine residues in the OsGSTU17 protein (the peak in the wavelength regions of 270–290 nm disappeared). The replacement in the conserved amino acid residue Glu78 led to the disappearance of the peak in the near-UV CD spectrum of the protein within the wavelength range of 280–300 nm, which represents the spatial conformation of tryptophan.

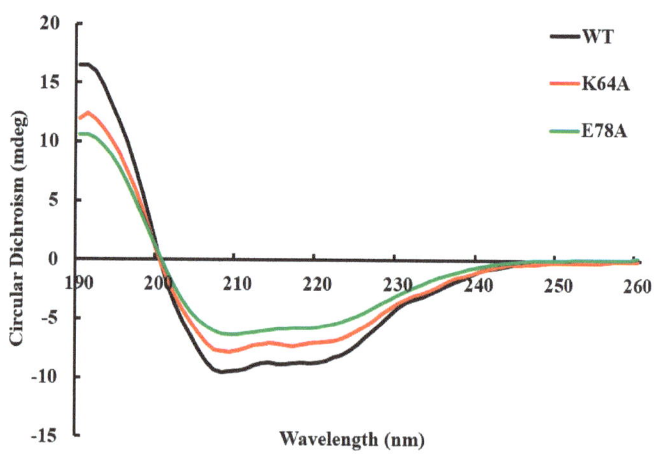

Figure 4. Far-UV CD spectra of the wild-type and mutants of OsGSTU17. The values of the wild-type were from our previous study [28].

ANS (1-aniline-8-naphthalene sulfonate), a fluorescent dye with low fluorescence intensity in a polar environment, when noncovalently bound with hydrophobic surfaces of proteins, its fluorescence intensity would increase significantly, accompanied by a maximum blue shift emission [29,30]. The fluorescent intensity reflects the scale of hydrophobicity of protein surfaces, which provides a clue as to what variations occurred in protein tertiary structures. The maximum fluorescent intensities of mutants K64A and E78A were almost identical (about 250 nm) but higher than that of wild-type, which indicated that the hydrophobic areas of protein OsGSTU17 were enlarged to a similar extent after the residues Lys64 and Glu78 were substituted with alanine (Figure 6).

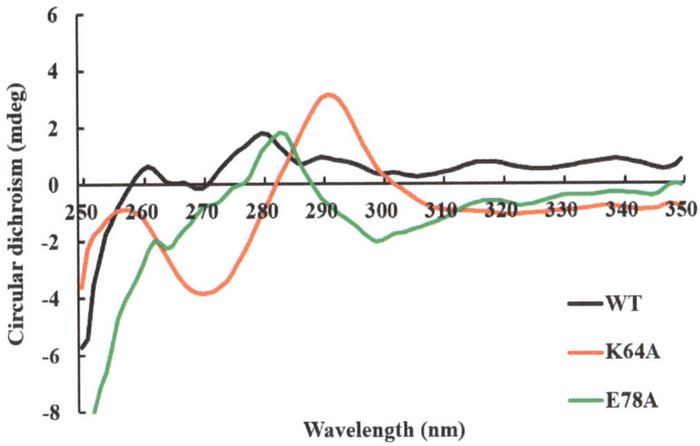

Figure 5. Near-UV CD spectra of the wild-type and mutants of OsGSTU17. The values of the wild-type were from our previous study [28].

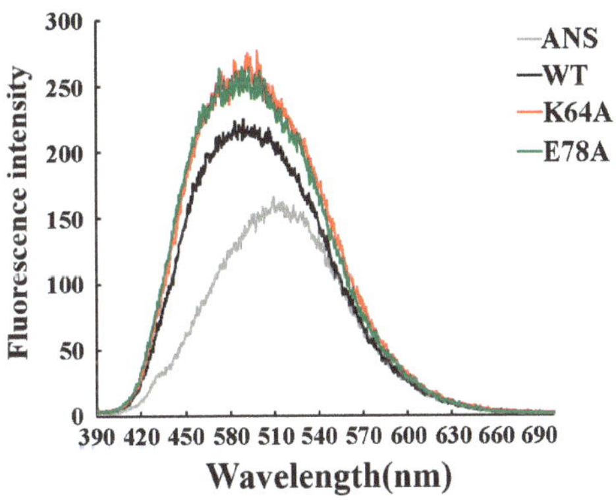

Figure 6. ANS fluorescence intensity of mutants and wild-type OsGSTU17. ANS represents 1-aniline-8-naphthalene sulfonate. The values of the wild-type were from our previous study [28].

The results of Native PAGE used for estimation of the molecular weight of mutants and wild-type could indicate whether the quaternary structure of OsGSTU17 has changed after amino acid substitution (Figure 7). The molecular weights of both the mutants K64A and E78A were similar to those of the dimeric wild-type OsGSTU17, approximately 52.0 kDa [21]. The above results, combined with the SDS-PAGE results of two denatured mutants, indicate that mutant K64A and E78A existed in the form of dimers, that is, replacing Lys64 and Glu78 with Ala will not cause the disruption of the dimeric interactions in the mutant proteins.

The substitutions of Lys64 and Glu78 with alanine decreased the thermal stability of OsGSTU17 to different extents, respectively (Figure 8). Compared with the wild-type, the enzyme activity of mutant E78A showed poor stability at 0 °C, with a 30% decrease in enzyme activity within 2 h at 0 °C. Therefore, reliable activity data for this enzyme could not be obtained within the range of 25 °C to 65 °C. The enzyme activity of mutant K64A remained 70% of its activity after heating at 40 °C for 15 min. Until the temperature increased to 60 °C, the enzyme activity of mutant K64A was completely abolished after being kept for 15 min. However, the wild-type OsGSTU17 still retained 80% of its maximum activity at 50 °C, while the mutant K64A only retained less than 20% activity at this temperature. The above results indicate that Glu78 is more important for the stability of OsGSTU17 than Lys64.

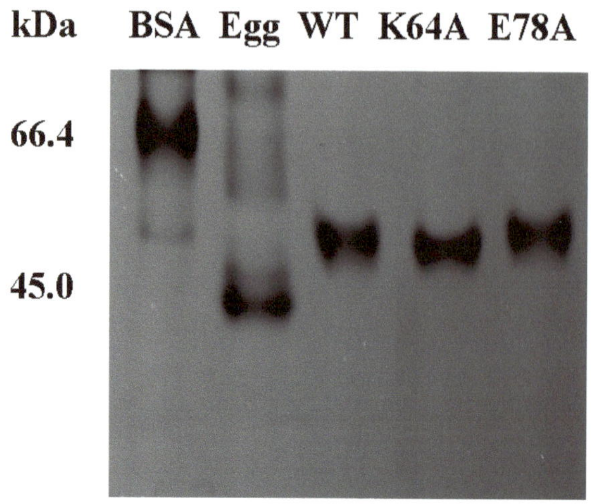

Figure 7. Native polyacrylamide gel electrophoresis of the purified mutant proteins and wild-type. BSA represents bovine albumin; egg represents ovalbumin.

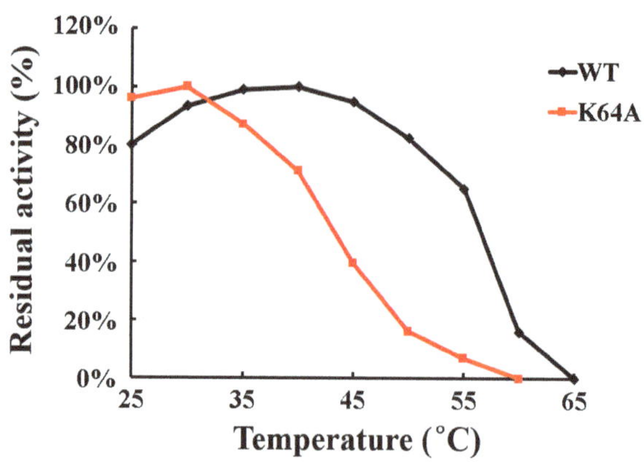

Figure 8. Thermodynamic stability of mutant K64A and wild-type. The values of the wild-type were from our previous study [21].

3. Discussion

Glutathione transferases (GSTs) are a large family of dimeric proteins capable of conjugating glutathione with a variety of compounds containing electrophilic centers to exert their detoxification and antioxidant functions, which rely on the active sites in each subunit as well as the dimeric structure between subunits [31]. Compared with plant GSTs, animal GST structures were more extensively studied due to their roles in drug metabolism [32,33], and it was evidenced that a unique hydrophobic lock-and-key motif and surrounding residues at the subunit interface were essential for molecular recognition in the alpha, mu, pi, epsilon, and delta classes of animal GSTs [12–15,31]. To date, though several plant GST protein structures have been resolved, which sheds light on how they bind with and catalyze substrates, less is known about their dimerization requirements, which are considered crucial for their enzymatic functions.

Tau is one class of plant GSTs that has been widely studied, especially in crops, due to its function of herbicide detoxification; however, its underlying mechanism of dimerization is still unclear. In this study, a series of conserved residues at the interface of tau class GST subunits were identified (Figure 1). Five of them, Pro65, Glu78, Tyr93, Arg95, and Arg99 (alignment positions of 88, 101, 116, 118, and 122, respectively, in Figure 1), were confirmed to be responsible for the formation of hydrogen and/or ionic bonds in TaGSTU4 and involved in the interactions between subunits [16]. The crystal structure of wheat TaGSTU4 (PDB: 1GWC) showed that a hydrogen bond was formed between the carboxyl group of the main chain of Pro65 and the hydroxyl group of the side chain of Tyr93 from another subunit [16]. In the wild-type OsGSTU17 protein, the conserved aromatic residue Tyr92 was naturally replaced with Phe92 (Figure 1), which might result in its inability to form a hydrogen bond between the subunits. However, OsGSTU17 did exert its function in the form of a dimer [21]. The conjecture that this hydrogen bond interaction was not essential for dimerization to occur in OsGSTU17 was consistent with the study of Wang et al. [20]. Also, they found that the mutant Y93F of TaGSTU4 virtually lost most of its detoxification activity, with 13% retained towards CDNB. However, the substrate specificity of wild-type OsGSTU17 (equivalent to mutant Y92F) was shown to be comparable to that of wild-type TaGSTU4. Although its activity towards CDNB was similar to that of mutant Y93F of TaGSTU4, its activity towards NBD-Cl was shown to be similar to that of the wild-type TaGSTU4. Moreover, significantly stronger activity was exhibited towards NBC by OsGSTU17 than that of TaGSTU4 (Table S2). This indicated that the effect of the substitution of Tyr92 with Phe on the enzymatic activity of OsGSTU17 was not as profound as that shown in TaGSTU4. Additionally, higher thermal stability was presented by the wild-type OsGSTU17, which retained 60% of its maximum activity at 55 °C, compared to the mutant Y93F of TaGSTU4, which was inactivated at 45 °C and 55 °C for its wild-type [20], suggesting the less important role of Tyr92 in maintaining the thermal stability of OsGSTU17. By further analysis of the protein alignment in Figure 1, the polarity of Ala64 of TaGSTU4 was opposite to that of a conserved lysine residue in the corresponding position across other tau class GSTs, including OsGSTU17, and this Ala64 was adjacent to Pro65, which formed a hydrogen bond with Tyr93 in a second subunit of TaGSTU4. We hypothesized that other differences in residues on the subunit surface may likely compensate for the loss of the hydrogen bond between Pro65 and Phe92 in OsGSTU17. We constructed a mutant K64A of OsGSTU17 so that the functional structures would be closer to those of mutant Y93F of TaGSTU4 to verify our hypothesis. Compared with the wild-type OsGSTU17, the detoxification ability of mutant K64A towards NBD-Cl and NBC decreased significantly; in particular, it lost almost its entire activity towards NBC (around a 96% reduction). And the thermodynamic stability of the mutant K64A of OsGSTU17 was shown to be more similar to that of the wild-type TaGSTU4 than to that of its mutant Y93F (Figure S1). Lys64 was located at the beginning of the β4-sheet (Figure 1), one component of the core ββα motif in the N-terminal domain, which is highly conserved across all the cytosolic GSTs, facilitating the recognition of the γ-glutamyl portion of GSH [6]. Kinetic analysis showed that the substitution of Lys64 with Ala significantly disrupted the affinity of OsGSTU17 to GSH (Table 2). Meanwhile, the near-UV CD spectroscopy results showed significant changes in the spatial conformation of aromatic amino acids in the mutant K64A (Figure 5), but no aromatic amino acids were found next to Lys64 in the structure of OsGSTU17. In the primary structure, the closest aromatic amino acid to Lys64, tyrosine (Tyr75), was located in the α3-helix (Figure 1). Since α3-helix is adjacent to the β4-sheet, the substitution of the Lys64 located at the starting position of the β4-sheet may cause a shift in the spatial conformation of the α3-helix, which explains the disappearance of the tyrosine peak in the near-UV CD spectrum of mutant K64A. Meanwhile, due to this substitution, the spatial conformation of the α3-helix in the core ββα motif has changed, suggesting that the interaction force between amino acids on the subunit surface of mutant K64A has decreased. When external energy (increasing temperature) is applied to wild-type and mutant K64A, the structural instability of mutant K64A leads to a decrease in activity, resulting in lower thermal stability than wild-type. But this change did not result in the depolymerization of the protein. However, the effects of the substitution of Lys64 residue on the glutathione peroxidase (GPOX) activity of OsGSTU17 were shown to be contrary to those on its detoxification capacity. The activity of mutant K64A towards Cum-OOH were increased about 2-fold (Table 1). This suggested that different catalytic mechanisms may be adopted by OsGSTU17 towards Cum-OOH and hydrophobic toxic substances, respectively, which was consistent with our previous finding that variations in residues at the H-site could result in inverse consequences in the detoxification capabilities and GPOX activities of OsGSTU17 [8].

Salt bridges have been proven to play an important role in protein folding, stability, and catalysis [34–36]. In addition to hydrogen bonds, the salt bridge interactions between glutamate of one subunit and two arginines of the other subunit were observed in both dimeric crystal structures of GmGSTU4-4 [17] and TaGSTU4 [16]. The three conserved residues, according to the alignment sites, Glu101, Arg118, and Arg122 in Figure 1, involved in the salt bridge formation were also identified in OsGSTU17. These residues were superimposed on the similar spatial conformation of the crystal structure of TaGSTU4 (Figure 2), suggesting that there may be similar electrostatic interactions between Glu78, Arg94, and Arg98 in OsGSTU17. When residues involved in ionic bonds were replaced with non-polar amino acids, TaGSTU4 protein either existed in insoluble form or abolished its entire activity. However, when these residues were substituted with an amino acid of the same polarity, the protein restored its dimeric form expressed in soluble form, and except for Cum-OOH, their activities towards other substrates were significantly reduced or lost entirely [20]. This indicated that the ionic bonds were critical to the folding, stability, and substrate activity of TaGSTU4. However, this study showed the ionic bonds seemed to be less important for OsGSTU17. The conserved amino acid residue Glu78 is located near the end of the α3-helix, and the substitution of it with alanine may shorten the length of the α3-helix (the proportion of α-helix decreases, Table S1, and the degree of reduction is greater than that of mutant K64A), thereby affecting the spatial conformation of tryptophan (Trp80), located one amino acid away from Glu78. This structural change resulted in the disappearance of the peak of mutant E78A in the near-UV CD spectrum at 280–300 nm. At the same time, the shortening of the α3-helix may lead to a decrease in the intermolecular forces between subunits, resulting in poorer stability of mutant E78A under low-temperature conditions, and its impact on protein stability is greater than that of mutant K64A. Even though the mutant E78A of OsGSTU17 exhibited poor thermal stability, it existed in soluble form, which differs from the mutant E78A of TaGSTU4. Meanwhile, compared with the wild-type, the mutant E78A of OsGSTU17 reserved its substrate specificity and slightly increased its activity towards CDNB. Thus, we speculated that these salt bridges were not a necessary force for maintaining the dimer of OsGSTU17, but they might facilitate the enhancement of the protein's thermal stability [37]. When the Glu78 was replaced by Ala, we found that the protein remained in dimer form even though it had lost both hydrogen and ionic bonds (Figures 3 and 7), which suggested that there were likely other interactions that existed in OsGSTU17 to assist in the formation of dimers. From the alignment of protein sequences, DHAR class GSTs also contained the conserved structural basis for the hydrogen bond and/or salt bridges (Figure 1), but DHAR class GSTs function in the form of monomers, which indicates that the hydrogen bond or salt bond might not be the main force to form dimers, and there might be other structural basis or forces that exist for assisting in the maintenance of dimers of tau class GSTs. It was speculated that hydrophobic interactions might be the primary force for the dimerization of OsGSTU17. Numerous studies have suggested that hydrophobic interactions were stronger than salt bridges in protein folding, and unique hydrophobic interactions existed between subunits of GSTs [12–15,31,34,36,38–40]. According to the sequence analysis by Thom et al. [16], the conserved residues residing at the dimer interface account for a large proportion of the conserved residues in the tau class GSTs. However, whether these amino acids interact with each other and whether these interactions contribute to protein dimerization needs to be further explored.

4. Materials and Methods

4.1. Alignment of Protein Sequences

The protein sequences of tau and DHAR class GSTs with crystal structure information were aligned with OsGSTU17 using BioEdit 7.0.5.3 [41]. The sequences involved include OsGSTU17 (*Oryza sativa*, AF402804), TaGSTU4 (*Triticum tauschii*, 1GWC), GmGSTU4-4 (*Glycine max*, 2VO4), GmGSTU10 (*Glycine max*, 4CHS), AtGSTU23 (*Arabidopsis thaliana*, 6EP6), MiGSTU (*Mangifera indica*, 5G5F), PtGSTU30 (*Pinus tabulaeformis*, 5J4U), PvGmGSTUG (*Phaseolus vulgaris*, 6GHF), OsDHAR1 (*Oryza sativa*, 5D9T), PgDHAR1 (*Pennisetum glaucum*, 5EVO), AtDHAR2 (*Arabidopsis thaliana*, 5LOL), and PtDHAR3 (*Populus trichocarpa*, 2N5F).

4.2. Three-Dimensional Structural Prediction of OsGSTU17 Protein

Taking 2VO4 (Protein Data Bank, PDB) as the template, the three-dimensional structure model of OsGSTU17 was constructed by homologous modeling using Insight II 2005 (Accelrys Software Inc., San Diego, CA, USA). The three-dimensional structural comparison and analysis of OsGSTU17 and TaGSTU4 were completed using Accelrys Discovery Studio version 1.6 (Accelrys Software Inc., San Diego, CA, USA).

4.3. Construction of Mutants

The mutant sequences of OsGSTU17 were generated by overlap extension using two-round polymerase chain reactions (PCR) described by Ho et al. [42]. Briefly, in the first round of PCR, two overlapped fragments (left and right) were obtained by two separate PCR reactions using the plasmid containing the cDNA of OsGSTU17 as the template. The left fragment containing the mutation in the 3′ end was obtained by PCR using R and Ex1 as primers; the right fragment was obtained by PCR using F and Ex2 as primers. In the second round of PCR, both the purified left and right overlapped fragments were used as the template, and primers Ex1 and Ex2 with restriction sites were used to assemble the sequences containing mutations of interest. The two rounds of PCRs were performed using *Pyrobest*® DNA Polymerase (TaKaRa, Dalian, China). The primer sequences are shown in Table S3. The purified PCR product was digested by *Eco*RI and *Hind*III (New England Biolabs Ltd., Beijing, China) and subcloned into the modified prokaryotic expression vector pET30a, which carried an 18-nucleotide fragment encoding 6×His-tag between the start codon and multiple cloning site (MCS) to avoid the influence of non-OsGSTU17 protein amino acids on its folding [21]. The recombinant plasmids ΔpET30a/Mutant were transformed into *E. coli* BL21 (DE3). The mutated sequences were confirmed by Sanger sequencing (Beijing Genomics Institute, Shenzhen, China).

4.4. Expression and Purification of Mutant Recombinant Proteins

The expression, purification, and desalination of proteins were carried out according to the description of Yang et al. [21]. The overnight culture of *E. coli* BL21 (DE3) containing the recombinant plasmids ΔpET30a/Mutant was diluted at 1:100 and cultured to an OD value of 0.6, then the inducer of isopropylthio-β-D-galactopyranoside (IPTG) was added to a final concentration of 0.1 mM. The bacteria were harvested after a continuous culture for 12 h at 37 °C and then resuspended with binding buffer (20 mM Na_3PO_4, 0.5 M NaCl, 20 mM imidazole, pH = 7.4). The resuspended solution mixture was sonicated at 0 °C for 10 min with a 5 min interval. The supernatant containing the target protein was then loaded onto a pre-equilibrated Ni Sepharose High Performance column (Amersham Pharmacia Biotech, GE Healthcare, Chicago, IL, USA). The precipitate was dissolved in binding buffer and then detected with 12% SDS—polypropylene gel electrophoresis (SDS-PAGE). The mutant proteins with 6×His-tag were supposed to be bound to the nickel ion chelated on Sepharose, then eluted with elution buffer (20 mM Na_3PO_4, 0.5 M NaCl, 500 mM imidazole, pH = 7.4) and collected. The purified target proteins were then loaded onto a PD-10 desalting column (Amersham Pharmacia Biotech, GE Healthcare, Chicago, IL, USA) for buffer exchange. Finally, the mutant proteins were dissolved in 10 mM Tris-HCl buffer (pH = 7.4). The purity of the purified protein was detected by SDS-PAGE, and the concentration of the protein was determined by its absorbance value at 280 nm, measured by a Thermo Scientific™ Evolution™ 300 (Waltham, MA, USA) ultraviolet spectrophotometer.

4.5. Detection of Enzymatic Properties

Substrates NBD-Cl, CDNB, NBC, DCNB, ECA, 4-NPA, and Cum-OOH (Merck, Germany), which are typical substrates for detecting detoxification and GPOX activities of tau class GSTs, were used to investigate the enzyme activities of wild-type and mutants. The activity towards NBD-Cl was determined according to the method described by Ricci et al. [43], the activity towards CDNB, NBC, DCNB, and ECA was determined according to the methods described by Habig et al. [44], and the activity towards Cum-OOH was determined according to the method described by Edwards and Dixon [45]. Taking NBD-Cl and GSH as substrates (Merck, Germany), the kinetic parameters of mutant enzymes were monitored according to the description of Yang et al. [21]. The K_m and V_{max} values for GSH were obtained by fixing the NBD-Cl concentration at 1.0 mM and detecting the enzyme activity with a GSH concentration in the range of 0.2 to 1.0 mM. The K_m and V_{max} values for NBD-Cl were obtained by fixing the GSH concentration at 1.0 mM and detecting the enzyme activity with NBD-Cl concentrations ranging from 0.2 to 1.0 mM. Non-linear regression analysis of Graphpad Prism 10.0 (Graphpad Software Inc., San Diego, CA, USA) was applied to the data to obtain the kinetic parameters of mutants. The experimental data were represented as the average of three independent experiments.

4.6. Detection of Protein Structure Changes

The circular dichroism (CD) spectra of near- and far-UV absorbance of the mutants and wild-type OsGSYU17 were measured by Chirascan-plus Dynamic Multimode Spectroscopy (DMS) (Applied Photophysics Inc., Leatherhead, UK). The final absorbent values were determined by subtracting the buffer-blank spectrum from that acquired for the protein samples, and the reported spectra were presented as the average of three scans for each sample. The near-UV CD data were obtained by dissolving the proteins in elution buffer without imidazole (20 mM Na_3PO_4, 0.5 M NaCl, pH 7.4) to a concentration of 0.91 mg/mL in a 0.5 mm pathlength cell at 25 °C. The far-UV CD data of proteins were obtained by dissolving the proteins in 10 mM Tris-HCl (pH 7.4) buffer to a concentration of 0.24 mg/mL in a 10 mm pathlength cell at 25 °C.

For the ANS binding assay, the fluorescence intensity of ANS was detected by the HITACHI F-4500 FL Spectrophotometer with a 1 mL reaction mixture (PBS buffer, pH = 7.4) containing 200 μM ANS and enzymes with a final concentration of 0.05 mg/mL. The parameters of excitation wavelength, receiving range of emission spectrum, scanning speed, excitation and emission slit, and PMT voltage were determined according to Wang et al. [20]. Triplicate scanning data of both the blank and sample were recorded, respectively. The reported spectra were represented as the average of three independent experiments.

Native PAGE was performed according to the method described by Nijtmans et al. [46]. To remove the monomers, dimers, and polymerization initiator from the gel and improve its resolution, pre-electrophoresis of the prepared gel was performed for 30–60 min before the samples were loaded. In the process of electrophoresis, the voltage was set between 100 and 200 V, and the ambient temperature was kept at 0–4 °C to avoid protein denaturation due to heat generated by high voltage.

The thermal stability curves of mutants were plotted based on the activity of the mutant towards CDNB. The activities towards the substrate were monitored after the enzyme was incubated at 25 to 55 °C for 15 min at intervals of 5 °C, respectively. The experimental results represented the average of three independent experiments.

5. Conclusions

In conclusion, the five conserved residues, Pro65, Glu78, Tyr93, Arg95, and Arg99, identified in TaGSTU4 as responsible for the formation of hydrogen bonds and ionic bonds between subunits, played an important role in its folding and enzymatic activities. However, the corresponding residues identified in OsGSTU17 displayed functional differences compared to the residues in TaGSTU4. The conserved residue Glu78 in OsGSTU17 was revealed to be essential for the protein's thermal stability, while the hydrogen bonds formed by it on the subunit surface did not play a crucial role in protein dimerization. But interestingly, the hydrogen bonds formed by Glu78 were indispensable in the dimerization of the TaGSTU4 protein. The conserved Lys64 residue at the subunit interface did not participate in the formation of hydrogen bonds and had no effect on protein dimerization. However, the change in this residue was not conducive to the thermal stability of the OsGSTU17 protein or its binding with GSH. Therefore, a question is raised: "Why does TaGSTU4 still retain alanine at this position during evolution?" The relationship between amino acids and structure still need further research and exploration. It is inferred that the interactions between other amino acids on the surface of protein subunits, such as hydrophobic interactions, might be the primary force for maintaining protein dimerization.

Supplementary Materials: The supporting information can be downloaded at: https://www.mdpi.com/article/10.3390/ijms25010398/s1.

Author Contributions: Conceptualization, X.Y. and J.G.; validation, X.Y., Z.Z., L.W. and M.Y.; writing—original draft preparation, X.Y.; writing—review and editing, X.Y., J.G. and S.L.; funding acquisition, X.Y. and S.L. All authors have read and agreed to the published version of the manuscript.

Funding: This research was funded by the Department of Education of Jilin Province, grant numbers JJKH20230379CY and JJKH20230376KJ; and the Department of Science and Technology of Jilin Province, grant number 20210101370JC.

Institutional Review Board Statement: Not applicable.

Informed Consent Statement: Not applicable.

Data Availability Statement: Data is contained within the article or Supplementary Material.

Acknowledgments: We acknowledge the Technology Center of Protein Research, Tsinghua University, and Xuemin Han of the Chinese Academy of Forestry for their help in protein circular dichroism spectra. We also acknowledge Bintao Cui of RMIT University of Australia for providing language help.

Conflicts of Interest: The authors declare no conflicts of interest.

Abbreviations

GSTs	Glutathione transferases
NBD-Cl	7-chloro-4-nitrobenzo-2-oxa-1,3-diazole
CDNB	1-chloro-2,4-dinitrobenzene
NBC	4-nitrobenzyl chloride
Cum-OOH	Cumene hydroperoxide
DCNB	1,2-dichloro-4-nitrobenzene
4-NPA	4-nitrophenyl acetate
ECA	Ethacrynic Acid
ANS	1-aniline-8-naphthalene sulfonate
CD	Circular Dichroism

References

1. Kao, C.W.; Bakshi, M.; Sherameti, I.; Dong, S.; Reichelt, M.; Oelmüller, R.; Yeh, K.W. A Chinese cabbage (*Brassica campetris* subsp. Chinensis) tau-type glutathione-S-transferase stimulates Arabidopsis development and primes against abiotic and biotic stress. *Plant Mol. Biol.* **2016**, *92*, 643–659. [CrossRef] [PubMed]

2. Srivastava, D.; Verma, G.; Chauhan, A.S.; Pande, V.; Chakrabarty, D. Rice (*Oryza sativa* L.) tau class glutathione S-transferase (OsGSTU30) overexpression in Arabidopsis thaliana modulates a regulatory network leading to heavy metal and drought stress tolerance. *Met. Integr. Biomet. Sci.* **2019**, *11*, 375–389. [CrossRef] [PubMed]
3. Sylvestre-Gonon, E.; Law, S.R.; Schwartz, M.; Robe, K.; Keech, O.; Didierjean, C.; Dubos, C.; Rouhier, N.; Hecker, A. Hecker, Functional, Structural and Biochemical Features of Plant Serinyl-Glutathione Transferases. *Front. Plant Sci.* **2019**, *10*, 608. [CrossRef] [PubMed]
4. Morel, M.; Meux, E.; Mathieu, Y.; Thuillier, A.; Chibani, K.; Harvengt, L.; Jacquot, J.; Gelhaye, E. Xenomic networks variability and adaptation traits in wood decaying fungi. *Microb. Biotechnol.* **2013**, *6*, 248–263. [CrossRef]
5. Munyampundu, J.-P.; Xu, Y.-P.; Cai, X.-Z. Phi Class of Glutathione S-transferase Gene Superfamily Widely Exists in Nonplant Taxonomic Groups. *Evol. Bioinform. Online* **2016**, *12*, 59–71. [CrossRef]
6. Armstrong, R.N. Structure, catalytic mechanism, and evolution of the glutathione transferases. *Chem. Res. Toxicol.* **1997**, *10*, 2–18. [CrossRef] [PubMed]
7. Sommer, A.; Peter, B. Characterization of Recombinant Corn GlutathioneS-Transferase Isoforms I, II, III, and IV. *Pestic. Biochem. Physiol.* **1999**, *63*, 127–138. [CrossRef]
8. Yang, X.; Wei, J.; Wu, Z.; Gao, J. Effects of Substrate-Binding Site Residues on the Biochemical Properties of a Tau Class Glutathione S-Transferase from *Oryza sativa*. *Genes* **2019**, *11*, 25. [CrossRef]
9. Nianiou-Obeidat, I.; Madesis, P.; Kissoudis, C.; Voulgari, G.; Chronopoulou, E.; Tsaftaris, A.; Labrou, N.E. Plant glutathione transferase-mediated stress tolerance: Functions and biotechnological applications. *Plant Cell Rep.* **2017**, *36*, 791–805. [CrossRef]
10. Hayes, J.D.; Flanagan, J.U.; Jowsey, I.R. Glutathione transferases. *Annu. Rev. Pharmacol. Toxicol.* **2005**, *45*, 51–88. [CrossRef]
11. Lallement, P.A.; Meux, E.; Gualberto, J.M.; Prosper, P.; Didierjean, C.; Saul, F.; Haouz, A.; Rouhier, N.; Hecker, A. Structural and enzymatic insights into Lambda glutathione transferases from *Populus trichocarpa*, monomeric enzymes constituting an early divergent class specific to terrestrial plants. *Biochem. J.* **2014**, *462*, 39–52. [CrossRef] [PubMed]
12. UHegazy, M.; Mannervik, B.; Stenberg, G. Functional role of the lock and key motif at the subunit interface of glutathione transferase p1-1. *J. Biol. Chem.* **2004**, *279*, 9586–9596. [CrossRef] [PubMed]
13. Wongsantichon, J.; Ketterman, A.J. An intersubunit lock-and-key 'clasp' motif in the dimer interface of Delta class glutathione transferase. *Biochem. J.* **2006**, *394*, 135–144. [CrossRef] [PubMed]
14. Wongsantichon, J.; Robinson, R.C.; Ketterman, A.J. Epsilon glutathione transferases possess a unique class-conserved subunit interface motif that directly interacts with glutathione in the active site. *Biosci. Rep.* **2015**, *35*, e00272. [CrossRef] [PubMed]
15. Hornby, J.A.; Codreanu, S.G.; Armstrong, R.N.; Dirr, H.W. Molecular recognition at the dimer interface of a class mu glutathione transferase: Role of a hydrophobic interaction motif in dimer stability and protein function. *Biochemistry* **2002**, *41*, 14238–14247. [CrossRef]
16. Thom, R.; Cummins, I.; Dixon, D.P.; Edwards, R.; Cole, D.J.; Lapthorn, A.J. Structure of a tau class glutathione S-transferase from wheat active in herbicide detoxification. *Biochemistry* **2002**, *41*, 7008–7020. [CrossRef] [PubMed]
17. Axarli, I.; Dhavala, P.; Papageorgiou, A.C.; Labrou, N.E. Crystallographic and functional characterization of the fluorodifen-inducible glutathione transferase from Glycine max reveals an active site topography suited for diphenylether herbicides and a novel L-site. *J. Mol. Biol.* **2009**, *385*, 984–1002. [CrossRef]
18. Valenzuela-Chavira, I.; Contreras-Vergara, C.A.; Arvizu-Flores, A.A.; Serrano-Posada, H.; Lopez-Zavala, A.A.; García-Orozco, K.D.; Hernandez-Paredes, J.; Rudiño-Piñera, E.; Stojanoff, V.; Sotelo-Mundo, R.R.; et al. Insights into ligand binding to a glutathione S-transferase from mango: Structure, thermodynamics and kinetics. *Biochimie* **2017**, *135*, 35–45. [CrossRef]
19. Chronopoulou, E.G.; Papageorgiou, A.C.; Ataya, F.; Nianiou-Obeidat, I.; Madesis, P.; Labrou, N.E. Expanding the Plant GSTome Through Directed Evolution: DNA Shuffling for the Generation of New Synthetic Enzymes With Engineered Catalytic and Binding Properties. *Front. Plant Sci.* **2018**, *9*, 1737. [CrossRef]
20. Wang, C.L.; Yang, H.L. Conserved residues in the subunit interface of tau glutathione s-transferase affect catalytic and structural functions. *J. Integr. Plant Biol.* **2011**, *53*, 35–43. [CrossRef]
21. Yang, X.; Sun, W.; Liu, J.P.; Liu, Y.J.; Zeng, Q.Y. Biochemical and physiological characterization of a tau class glutathione transferase from rice (*Oryza sativa*). *Plant Physiol. Biochem.* **2009**, *47*, 1061–1068. [CrossRef] [PubMed]
22. Li, J.; Meng, L.; Ren, S.; Jia, C.; Liu, R.; Jiang, H.; Chen, J. OsGSTU17, a Tau Class Glutathione S-Transferase Gene, Positively Regulates Drought Stress Tolerance in *Oryza sativa*. *Plants* **2023**, *12*, 3166. [CrossRef]
23. Greenfield, N.J. Using circular dichroism spectra to estimate protein secondary structure. *Nat. Protoc.* **2006**, *1*, 2876–2890. [CrossRef] [PubMed]
24. Whitmore, L.; Wallace, B.A. Protein secondary structure analyses from circular dichroism spectroscopy: Methods and reference databases. *Biopolymers* **2008**, *89*, 392–400. [CrossRef] [PubMed]
25. Jin, Y.; Sakurai, H.; Nagai, Y.; Nagai, M. Changes of near-UV CD spectrum of human hemoglobin upon oxygen binding: A study of mutants at alpha 42, alpha 140, beta 145 tyrosine or beta 37 tryptophan. *Biopolymers* **2004**, *74*, 60–63. [CrossRef] [PubMed]
26. Strickland, E.H.; Beychok, S. Aromatic Contributions To Circular Dichroism Spectra Of Protein. *CRC Crit. Rev. Biochem.* **1974**, *2*, 113–175. [CrossRef] [PubMed]
27. Nagatomo, S.; Nagai, M.; Ogura, T.; Kitagawa, T. Near-UV circular dichroism and UV resonance Raman spectra of tryptophan residues as a structural marker of proteins. *J. Phys. Chem. B* **2013**, *117*, 9343–9353. [CrossRef] [PubMed]

28. Yang, X.; Wu, Z.; Gao, J. Effects of conserved Arg20, Glu74 and Asp77 on the structure and function of a tau class glutathione S-transferase in rice. *Plant Mol. Biol.* **2021**, *105*, 451–462. [CrossRef]
29. Schonbrunn, E.; Eschenburg, S.; Luger, K.; Kabsch, W.; Amrhein, N. Structural basis for the interaction of the fluorescence probe 8-anilino-1-naphthalene sulfonate (ANS) with the antibiotic target MurA. *Proc. Natl. Acad. Sci. USA* **2000**, *97*, 6345–6349. [CrossRef]
30. Piromjitpong, J.; Wongsantichon, J.; Ketterman, A.J. Differences in the subunit interface residues of alternatively spliced glutathione transferases affects catalytic and structural functions. *Biochem. J.* **2007**, *401*, 635–644. [CrossRef]
31. Sayed, Y.; Wallace, L.A.; Dirr, H.W. The hydrophobic lock-and-key intersubunit motif of glutathione transferase A1-1: Implications for catalysis, ligandin function and stability. *FEBS Lett.* **2000**, *465*, 169–172. [CrossRef]
32. Tew, K.D. Glutathione-associated Enzymes in Anticancer Drug Resistance. *Cancer Res.* **1994**, *54*, 4313. [CrossRef] [PubMed]
33. Townsend, D.M.; Tew, K.D. The role of glutathione-S-transferase in anti-cancer drug resistance. *Oncogene* **2003**, *22*, 7369–7375. [CrossRef] [PubMed]
34. Hendsch, Z.S.; Tidor, B. Do salt bridges stabilize proteins? A continuum electrostatic analysis. *Protein Sci. Publ. Protein Soc.* **1994**, *3*, 211–226. [CrossRef] [PubMed]
35. Xu, D.; Lin, S.L.; Nussinov, R. Protein binding versus protein folding: The role of hydrophilic bridges in protein associations. *J. Mol. Biol.* **1997**, *265*, 68–84. [CrossRef] [PubMed]
36. Sindelar, C.V.; Hendsch, Z.S.; Tidor, B. Effects of salt bridges on protein structure and design. *Protein Sci. Publ. Protein Soc.* **1998**, *7*, 1898–1914. [CrossRef] [PubMed]
37. Lee, C.W.; Wang, H.J.; Hwang, J.K.; Tseng, C.P. Protein thermal stability enhancement by designing salt bridges: A combined computational and experimental study. *PLoS ONE* **2014**, *9*, e112751. [CrossRef] [PubMed]
38. Waldburger, C.D.; Schildbach, J.F.; Sauer, R.T. Are buried salt bridges important for protein stability and conformational specificity? *Nat. Struct. Biol.* **1995**, *2*, 122–128. [CrossRef]
39. Wimley, W.C.; Gawrisch, K.; Creamer, T.P.; White, S.H. Direct measurement of salt-bridge solvation energies using a peptide model system: Implications for protein stability. *Proc. Natl. Acad. Sci. USA* **1996**, *93*, 2985–2990. [CrossRef]
40. Stenberg, G.; Abdalla, A.M.; Mannervik, B. Tyrosine 50 at the subunit interface of dimeric human glutathione transferase P1-1 is a structural key residue for modulating protein stability and catalytic function. *Biochem. Biophys. Res. Commun.* **2000**, *271*, 59–63. [CrossRef]
41. Thomas, H. BioEdit: A user-friendly biological sequence alignment program for Windows 95/98/NT. *Nucleic Acids Symp. Ser.* **1999**, *41*, 95–98.
42. Ho, S.N.; Hunt, H.D.; Horton, R.M.; Pullen, J.K.; Pease, L.R. Site-directed mutagenesis by overlap extension using the polymerase chain reaction. *Gene* **1989**, *77*, 51–59. [CrossRef] [PubMed]
43. Ricci, G.; Caccuri, A.M.; Bello, M.L.; Pastore, A.; Piemonte, F.; Federici, G. Colorimetric and fluorometric assays of glutathione transferase based on 7-chloro-4-nitrobenzo-2-oxa-1,3-diazole. *Anal. Biochem.* **1994**, *218*, 463–465. [CrossRef] [PubMed]
44. Habig, W.H.; Pabst, M.J.; Jakoby, W.B. Glutathione S-transferases. The first enzymatic step in mercapturic acid formation. *J. Biol. Chem.* **1974**, *249*, 7130–7139. [CrossRef]
45. Edwards, R.; Dixon, D. Plant Glutathione Transferases. *Methods Enzymol.* **2005**, *401*, 169–186.
46. Nijtmans, L.G.J.; Henderson, N.S.; Holt, I.J. Blue Native electrophoresis to study mitochondrial and other protein complexes. *Methods* **2002**, *26*, 327–334. [CrossRef]

International Journal of *Molecular Sciences*

Article

Overexpression of an ART1-Interacting Gene *OsNAC016* Improves Al Tolerance in Rice

Fuhang Liu, Dan Ma, Jinyu Yu, Ran Meng, Zhigang Wang, Baolei Zhang, Xingxiang Chen, Lin Zhang, Liyun Peng and Jixing Xia *

State Key Laboratory for Conservation and Utilization of Subtropical Agro-Bioresources, College of Life Science and Technology, Guangxi University, Nanning 530004, China; 1708404002@st.gxu.edu.cn (F.L.); 2208301045@st.gxu.edu.cn (D.M.); 2208301085@st.gxu.edu.cn (J.Y.); 2108301043@st.gxu.edu.cn (R.M.); wangzhigang816@163.com (Z.W.); 1908301074@st.gxu.edu.cn (B.Z.); 1908401001@st.gxu.edu.cn (X.C.); 2108301081@st.gxu.edu.cn (L.Z.); 2008401020@st.gxu.edu.cn (L.P.)

* Correspondence: xiajx@gxu.edu.cn; Tel.: +86-0771-323-6616

Abstract: Rice (*Oryza sativa*) exhibits tremendous aluminum (Al)-tolerance. The C2H2-transcription factor (TF) ART1 critically regulates rice Al tolerance via modulation of specific gene expression. However, little is known about the posttranscriptional ART1 regulation. Here, we identified an ART1-interacted gene *OsNAC016* via a yeast two-hybrid (Y2H) assay. *OsNAC016* was primarily expressed in roots and weakly induced by Al. Immunostaining showed that OsNAC016 was a nuclear protein and localized in all root cells. Knockout of *OsNAC016* did not alter Al sensitivity. Overexpression of *OsNAC016* resulted in less Al aggregation within roots and enhanced Al tolerance in rice. Based on transcriptomic and qRT-PCR evaluations, certain cell-wall-related or ART-regulated gene expressions such as *OsMYB30* and *OsFRDL4* were altered in *OsNAC016*-overexpressing plants. These results indicated that OsNAC016 interacts with ART1 to cooperatively regulate some Al-tolerance genes and is a critical regulatory factor in rice Al tolerance.

Keywords: Al tolerance; ART1; NAC; transcription factor; rice

check for updates

Citation: Liu, F.; Ma, D.; Yu, J.; Meng, R.; Wang, Z.; Zhang, B.; Chen, X.; Zhang, L.; Peng, L.; Xia, J. Overexpression of an ART1-Interacting Gene *OsNAC016* Improves Al Tolerance in Rice. *Int. J. Mol. Sci.* **2023**, *24*, 17036. https://doi.org/10.3390/ijms242317036

Academic Editor: Ki-Hong Jung

Received: 6 November 2023
Revised: 27 November 2023
Accepted: 29 November 2023
Published: 1 December 2023

1. Introduction

Aluminum (Al) is ubiquitously found within soil. In acidic soil, Al takes up an ionic form, which inhibits root growth and mineral absorption, thereby producing dwarf roots with impaired ability to absorb water and nutrients, thus, enhancing plant vulnerability to numerous stressors, including drought [1,2]. Given these factors, Al toxicity substantially restricts crop production in acidic soils.

To remain viable under toxic Al or acidic soil conditions, plants established multiple approaches to internally and/or externally stimulate Al detoxification [3,4]. To internally detoxify Al, plants sequestrate Al within vacuoles, whereby, it is combined with organic acids or other chelators, namely, citrate and oxalate. Alternately, well-established pathways of external Al detoxification involve root-based release of organic acid anions (OAA), including malate, citrate and oxalate release [2–4]. These OAAs strongly sequester Al and remove them from the rhizosphere. Recently, the genes encoding malate- and citrate transporters were recognized, for example, Al-activated malate transporter (*ALMT1*) in wheat, and citrate transporters (*HvAACT1* and *SbMATE*) in barley and sorghum [5–7].

Rice *Japonica* cultivars demonstrate very high Al tolerance in hydroponic conditions and in soil [8]. Dissimilar to other cereal crops such as barley and wheat, root OAAs release was not one main mechanism for rice Al tolerance. Recent advances in rice Al tolerance research revealed that several kinds of transcription factors (TFs) play critical roles in regulating the expression of Al tolerance genes such as zinc finger protein, WRKY, and MYB [4]. A C2H2 zinc finger TF ART1 (*Al resistance TF1*) is reported to regulate at least 31 down-stream genes that participate in rice Al tolerance [9]. Among them, researchers have functionally characterized ten genes (*STAR1*, *STAR2*, *Nrat1*, *OsALS1*, *OsCDT3*, *OsFRDL2*, *OsFRDL4*, *OsMGT1*, *OsEXPA10*, and *OsART2*) using mutant approach [10–18]. For example, *STAR1* and *STAR2* associate to form an ABC transporter, which exports UDP-glucose and critically regulates Al detoxification through cell wall modification [16]. *Nrat1* produces a plasma-membrane-localized transporter protein that transports trivalent Al [10,17], whereas *OsALS1* produces a tonoplast-residing half-size Al-shuttling ABC transporter [18]. *Nrat1* and *OsALS1* work synergistically to internally detoxify Al within rice. On the other hand, *OsFRDL4* encoding a citrate transporter modulates root citrate release, which sequesters Al within the rhizosphere [14]. Moreover, a small cysteine-rich plasma membrane-localized peptide OsCDT3 sequesters Al and blocks its entry into cells [10]. Although the ART1-regulated Al-tolerance gene expressions were enhanced by Al stress, Al did not regulate the mRNA level of *ART1*. Based on these findings, ART1 posttranscriptional or posttranslational modulation might be required for the activation of ART1.

The plant-specific NAC (NAM, ATAF, CUC1/2) TF family dominates the TF population within plants [19]. At present, we recognize 117 Arabidopsis and 151 rice NAC proteins [19]. Generally, NAC proteins contain a conserved target-interacting domain (NAC domain) at the N-terminal region and a highly diverse transcriptional regulating C-terminal domain. NAC TFs are critical regulators of various stress responses such as cold, salt, drought and ABA stresses [20,21]. Recently, several studies have reported the involvement of NAC TFs in Al stress response within plants. Rice bean NAC TF VuNAR1, for example, critically regulates Al tolerance through the regulation of cell wall pectin metabolism in Arabidopsis [22]. A NAC TF SOG1 (GAMMA RESPONSE1 SUPPRESSOR) plays a central role in DNA damage response. Disruption of SOG1 showed hypersensitivity to high Al treatment [23]. More recently, ANAC017 was reported to participate in Al tolerance via modulation of xyloglucan transglucosylase/hydrolase 31 (XTH31) expression in Arabidopsis [24]. However, the significance of NAC TF in regulating rice Al tolerance remains unknown.

To gain insights into how ART1 is regulated at the posttranscriptional level, we screened for ART1-interacting proteins via yeast two-hybrid (Y2H) assay with truncated ART1 as bait and identified a NAC TF OsNAC016. We investigated the phenotypes of *OsNAC016*-knockout and overexpressing lines under normal and Al treatment conditions. We also analyzed the root transcriptomes of *OsNAC016*-overexpressing and wild-type (WT) plants with or without Al treatment. The presented evidence revealed that OsNAC016 is involved in rice Al tolerance via the modulation of certain cell wall related genes and partial ART1-regulated genes.

2. Results

2.1. Identification of ART1-Interacting Protein OsNAC016

A previous study showed that ART1 can transcriptionally activate target genes within yeast. Truncation analysis indicated that its activation domain resides within its C terminus (176–466 aa). To probe the molecular mechanism of ART1 activation by Al, we utilized truncated ART1 (1–175 aa) lacking an activation domain as bait to screen for ART1-interacting proteins via yeast two-hybrid (Y2H) assay and identified eight candidate proteins (Figure 1A; Supplemental Figure S1A,B; Table S1). As several NAC TFs were reported to be involved in regulating Al tolerance in plants [22,23], we selected a NAC TF OsNAC016 (LOC_Os01g01430) for further functional analysis. *OsNAC016* encodes a 340-amino-acid protein harboring an NAM domain at its N terminus. To verify a physical association between OsNAC016 and ART1, we conducted a bimolecular fluorescence complementation (BiFC) assessment in rice protoplast cells. We observed the YFP signal in rice protoplasts co-expressing OsNAC016-nYFP and ART1-cYFP or ART1-nYFP and OsNAC016-cYFP, but not in those co-expressing control constructs. Furthermore, YFP signal nicely overlapped with the RFP signal of the nuclear marker Ghd7 [25] (Figure 1B), confirming in vivo interaction of OsNAC016 and ART1 at the nucleus. We also performed co-immunoprecipitation

(Co-IP) experiment in *N. benthamiana* leaves co-expressing MYC-tagged OsNAC016 and FLAG-tagged ART1 and found that OsNAC016 was coimmunoprecipitated with ART1 in planta (Figure 1C). Taken together, the aforementioned evidence indicated the binding of OsNAC016 to ART1 within the nuclear compartment of plant cells.

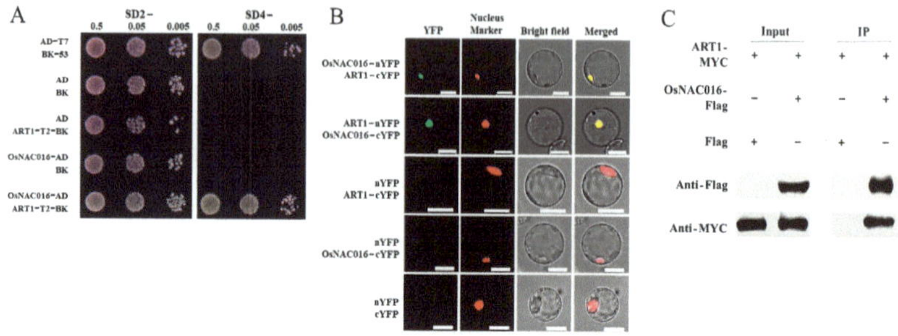

Figure 1. Interaction of OsNAC016 with ART1. (**A**) Interaction between OsNAC016 and truncated ART1 examined by Y2H assay. SD2− medium lacking leucine and tryptophan, SD4− medium lacking tryptophan, leucine, histidine, and adenine. (**B**) OsNAC016-ART1 interaction shown by a BiFC assay. Pair of plasmid constructs (positive or negative control, *OsNAC016-nYFP* and *ART1-cYFP*, *ART1-nYFP* and *OsNAC016-cYFP*) with Ghd7-RFP were transiently co-expressed in rice protoplasmic cells. YFP (green), RFP (red), and merged image (yellow) of YFP (green) and RFP (red) were shown. Ghd7-RFP was used as a nuclear marker. Bar = 10 μm. (**C**) Coimmunoprecipitation of OsNAC016 with ART1. *OsNAC016-Flag* and *ART1-MYC* were transiently co-expressed in *N. benthamiana* leaves. Crude proteins were immunoprecipitated using anti-Flag antibody-conjugated agarose beads and then detected using anti-Flag and anti-MYC antibodies, respectively.

2.2. Expression Pattern of OsNAC016

QRT-PCR assessment was utilized to determine the *OsNAC016* expression within roots, stems, leaf blades, leaf sheaths and panicles. OsNAC016 was primarily expressed in roots (Figure 2A), and its expression was higher in the root tip (0–1 cm) than in the mature root segments (1–2 and 2–3 cm) (Figure 2B). Dose-reliant and time course assessments revealed that the root OsNAC016 expression was weakly up-regulated by Al treatments (Figure 2C,D).

2.3. Subcellular and Cellular Localization of OsNAC016

To investigate the OsNAC016 subcellular localization, the OsNAC016-GFP fusion construct or GFP empty vector was co-introduced with the nuclear marker Ghd7-RFP into rice protoplast cells. The resulting OsNAC016-GFP fluorescence showed a perfect overlap with the red fluorescence produced by Ghd7-RFP while the green fluorescence of GFP alone were widely distributed in the nucleus and cytoplasm. Based on this evidence, OsNAC016 primarily resides in the nucleus (Figure 3A–H).

To determine the OsNAC016 cell-specific localization in rice roots, we conducted immunostaining of the transgenic rice expressing the *ProOsNAC016:OsNAC016-GFP* construct. GFP antibody co-staining with a nuclear stain DAPI revealed that red fluorescence was merged with the blue DAPI fluorescence in all the root cell layers (Figure 3M–T). No red fluorescence was evident in the WT root (Figure 3I–L). These results indicated that OsNAC016 locates to the nucleus of all root cells.

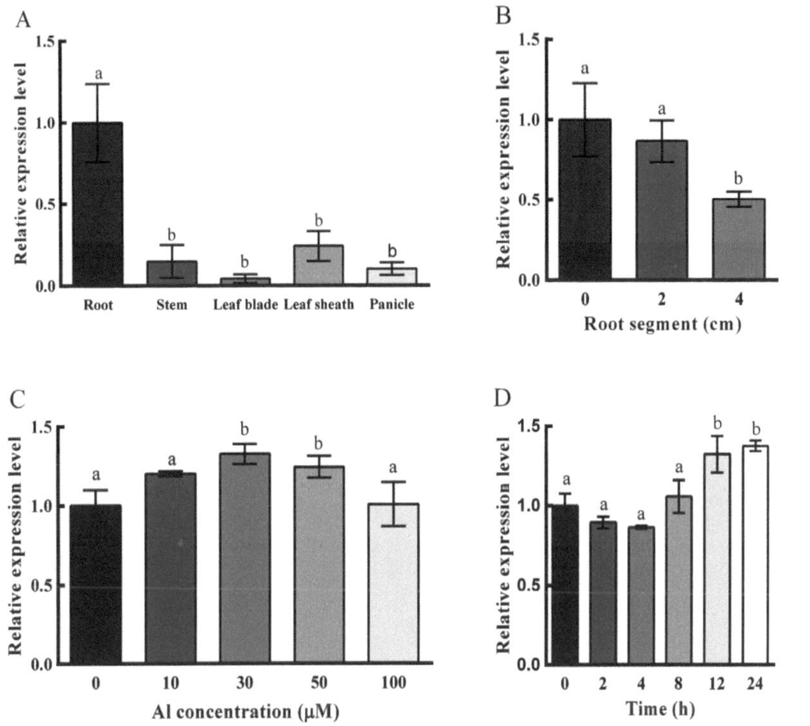

Figure 2. Expression pattern of *OsNAC016*. (**A**) Expression of *OsNAC016* in various tissues of rice cv. Nipponbare (WT). (**B**) Expression in different root segments. RNA was extracted from different root segments (0–1, 1–2, and 2–3 cm from root apex) of 5-d-old wild-type rice seedlings. (**C**) Dose-dependent expression of *OsNAC016* in rice roots with Al treatment. Seedlings were exposed to different concentrations of Al at pH 4.5 for 8 h. (**D**) Time-dependent expression of *OsNAC016* in rice roots with Al treatment. Seedlings were treated with 30 μM Al at pH 4.5 for different times. *Histone H3* was used as an internal standard. Data are means ± SD of three biological replicates. Significant differences were determined by Tukey's test and are labeled with different letters ($p < 0.05$).

2.4. Mutation of OsNAC016 Did Not Affect Al Tolerance in Rice

To test whether *OsNAC016* participated in rice Al-tolerance, we created two null mutants for *OsNAC016* via CRISPR/Cas9-mediated gene editing; one with a 1-bp insertion (*osnac016-1*) and another with a 307-bp deletion (*osnac016-2*), at varying sites within the first exon (Supplemental Figure S2). Additionally, the expression level of *OsNAC016* was significantly reduced in the two *osnac016* lines compared with WT (Supplemental Figure S3). First, we compared the *ART1* and some ART1-regulated gene expressions in roots between WT and two *OsNAC016*-knockout rice lines with or without Al treatment. Under Al-free conditions, the *ART1*, *OsASL1* and *OsNrat1* expressions were slightly elevated in the *OsNAC016*-knockout lines, relative to the WT rice (Figure 4A,B,D). However, following Al exposure, the gene expressions were similar between WT rice and two *OsNAC016*-knockout lines (Figure 4A–E). Next, we evaluated WT rice and two *OsNAC016*-knockout line root growth, and demonstrated similar root growth in the two *osnac016* mutants and WT rice both with and without Al (Figure 5A,B). Additionally, phylogenetic analysis showed that multiple homologs for OsNAC016 exist in the rice genome such as OsNAC103, OsNAC58, OsNAC122, and OsNAC131 (Supplemental Figure S4). Like *OsNAC016*, the expression levels of *OsNAC58*, *OsNAC103*, *OsNAC122*, and *OsNAC131* in roots were weakly affected by Al treatment (Supplemental Figure S5). The aforementioned results indicated that no differences in *ART1* and some ART1-regulated gene expressions as well as Al tolerance between the WT and mutant lines under Al stress might be explained by the emergence of redundant genes with comparable activities to OsNAC016 in rice.

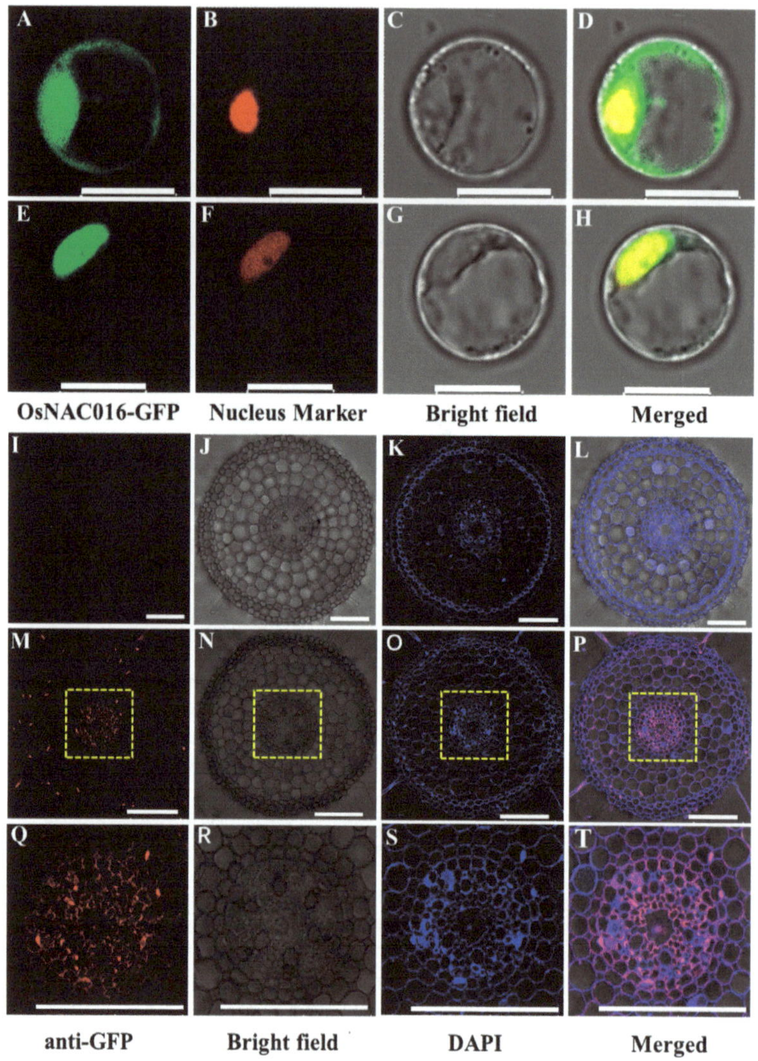

Figure 3. Subcellular and cellular localization of OsNAC016. (**A–H**) Subcellular localization of OsNAC016. *GFP* (**A–D**) or *OsNAC016-GFP* (**E–H**) plus *Ghd7-RFP* was separately transformed into rice protoplast cells. GFP (green), RFP (red), and merged image (yellow) of YFP (green) and RFP (red) were shown. Bars = 10 μm. (**I–T**) Cellular localization of OsNAC016. Immunostaining of the roots of Nipponbare (**I–L**) and *ProOsNAC016:OsNAC016-GFP* transgenic plants (**M–P**) was performed using the anti-GFP antibody. (**Q–T**) Magnified images of the yellow boxed areas in (**M–P**), respectively. Red shows the fluorescent signal of the anti-GFP antibody, blue shows autofluorescence of cell wall and nuclei stained with DAPI. Scale bar = 100 μm.

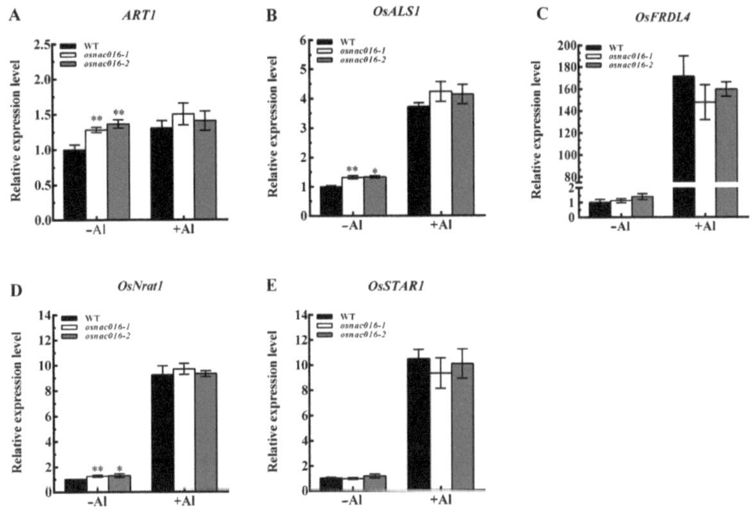

Figure 4. Effect of knockout of *OsNAC016* on the expression of *ART1* and some ART1-regulated genes in roots. *ART1* (**A**), *OsALS1* (**B**), *OsFRDL4* (**C**), *OsNrat1* (**D**), *OsSTAR1* (**E**). Both wild-type rice and two *OsNAC016*-knockout lines were exposed to 0 or 50 μM Al for 8 h. *Histone H3* was used as an internal standard. Data are means ± SD (n = 3). Asterisks indicate significant differences between WT and the knockout lines (Tukey's test, ** $p < 0.01$, * $p < 0.05$).

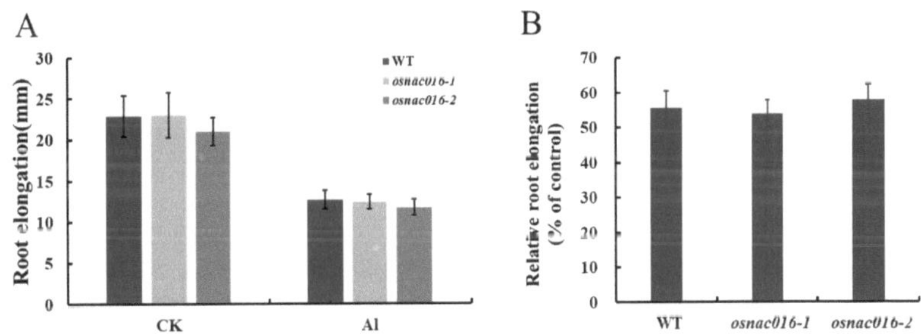

Figure 5. Sensitivity analysis of *OsNAC016* knockout lines to Al. (**A**) Effect of Al on root length. (**B**) Effect of Al on root elongation. Five-day-old seedlings of the WT and two *OsNAC016*-knockout lines were exposed to a 0.5 mM $CaCl_2$ solution, pH 4.5, containing 0 or 50 μM Al for 24 h. Root length was measured before and after the treatment. Relative root elongation refers to (root elongation with Al)/(root elongation without Al) × 100. Data are means ± SD (n = 8).

2.5. Overexpression of OsNAC016 Improves Al Tolerance in Rice

To further elucidate the *OsNAC016*-mediated regulation of rice Al-tolerance, we generated *OsNAC016*-overexpressing lines in which *OsNAC016* is regulated by the maize *ubiquitin 1* promoter. The root *OsNAC016* expression within two independent transgenic lines was increased by more than 70 times (Supplemental Figure S6). Under Al-free conditions, root growth of the two *OsNAC016*-overexpressing line was comparable to WT rice. Alternately, with 20 μM Al exposure, WT root elongation was strongly suppressed, compared to the two overexpression lines of *OsNAC016* (Figure 6A–D). Therefore, *OsNAC016* overexpression can enhance Al tolerance in rice.

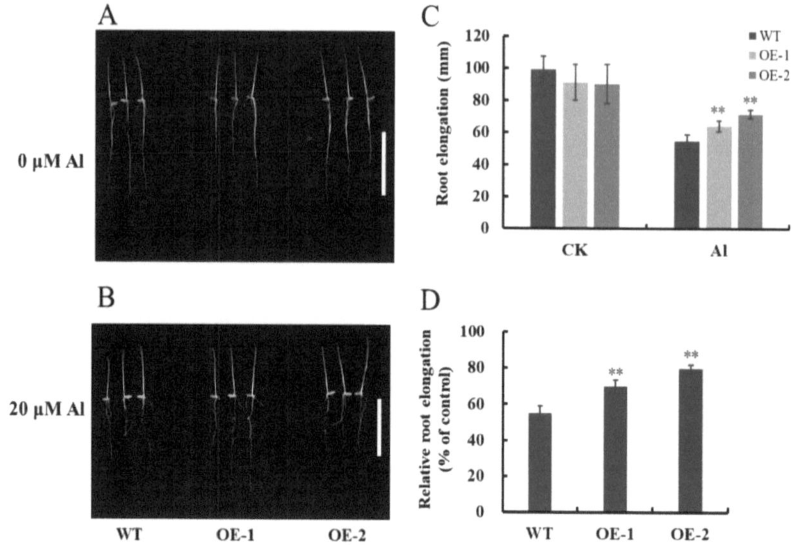

Figure 6. Al sensitivity of the WT and *OsNAC016*-overexpressing lines. (**A**,**B**) Phenotypes of the WT and *OsNAC016*-overexpressing lines after the treatment with 0 (**A**) and 20 μM Al (**B**). (**C**) Effect of Al on root length. (**D**) Effect of Al on root elongation. Germinated seeds of each line were exposed to a 0.5 mM $CaCl_2$ solution (pH 4.5) containing 0 or 20 μM Al for 5 d. Scale bar = 5 cm. Root length was measured before and after the treatment. Relative root elongation refers to (root elongation with Al)/(root elongation without Al) × 100. Data are means ± SD of biological replication (n = 8). Asterisks indicates significant differences ($p < 0.01$, Tukey's test).

Moreover, we compared the root Al levels within WT rice and two *OsNAC016*-overexpressing lines. After a short-term (8 h) treatment with 50 μM Al, Al levels within roots and the cell walls within root tips (0 to 1 cm from the apex) were drastically reduced in the overexpression lines, relative to the WT (Figure 7A,B). However, Al concentration within root cell sap among all lines was comparable (Figure 7C). Collectively, this suggested that *OsNAC016* overexpression reduces Al accumulation within root tip cell walls.

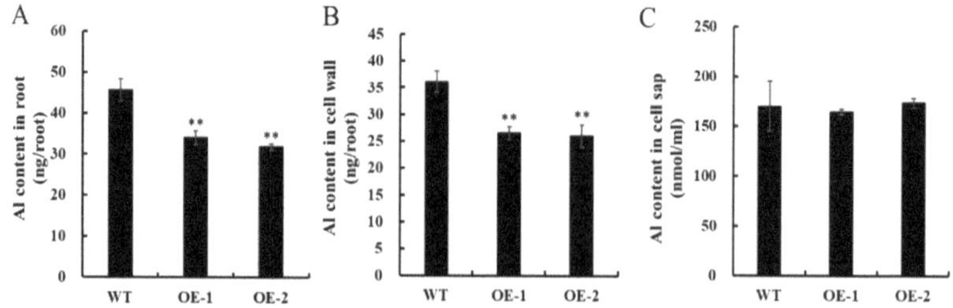

Figure 7. Al accumulation in the root tips. (**A**) Al content in the root tips. (**B**) Al content in the cell wall of root tips. (**C**) Al concentration in the cell sap of root tips. Seedlings (5 days old) of the WT and two *OsNAC016*-overexpressing lines (OE-1, OE-2) were exposed to a solution (0.5 mM $CaCl_2$, 50 μM $AlCl_3$, pH = 4.5) for 8 h. Root tips (0–1 cm) were excised and Al content in the root tips, cell wall, and cell sap was measured by ICP-MS. Data were means ± SD (n = 3). Asterisks indicates significant differences ($p < 0.01$, Tukey's test).

2.6. Overexpression of OsNAC016 Modulates Multiple Gene Expressions in Rice

To help understand the possible mechanisms of the enhanced Al tolerance, we analyzed the transcriptomes of Nipponbare (WT) and *OsNAC016*-overexpressed line (OE-1) rice roots treated with 0 or 40 μM Al for 8 h via an RNA-seq approach. Based on a threshold of twofold change, under normal conditions, we identified 1430 differentially expressed genes (DEGs) between the WT and OE line, and following Al exposure, 1441 DEGs between the lines (Figure 8A; Table S2). Among them, there were 718 down-expressed DEGs and 105 up-expressed DEGs under both two conditions (Table S2). These 823 genes were considered as the OsNAC016-regulated genes under Al-treated conditions. Based on gene ontology (GO) enrichment analysis, the up-expressed genes were primarily related to oxidative stress response and fatty acid metabolic process (Figure 8B), while the down-expressed genes were primarily involved in stress response and transmembrane transport (Figure 8C). In the cellular component category, some up- and down-regulated genes were enriched in the cell wall group. To further validate the RNA-seq data reliability, we chose seven stress- or cell-wall-related genes for RT-qPCR analysis, including *OsPAL6* (Os04g0518400), *OsEXPA10* (Os04g0583500), *OsMYB30* (Os02g0624300), *OsCesA7* (Os10g0467800), *OsBC1* (Os03g0416200), *OsUgp1* (Os09g0553200) and *OsUgp2* (Os02g0117700). Our RT-qPCR results corroborated our RNA-seq information, confirming the RNA-seq data reliability (Figure 9A–H). Therefore, based on these findings, OsNAC016 likely modulated both stress- and cell-wall-associated genes to induce Al stress resistance within rice.

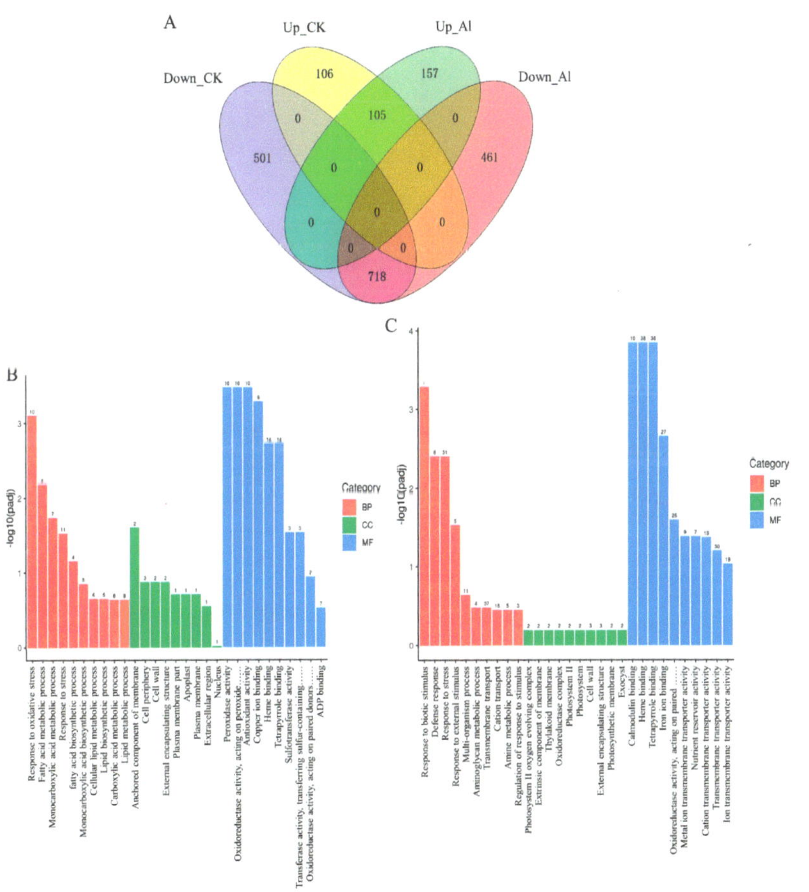

Figure 8. Transcriptome analysis of OsNAC016-regulated genes. (**A**): Venn diagram of the numbers of OsNAC016-regulated genes in the OE and WT lines. UP_CK: the up-regulated genes in the OE compared with the WT under normal conditions (fold-change > 2). UP_Al: the up-regulated genes in

the OE compared with the WT under Al treatments (fold-change > 2). DOWN_Al: the down-regulated genes in the OE compared with the WT under Al treatments (fold-change < 0.5). DOWN_CK: the down-regulated genes in the OE compared with the WT under normal conditions (fold-change < 0.5). (**B**,**C**) Classification of up-regulated genes (**B**) and down-regulated genes (**C**) in OE line compared with the WT under Al treatments. BP: biological process. CC: cellular component. MF: molecular function. Number represents the numbers of the genes. Three biological replicates (n = 3) were performed for each treatment.

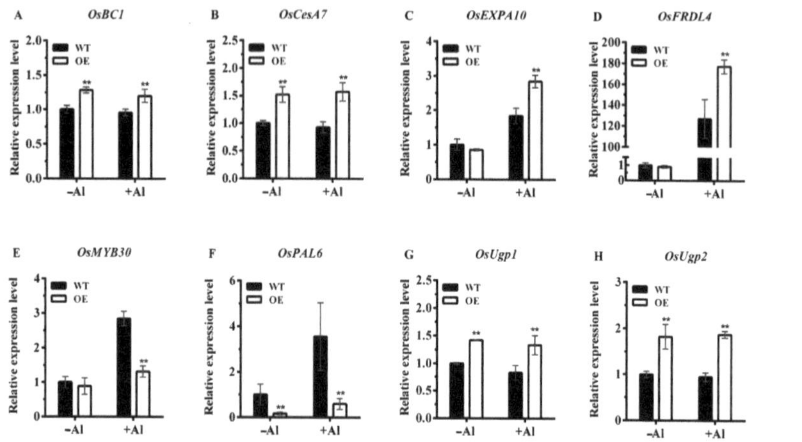

Figure 9. The expression level of eight candidate DEGs in WT and OE (*OsNAC016* overexpression) roots under normal conditions and Al treatments. *OsBC1* (**A**), *OsCesA7* (**B**), *OsEXPA10* (**C**), *OsFRDL4* (**D**), *OsMYB30* (**E**), *OsPAL6* (**F**), *OsUgp1* (**G**), *OsUgp2* (**H**). −Al represents WT and OE lines under normal conditions. +Al represents WT and OE lines under Al treatment conditions (0.5 mM $CaCl_2$, 40 μM $AlCl_3$, pH = 4.5) for 8 h. *Histone H3* was used as an internal standard. Data are means ± SD (n = 3). Asterisks indicates significant differences ($p < 0.01$, Tukey's test).

Since OsNAC016 interacted with ART1, overexpression of *OsNAC016* might influence the ART1-modulated gene expressions. To verify this hypothesis, we assessed the expressions of 31 ART1-regulated downstream genes between WT and OE line in RNA-seq data (Supplemental Table S3). Among these genes, most genes were similarly upregulated by Al treatment between two lines. However, four genes (Os02g0770800, Os01g0919200, Os01g0919100, Os01g0731600) showed largely elevated expression differences between OE line versus WT rice. Among them, only Os01g0919100 (OsFLRD4) was functionally characterized and was involved in rice Al tolerance. We conducted RT-qPCR analysis and confirmed that the *OsFLRD4* expression was substantially elevated in the *OsNAC016*-overexpressing lines, relative to WT rice (Figure 9D). Therefore, *OsNAC016* overexpression strongly modulates the expressions of partial ART1-regulated genes.

3. Discussion

In this study, we obtained an ART1-interacted gene *OsNAC016* using a yeast two-hybrid (Y2H) assay. A physical association between ART1 and OsNAC016 was further proved via BiFC and Co-IP experiments. *OsNAC016* was primarily expressed within roots and slightly modulated via Al. Immunostaining showed that OsNAC016 was a nuclear protein and localized in all root cells. Knockout of *OsNAC016* has a weak or no effect on the expressions of *ART1* and ART1-modulated Al-tolerant genes under normal and Al stress. Furthermore, we observed no marked alterations in Al tolerance between WT and *osnac016* mutants. One possibility is that there are redundant genes with similar function to OsNAC016, leading to no obvious Al sensitive phenotype in *osnac016* mutants. This notion is supported by the presence of multiple homologs for OsNAC016 in the rice genome such as OsNAC58, OsNAC103, OsNAC122, and OsNAC131. On the other hand, overexpression of *OsNAC016* resulted in less Al root aggregation and enhanced rice tolerance to Al stress. Based on transcriptomic sequencing and qRT-PCR assessment, the expressions of some genes related to cell wall or Al stress were altered in *OsNAC016*-overexpressing plants. These findings suggested that OsNAC016 contributes to the regulation of Al tolerance within rice.

The cell wall acts as an initial physical barrier that protects plant cells from harmful external environmental factors, such as, Al toxicity. Among the main Al-resistance strategies in plants is to decrease Al accumulation in the cell wall through modifying cell wall properties to alter its Al-binding capacity [26–29]. Normally, high Al tolerance in plants is associated with low Al accumulation in the cell wall, as described in rice [27], *Secale sylvestre* [29], and *Arabidopsis* [28]. Interestingly, *OsNAC016* overexpressing lines showed reduced cell wall Al content and enhanced Al tolerance relative to the WT rice. The RNA-seq and qRT-PCR analyses demonstrated that the expression levels of some genes related to cell wall modification in the *OsNAC016*-overexpressing lines were significantly downregulated or upregulated relative to the WT rice under Al stress, including *OsPAL6, OsMYB30, OsEXPA10, OsBC1, OsCesA7, OsUgp1* and *OsUgp2*. Among them, *OsPAL6* and *OsMYB30* are down-regulated while the other five genes are up-regulated in the *OsNAC016*-overexpressing lines. OsPAL6, a phenylalanine ammonia-lyase, forms the first major enzyme in the phenylpropanoid pathway, participating in lignin biosynthesis [30]. Several studies have showed that down-regulation or knockout of lignin biosynthesis-related genes impact the Al interaction with the cell wall and Al tolerance such as *4-coumarate: coenzyme A ligase 4CL4* and *4CL5* within rice [26,31]. The expression of *OsPAL6* and *4CL5* is positively regulated by OsMYB30 [26,30,31]. Knockout of *OsMYB30* resulted in a lower cell wall Al aggregation and an enhanced Al tolerance within rice [26]. *OsEXPA10* encoding a cell wall expansin participated in root cell elongation through the modulation of cell-wall loosening. Knocking out *OsEXPA10* led to a reduction in root length [11]. OsBC1 belonging to the classical dynamin-related protein family (OsDRPs), is homologous to OsBC3/OsDRP2B. Disruption of *OsBC3* caused a marked rise in cell wall pectin content [32]. Generally, pectic matrix is proposed to be the primary cell wall Al-binding site [33]. OsCesA7 is cellulose synthase A subunit 7 functioning in cellulose biosynthesis. Mutation of *OsCesA7* resulted in a reduced cellulose and an increased cell wall hemicellulose content [34]. Hemicellulose possesses Al-binding ability. Its content in the cell wall affects Al accumulation and tolerance [31]. Both *OsUgp1* and *OsUgp2* encode UDP-glucose pyrophosphorylase (UGPase), which accelerates the UTP and glucose-1-phosphate conversion to pyrophosphate and UDP-glucose [35]. UDP-glucose is reported to be used to modify the cell wall to alleviate Al stress-induced root growth inhibition through masking cell-wall-based Al-binding sites. Therefore, the reduced cell wall Al aggregation in roots found within *OsNAC016*-overexpressing lines may be explained via modulation of the above-mentioned gene expressions, relative to the WT rice.

ART1 is the central regulator for rice Al tolerance, which regulate ≥31 Al-detoxifying genes. Overexpression of *OsNAC016* significantly increases the partial ART1-modulated gene expressions such as *OsFRDL4*, Os02g0770800, Os01g0919200, and Os01g0731600. Among the previously characterized ART1-modulated genes, *OsFRDL4* expression was the most affected. OsFRDL4 is a plasma-membrane-localized citrate transporter that facilitates rice Al tolerance through releasing citrate from roots following Al stimuli. ART1 regulates *OsFRDL4* expression by directly interacting with its promoter's cis-element GGN(T/g/a/C)V(C/A/g)S(C/G) region [14]. A recent study showed that OsNAC016 can bind to the cis-element CACG-motif [36]. We performed an analysis of cis-acting elements and identified 11 CACG-motifs in the 1–1642 bp promoter region of *OsFRDL4* (Supplemental Figure S7). Some OsNAC016-binding motifs are adjacent to ART1-interacting cis-element. These suggested that overexpression of *OsNAC016* might cause more OsNAC016-ART1 complexes to interact with the *OsFRDL4* promoter region to promote its expression and citrate secretion under Al stress conditions, resulting in an enhanced Al tolerance within rice.

Additionally, transcriptome analysis revealed that overexpression of *OsNAC016* up-regulated many biological processes-related gene expressions such as peroxidase activity, antioxidant activity, and response to oxidative stress under normal conditions, implying that OsNAC016 may also play important regulatory roles in other stress responses. A recent study reported that OsNAC016 negatively regulated drought tolerance via down-regulating ABA- and drought-responsive genes in rice [37]. More recently, OsNAC016 was also shown to negatively modulate Pi-starvation tolerance though the activation of *OsSPX2* in rice [36]. These findings suggested that OsNAC016 acts as an important TF that regulates multiple stress responses in rice.

In conclusion, OsNAC016 is the first protein identified as ART1-interacting protein. Overexpression of *OsNAC016* can lower root cell wall Al aggregation through modulation of some cell-wall-related and ART1-modulated gene expressions and thus improve Al tolerance in rice.

4. Materials and Methods

4.1. Plant Materials and Growth Conditions

Herein, we used WT (*Oryza sativa* cv Nipponbare), as well as 2 *OsNAC016*-knockout lines, and 2 *OsNAC016* overexpression lines which were generated within the laboratory. Rice seeds underwent a 2-day submersion in deionized water without light at 28 °C. Germinated seeds were transferred to a floating net under a solution with 0.5 mM $CaCl_2$ in an illumination incubator and used for various experiments.

Roots, stems, leaf blades, leaf sheaths and panicles were extracted from WT plants at the heading stage for organ-specific expression analysis of *OsNAC016*. We examined the spatial *OsNAC016* expression by individually harvesting root tips (0–1 cm), basal roots (1–2 cm) and mature region (2–3 cm). To analyze the *OsNAC016*-mediated response to Al stress, rice seedlings (5 days old) were treated to varying Al concentrations (0, 10, 30, 50, 100 μM) at pH 4.5 for 8 h or 30 μM Al over a wide range of experiment durations (0, 2, 4, 8, 12 and 24 h).

4.2. Analysis of Al Tolerance

For short-term Al treatment, seedlings (5-d-old) from individual lines underwent a 24-h treatment with a 0.5 mM $CaCl_2$ and 50 μM Al solution (pH 4.5). For long-term Al treatment, seedlings (2-d-old) from individual lines underwent a 5-d treatment with a 0.5 mM $CaCl_2$ and 20 μM Al solution (pH 4.5). Root lengths were recorded both prior to and after treatments using a ruler. Relative root elongation was computed using the following formular: (Al-treated root elongation)/(Al-untreated root elongation) × 100.

4.3. Generation of Transgenic Rice Plants

To construct the overexpression vector for *OsNAC016*, *OsNAC016* cDNA was amplified from the Nipponbare root cDNA via PCR and specific primers. The amplified cDNA fragment was ligated to the pCAMBIA1300-Ubi vector between the maize *Ubiquitin 1* promoter and nopaline synthase terminator, generating the pCAMBIA1300-*Ubi1:OsNAC016* construct. To create the CRISPR construct for *OsNAC016*, two sgRNA expression cassettes containing two different specific target sites for OsNAC016 were digested by *Bsa*I and cloned into a pYLCRISPR/Cas9 vector [38], producing the pYLCRISPR-*OsNAC016* construct. These constructs were next introduced into rice cv. Nipponbare via *Agrobacterium tumefaciens* strain EHA105-mediated transformation. The employed primer sequences are summarized in Table S4.

4.4. RNA Extraction and Gene Expression Analysis

The Trizol reagent kit (Invitrogen, Carlsbad, CA, USA) was utilized for total RNA extraction as per kit directions. The Hiscript II Q RT SuperMix Kit (Vazyme, Nanjing, China) was then employed to eliminate genomic DNA and perform reverse transcription reactions. For qRT-PCR, ChanQTM SYBR Color qPCR Master Mix (Vazyme, Nanjing, China) was used in a StepOnePlus RT-PCR System (Analytik Jena AG). An internal standard of *Histone H3* was utilized. Relative gene expression employed the $2^{-\Delta\Delta CT}$ calculation method. The primers for qRT-PCR were shown in Table S4.

4.5. OsNAC016 Subcellular Localization

To assess OsNAC016 intracellular distribution, we inserted the coding sequence of *OsNAC016* into the pYL322-GFP vector upstream of GFP to create the pYL322-*OsNAC016-GFP* construct. Rice protoplasts were co-transfected with either pYL322-*OsNAC016-GFP* or a control vector along with a nuclear marker (OsGhd7-RFP), following the protocols described by Chen et al. [39]. After 12–16 h incubation, GFP or RFP fluorescence was observed using a confocal laser scanning microscope (TCS SP8; Leica, Weztlar, Germany).

4.6. OsNAC016 Cellular Distribution

The promoter (2000 bp) and CDS (1023 bp) sequences of *OsNAC016* were amplified using specific primers Pro-NAC016-F/Pro-NAC016-R and 1300-NAC016-F/1300-NAC016-R, respectively. Subsequently, the promoter and CDS fragments were cloned into the *Hind*III and *BamHI* sites of the pCAMBIA1300-*GFP* vector. The subsequent construct (pCAMBIA1300-*ProOsNAC016:OsNAC016-GFP*) was introduced into Nipponbare to generate the transgenic lines harboring *ProOsNAC016-OsNAC016-GFP*. The employed primer sequences are summarized in Supplemental Table S4.

To examine the OsNAC016 cellular localization, immunostaining was carried out using a GFP antibody following the protocol described by Yamaji and Ma in 2007 [40]. In brief, roots from both WT (WT) plants and transgenic lines expressing *ProOsNAC016:OsNAC016-GFP* were embedded in a 5% agarose and sliced into sections (100-μm thickness) using a microslicer (VT1000 S Leica), which were then transferred to microscope slides for incubation with rabbit anti-GFP polyclonal antibodies, with subsequent incubation with secondary antibodies (Alexa Fluor 555 goat anti-rabbit IgG; Molecular Probes) at room temperature (RT). Cell nucleus was stained with DAPI. Fluorescent images were then taken via a confocal laser scanning microscope (Leica model TCS SP8, Weztlar, Germany).

4.7. Y2H Assy

We screened a yeast cDNA library, prepared from the 12-h 30 μM Al-treated roots of 5-day-old WT rice seedling using truncated ART1 CDS (1–525 bp) as bait and the Matchmaker Yeast Two-Hybrid System (Clontech, Shiga, Japan). To verify the truncated ART1 and target protein associations, the truncated ART1 and target gene coding sequences were inserted into pGBKT7 and pGADT7 vectors, respectively. The resulting bait and prey vectors were co-introduced into the yeast strain AH109 and cultured on an SD medium without Leu and Trp or lacking Leu, Trp, Ade, and His. The employed primer sequences are summarized in Supplemental Data.

4.8. BiFC Assay

To prepare the BiFC constructs, the *OsNAC016* and *ART1* coding sequences were amplified from the Nipponbare root cDNA using corresponding primers, prior to cloning. The amplified products were cloned into the BiFC vectors pUC19-35S-Vn and pUC19-35S-Vc, producing the OsNAC016-cYFP, OsNAC016-nYFP, OsART1-cYFP and ART1-nYFP constructs. We prepared rice protoplast and conducted plasmid transformation into cells as reported previously [39]. Following plasmid transformation, cells were maintained without light at 28 °C for 12–15 h. Fluorescent images were captured via confocal laser scanning microscope (TCS SP8; Leica, Weztlar, Germany). The employed primer sequences are summarized in Table S4.

4.9. Co-IP Assay

The *OsNAC016* and *ART1* full-length coding sequences were ligated into the *Kpn*I and *BamHI* sites of pCAMBIA1300-35S-FLAG and pCAMBIA1300-35S-MYC vectors, respectively. The pCAMBIA1300-*35S:OsNAC016-FLAG* and pCAMBIA1300-*35S:ART1-MYC* recombinant constructs were introduced into *Agrobacterium* strain (GV3101), prior to transformation into 4-week-old *N. benthamiana* leaves. Approximately 24 h later, total protein isolation was completed from infected leaves by Lysis extract NB1 buffer (50 mM Tris-MES [pH 7.3], 500 mM Sucrose, 1 mM $MgCl_2$, 10 mM EDTA, 5 mM DTT, 1 mM PMSF, 100× Cocktail). Protein co-IP experiment was performed as reported by Xu et al. (2015) [41]. Proteins were detected with anti-FLAG (M2008S, Abmart, Shanghai, China, 1:5000) and anti-MYC (M2002S, Abmart, Shanghai, China, 1:5000) antibodies.

4.10. Evaluation of Root Tip Al Content

Five-day-old WT and two *OsNAC016*-overexpressing line seedlings were exposed for 8 h to a solution (pH 4.5) with 0.5 mM $CaCl_2$ and 50 μM $AlCl_3$. After 3 rinses in 0.5 mM $CaCl_2$, 20 root segments (0–1 cm from the root tip) each replicate were extracted and then transferred to a 2 mL tube or ultrafree-MC centrifugal filter unit (Millipore, Billerica, MA, USA). To obtain root cell sap, we centrifuged root-containing units for 10 min at 3000× *g* at 4 °C. After that, the units were frozen overnight at −80 °C. The root cell sap solution was obtained via centrifugation of samples at 20,400× *g* for 10 min at RT. The residues in the units were washed three times with alcohol (70%) by a vortex. The cell sap was diluted with 5% nitric acid. Both cell walls and whole roots were digested by 65% nitric acid. The Al contents of cell sap, cell wall and whole root tip were assessed using ICP-MS (Plasma Quant MS; Analytik Jena AG).

4.11. RNA-Seq Assay

Five-day-old WT and *OsNAC016*-overexpression line (OE) seedlings were subjected to a 0.5 mM $CaCl_2$ solution (pH 4.5) with 0 or 50 μM $AlCl_3$ for 8 h. The seedling roots were sampled for RNA-seq experiments. Following total root RNA extraction, we synthesized corresponding cDNA, and conducted cDNA library generation. cDNA library sequencing was conducted with the Illumina NovaSeq 6000 platform. Deseq2 was employed for gene expression comparison analysis between WT and OE lines. DEGs were identified using the following criteria: $|\log2(\text{FoldChange})| \geq 1$ and padj ≤ 0.05. DEGs GO (Gene Ontology; http://geneontology.org/, accessed on 1 October 2022) analysis was done using hypergeometric tests, and *p*-values were utilized for determining significant enrichment of corresponding categories.

Supplementary Materials: The following supporting information can be downloaded at: https://www.mdpi.com/article/10.3390/ijms242317036/s1.

Author Contributions: J.X. conceived and designed the experiment. F.L. performed the experiments. D.M., J.Y., R.M., Z.W., B.Z., X.C., L.Z. and L.P. participated in the research. J.X. and F.L. analyzed the data. F.L. and J.X. wrote the paper. All authors have read and agreed to the published version of the manuscript.

Funding: Guangxi Natural Science Foundation (2021GXNSFDA220002) and the Hundred Talents Program of Guangxi (2014).

Institutional Review Board Statement: Not applicable.

Informed Consent Statement: Not applicable.

Data Availability Statement: All the data supporting the conclusions of this article are provided within the article and in its additional files. All data and materials are available upon reasonable request from the corresponding author.

Conflicts of Interest: The authors declare no conflict of interest.

References

1. Kochian, L.V.; Pineros, M.A.; Hoekenga, O.A. The physiology, genetics and molecular biology of plant aluminum resistance and toxicity. *Plant Soil* **2005**, *274*, 175–195. [CrossRef]
2. Ma, J.F. Syndrome of aluminum toxicity and diversity of aluminum resistance in higher plants. *Int. Rev. Cytol.* **2007**, *264*, 225–252. [PubMed]
3. Kochian, L.V.; Piñeros, M.A.; Liu, J.; Magalhaes, J.V. Plant adaptation to acid soils: The molecular basis for crop aluminum resistance. *Annu. Rev. Plant Biol.* **2015**, *66*, 571–598. [CrossRef] [PubMed]
4. Zhang, X.; Long, Y.; Huang, J.; Xia, J. Molecular mechanisms for coping with Al toxicity in plants. *Int. J. Mol. Sci.* **2019**, *20*, 1551. [CrossRef] [PubMed]
5. Sasaki, T.; Yamamoto, Y.; Ezaki, B.; Katsuhara, M.; Ahn, S.J.; Ryan, P.R.; Delhaize, E.; Matsumoto, H. A wheat gene encoding an aluminum-activated malate transporter. *Plant J.* **2004**, *37*, 645–653. [CrossRef] [PubMed]
6. Magalhaes, J.V.; Liu, J.; Guimaraes, C.T.; Lana, U.G.; Alves, V.M.; Wang, Y.H.; Schaffert, R.E.; Hoekenga, O.A.; Pineros, M.A.; Shaff, J.E.; et al. A gene in the multidrug and toxic compound extrusion (MATE) family confers aluminum tolerance in sorghum. *Nat. Genet.* **2007**, *39*, 1156–1161. [CrossRef] [PubMed]
7. Furukawa, J.; Yamaji, N.; Wang, H.; Mitani, N.; Murata, Y.; Sato, K.; Katsuhara, M.; Takeda, K.; Ma, J.F. An aluminum-activated citrate transporter in barley. *Plant Cell Physiol.* **2007**, *48*, 1081–1091. [CrossRef]
8. Ma, J.F.; Shen, R.; Zhao, Z.; Wissuwa, M.; Takeuchi, Y.; Ebitani, T.; Yano, M. Response of rice to Al stress and identification of quantitative trait loci for Al tolerance. *Plant Cell Physiol.* **2002**, *43*, 652–659. [CrossRef]
9. Yamaji, N.; Huang, C.F.; Nagao, S.; Yano, M.; Sato, Y.; Nagamura, Y.; Ma, J.F. A zinc finger transcription factor ART1 regulates multiple genes implicated in aluminum tolerance in rice. *Plant Cell* **2009**, *21*, 3339–3349. [CrossRef]
10. Xia, J.X.; Yamaji, N.; Ma, J.F. A plasma membrane-localized small peptide is involved in rice aluminum tolerance. *Plant J.* **2013**, *76*, 345–355. [CrossRef]
11. Che, J.; Yamaji, N.; Shen, R.F.; Ma, J.F. An Al-inducible expansin gene, OsEXPA10 is involved in root cell elongation of rice. *Plant J.* **2016**, *88*, 132–142. [CrossRef]
12. Che, J.; Tsutsui, T.; Yokosho, K.; Yamaji, N.; Ma, J.F. Functional characterization of an aluminum (Al)-inducible transcription factor, ART2, revealed a different pathway for Al tolerance in rice. *New Phytol.* **2018**, *220*, 209–218. [CrossRef]
13. Yokosho, K.; Yamaji, N.; Fujii-Kashino, M.; Ma, J.F. Functional analysis of a MATE gene OsFRDL2 revealed its involvement in Al-induced secretion of citrate, but a lower contribution to Al tolerance in rice. *Plant Cell Physiol.* **2016**, *57*, 976–985. [CrossRef] [PubMed]
14. Yokosho, K.; Yamaji, N.; Kashino-Fujii, M.; Ma, J.F. Retrotransposon-mediated aluminum tolerance through enhanced expression of the citrate transporter OsFRDL4. *Plant Physiol.* **2016**, *172*, 2327–2336. [CrossRef]
15. Chen, Z.C.; Yamaji, N.; Motoyama, R.; Nagamura, Y.; Ma, J.F. Up-regulation of a magnesium transporter gene OsMGT1 is required for conferring aluminum tolerance in rice. *Plant Physiol.* **2012**, *159*, 1624–1633. [CrossRef] [PubMed]
16. Huang, C.F.; Yamaji, N.; Mitani, N.; Yano, M.; Nagamura, Y.; Ma, J.F. A bacterial-type ABC transporter is involved in aluminum tolerance in rice. *Plant Cell* **2009**, *21*, 655–667. [CrossRef]
17. Xia, J.; Yamaji, N.; Kasai, T.; Ma, J.F. Plasma membrane-localized transporter for aluminum in rice. *Proc. Natl. Acad. Sci. USA* **2010**, *107*, 18381–18385. [CrossRef]
18. Huang, C.F.; Yamaji, N.; Chen, Z.; Ma, J.F. A tonoplast-localized half-size ABC transporter is required for internal detoxification of aluminum in rice. *Plant J.* **2012**, *69*, 857–867. [CrossRef]
19. Nuruzzaman, M.; Manimekalai, R.; Sharoni, A.M.; Satoh, K.; Kondoh, H.; Ooka, H.; Kikuchi, S. Genome-wide analysis of NAC transcription factor family in rice. *Gene* **2010**, *465*, 30–44. [CrossRef] [PubMed]
20. Hu, H.; You, J.; Fang, Y.; Zhu, X.; Qi, Z.; Xiong, L. Characterization of transcription factor gene SNAC2 conferring cold and salt tolerance in rice. *Plant Mol. Biol.* **2008**, *67*, 169–181. [CrossRef]
21. Zhang, X.; Long, Y.; Huang, J.; Xia, J. OsNAC45 is involved in ABA response and salt tolerance in rice. *Rice* **2020**, *13*, 79. [CrossRef]
22. Lou, H.Q.; Fan, W.; Jin, J.F.; Xu, J.M.; Chen, W.W.; Yang, J.L.; Zheng, S.J. A NAC-type transcription factor confers aluminium resistance by regulating cell wall-associated receptor kinase 1 and cell wall pectin. *Plant Cell Environ.* **2020**, *43*, 463–478. [CrossRef] [PubMed]
23. Sjogren, C.A.; Bolaris, S.C.; Larsen, P.B. Aluminum-dependent terminal differentiation of the Arabidopsis root tip is mediated through an ATR-, ALT2-, and SOG1-regulated transcriptional response. *Plant Cell* **2015**, *27*, 2501–2515. [CrossRef] [PubMed]
24. Tao, Y.; Wan, J.X.; Liu, Y.S.; Yang, X.Z.; Shen, R.F.; Zhu, X.F. The NAC transcription factor ANAC017 regulates aluminum tolerance by regulating the cell wall-modifying genes. *Plant Physiol.* **2022**, *189*, 2517–2534. [CrossRef] [PubMed]
25. Wang, Z.; Zhang, B.; Chen, Z.; Wu, M.; Chao, D.; Wei, Q.; Xin, Y.; Li, L.; Ming, Z.; Xia, J. Three OsMYB36 members redundantly regulate Casparian strip formation at the root endodermis. *Plant Cell* **2022**, *34*, 2948–2968. [CrossRef] [PubMed]
26. Gao, L.J.; Liu, X.P.; Gao, K.K.; Cui, M.Q.; Zhu, H.H.; Li, G.X.; Yan, J.Y.; Wu, Y.R.; Ding, Z.J.; Chen, X.W.; et al. ART1 and putrescine contribute to rice aluminum resistance via OsMYB30 in cell wall modification. *J. Integr. Plant Biol.* **2023**, *65*, 934–949. [CrossRef]
27. Yang, J.L.; Li, Y.Y.; Zhang, Y.J.; Zhang, S.S.; Wu, Y.R.; Wu, P.; Zheng, S.J. Cell wall polysaccharides are specifically involved in the exclusion of aluminum from the rice root apex. *Plant Physiol.* **2008**, *146*, 602–611. [CrossRef]

28. Zhu, X.F.; Shi, Y.Z.; Lei, G.J.; Fry, S.C.; Zhang, B.C.; Zhou, Y.H.; Braam, J.; Jiang, T.; Xu, X.Y.; Mao, C.Z.; et al. XTH31, encoding an in vitro XEH/XET-active enzyme, regulates Al sensitivity by modulating in vivo XET action, cell wall xyloglucan content, and Al binding capacity in Arabidopsis. *Plant Cell* **2012**, *24*, 4731–4747. [CrossRef]
29. Liu, Q.; Yang, J.L.; He, L.S.; Li, Y.Y.; Zheng, S.J. Effect of aluminum on cell wall, plasma membrane, antioxidants and root elongation in triticale. *Biol. Plant.* **2008**, *52*, 87–92. [CrossRef]
30. He, J.; Liu, Y.; Yuan, D.; Duan, M.; Liu, Y.; Shen, Z.; Yang, C.; Qiu, Z.; Liu, D.; Wen, P.; et al. An R2R3 MYB transcription factor confers brown planthopper resistance by regulating the phenylalanine ammonia-lyase pathway in rice. *Proc. Natl. Acad. Sci. USA* **2020**, *117*, 271–277. [CrossRef]
31. Liu, S.; Zhao, L.; Liao, Y.H.; Luo, Z.L.; Wang, H.; Wang, P.; Zhao, H.; Xia, J.X.; Huang, C.F. Dysfunction of the 4-coumarate: Coenzyme a ligase 4CL4 impacts aluminum resistance and lignin accumulation in rice. *Plant J.* **2020**, *104*, 1233–1250. [CrossRef] [PubMed]
32. Li, R.; Xiong, G.; Zhou, Y. Membrane trafficking mediated by OsDRP2B is specific for cellulose biosynthesis. *Plant Signal. Behav.* **2010**, *5*, 1483–1486. [CrossRef] [PubMed]
33. Yang, J.L.; Zhu, X.F.; Zheng, C.; Zhang, Y.J.; Zheng, S.J. Genotypic differences in Al resistance and the role of cell-wall pectin in Al exclusion from the root apex in *Fagopyrum tataricum*. *Ann. Bot.* **2011**, *107*, 371–378. [CrossRef] [PubMed]
34. Wang, D.; Qin, Y.; Fang, J.; Yuan, S.; Peng, L.; Zhao, J.; Li, X. A missense mutation in the zinc finger domain of OsCESA7 deleteriously affects cellulose biosynthesis and plant growth in rice. *PLoS ONE* **2016**, *11*, e0153993. [CrossRef] [PubMed]
35. Chen, R.Z.; Zhao, X.; Shao, Z.; Wei, Z.; Wang, Y.Y.; Zhu, L.L.; Zhao, J.; Sun, M.X.; He, R.F.; He, G.C. Rice UDP-glucose pyrophosphorylase1 is essential for pollen callose deposition and its cosuppression results in a new type of thermosensitive genic male sterility. *Plant Cell* **2007**, *19*, 847–861. [CrossRef]
36. Sun, Y.; Wu, Q.; Xie, Z.; Huang, J. Transcription factor OsNAC016 negatively regulates phosphate-starvation response in rice. *Plant Sci.* **2023**, *329*, 111618. [CrossRef]
37. Wu, Q.; Liu, Y.; Xie, Z.; Yu, B.; Sun, Y.; Huang, J. OsNAC016 regulates plant architecture and drought tolerance by interacting with the kinases GSK2 and SAPK8. *Plant Physiol.* **2022**, *189*, 1296–1313. [CrossRef]
38. Ma, X.; Zhang, Q.; Zhu, Q.; Liu, W.; Chen, Y.; Qiu, R.; Wang, B.; Yang, Z.; Li, H.; Lin, Y.; et al. A robust CRISPR/Cas9 system for convenient, high-efficiency multiplex genome editing in monocot and dicot plants. *Mol. Plant* **2015**, *8*, 1274–1284. [CrossRef]
39. Chen, J.; Liu, Y.; Ni, J.; Wang, Y.; Bai, Y.; Shi, J.; Gan, J.; Wu, Z.; Wu, P. OsPHF1 regulates the plasma membrane localization of low- and high-affinity inorganic phosphate transporters and determines inorganic phosphate uptake and translocation in rice. *Plant Physiol.* **2011**, *157*, 269–278. [CrossRef]
40. Yamaji, N.; Ma, J.F. Spatial distribution and temporal variation of the rice silicon transporter Lsi1. *Plant Physiol.* **2007**, *143*, 1306–1313. [CrossRef]
41. Xu, F.; Copeland, C.; Li, X. Protein immunoprecipitation using *Nicotiana benthamiana* transient expression system. *Bio-protocol* **2015**, *5*, e1520. [CrossRef]

International Journal of *Molecular Sciences*

MDPI

Article

Insertion of Transposable Elements in *AVR-Pib* of *Magnaporthe oryzae* Leading to LOSS of the Avirulent Function

Jinbin Li [1,*,†], Lin Lu [2,†], Chengyun Li [3], Qun Wang [1] and Zhufeng Shi [1]

1 The Ministry of Agriculture and Rural Affairs International Joint Research Center for Agriculture, The Ministry of Agriculture and Rural Affairs Key Laboratory for Prevention and Control of Biological Invasions, Yunnan Key Laboratory of Green Prevention and Control of Agricultural Transboundary Pests, Agricultural Environment and Resource Research Institute, Yunnan Academy of Agricultural Sciences, Kunming 650205, China; qunwang70@163.com (Q.W.)
2 Flower Research Institute, Yunnan Academy of Agricultural Sciences, Kunming 650205, China; lu_lin2005@sina.com
3 The Ministry of Education Key Laboratory for Agricultural Biodiversity and Pest Management, Yunnan Agricultural University, Kunming 650200, China; licheng_yun@163.com
* Correspondence: lijinbin@yaas.org.cn
† These authors contributed equally to this work.

Abstract: Rice blast is a very serious disease caused by *Magnaporthe oryzae*, which threatens rice production and food supply throughout the world. The avirulence (*AVR*) genes of rice blast are perceived by the corresponding rice blast resistance (*R*) genes and prompt specific resistance. A mutation in *AVR* is a major force for new virulence. Exploring mutations in *AVR* among *M. oryzae* isolates from rice production fields could aid assessment of the efficacy and durability of *R* genes. We studied the probable molecular-evolutionary patterns of *AVR-Pib* alleles by assaying their DNA-sequence diversification and examining their avirulence to the corresponding *Pib* resistance gene under natural conditions in the extremely genetically diverse of rice resources of Yunnan, China. PCRs detected results from *M. oryzae* genomic DNA and revealed that 162 out of 366 isolates collected from Yunnan Province contained *AVR-Pib* alleles. Among them, 36.1–73.3% isolates from six different rice production areas of Yunnan contained *AVR-Pib* alleles. Furthermore, 36 (28.6%) out of 126 isolates had a transposable element (TE) insertion in *AVR-Pib*, which resulted in altered virulence. The TE insertion was identified in isolates from rice rather than from *Musa nana* Lour. Twelve *AVR-Pib* haplotypes encoding three novel *AVR-Pib* variants were identified among the remaining 90 isolates. *AVR-Pib* alleles evolved to virulent forms from avirulent forms by base substitution and TE insertion of Pot2 and Pot3 in the 5′ untranslated region of *AVR-Pib*. These findings support the hypothesis that functional *AVR-Pib* possesses varied sequence structures and can escape surveillance by hosts via multiple variation manners.

Keywords: *Magnaporthe oryzae*; rice blast; *AVR-Pib*; evolution; transposable element (TE)

Citation: Li, J.; Lu, L.; Li, C.; Wang, Q.; Shi, Z. Insertion of Transposable Elements in *AVR-Pib* of *Magnaporthe oryzae* Leading to LOSS of the Avirulent Function. *Int. J. Mol. Sci.* **2023**, *24*, 15542. https://doi.org/10.3390/ijms242115542

Academic Editor: Hikmet Budak

Received: 13 August 2023
Revised: 24 September 2023
Accepted: 28 September 2023
Published: 24 October 2023

1. Introduction

In the coevolution of plants and pathogens, the latter can adapt to the host and environment, and selection is the major evolutionary force. Up to now, the "arms race" and "trench warfare" hypotheses of coevolution between host resistance (*R*) genes and pathogen avirulence (*AVR*) genes have been proposed. In the principal hypothesis of the arms-race, the mutation of *R* genes and *AVR* genes is derived by directional selection. Contrarily, it is derived by undirectional selection in the trench-warfare hypothesis [1].

Rice blast is one of the most serious diseases in rice worldwide, which caused by the fungus *Magnaporthe oryzae*. The application of rice varieties with multiple resistant genes is the most important method controlling the disease with an economical, environmentally and friendly manner. The resistance of the sole resistance gene rice variety to *M. oryzae* can be lost quickly by the high variation of fungus. Up to now, more than 35 rice blast *R* genes have been cloned in rice: *Pita, PiCO39, Pish, Pi1, Pik, Pikp, Pikh/Pi54, Pikm, Pb1, Pid3, Pia, Pib, Pid2, Pit, Pizt, Pi2, Pi5, Pi9, pi21, Pi25, Pi36, Pi37, Pi56, Pi63, Pi35, Pid3-4A, Pi50, Pii, Pi54, Pike, Piks* [2,3].

The effectiveness of resistance of *Pib* has been examined in different rice production provinces in China. *Pib* exerts a high level of resistance to *M. oryzae* from Heilongjiang Province, and can be applied as a parent for resistance breeding in Heilongjiang Province [4]. *Pib* is moderately resistant in Fujian Province [5], but *Pib* exhibits partial resistance in Guangdong, Sichuan and Guizhou Provinces [6,7]. Different resistance spectra of *Pib* were detected in 282 blast isolates collected from *Indica* rice- and *Japonica* rice production regions in Yunnan Province [8]. Those results showed that the *Pib* gene exhibits different resistance to the Chinese rice blast from different rice-growing regions. There were 42 out of 54 varieties containing *Pib* in Chinese elite hybrid rice varieties, two haplotypes of *Pib* were identified and 9 different haplotypes of *AVR-Pib* were found among 27 *M. oryzae* [9]. The reactions of differential isolates on 54 rice varieties between the frequency of *AVR-Pib* haplotypes in the differential isolates showed a good correlation [9]. They showed that the *Pib* gene was widely distributed in rice varieties in China, and the adapt variation of *AVR-Pib* of *M. oryzae* occurred. The rice resistance gene of *Pib* is located on the long arm of chromosome 2 [10,11]. The cDNA length of the *Pib* gene contains 306 bp of 5′ untranslated regions (UTRs), 3753 bp of open reading frames (ORFs) (containing 3 exons) and 229 bp of 3′ UTRs, and encodes a NBS-LRR protein with 1251 amino acids [12]. In China, 16 out of 204 (7.8%) varieties have been detected hoarding the *Pib* gene in a mini-core collection of Chinese rice germplasm using their functional markers [13]. Moreover, 11 varieties were identified with the *Pib* gene among 58 leading rice cultivars or hybrid rice parents in China using functional DNA markers [14]. *Pi-b* was detected in 33 landraces among 176 landraces (18.8%) from Yunnan Province in China [15]. *Pib* gene homologs (87-bp deletion in exon 1 of *Pib*, leading to a loss of the resistance function of *Pib*) were identified in Yunnan Yuanjiang type of common wild rice (*Oryza rufipogon* Griff) [16]. In the Philippines, 32 out of 52 commercial rice varieties have been shown to contain *Pib* as detected by polymerase chain reaction (PCR) primers specific to *Pib* [17]. They showed that the *Pib* gene was widely used in rice breeding in China.

Based on the gene-for-gene theory, rice *R* gene(s) can discern the corresponding *AVR* of *M. oryzae* and trigger the defense response to prevent invasion. So far, 12 *AVR* genes have been cloned in *M. oryzae*: *AVR-Pib* [18], *AVR-Pi54* [19], *AVR-Pi9* [20], *AVR-Pia* [21], *AVR-Pik/km/kp* [21], *AVR-Pii* [21], *AVR-Pizt* [22], *ACE1* [23], *AVR1-CO39* [24], *AVR-Pita* [25], *PWL1* [26], and *PWL2* [27]. The *AVR-Pib* allele of *M. oryzae* predicts the resistance efficacy of the rice *R* gene *Pib*. *AVR-Pib* encodes a putative secreted protein with 75 amino acids. *AVR-Pib* is perceived by the host as a Pib resistance protein and prompts the innate immune response [18]. Transposable element (TE) insertion, segmental deletion, absence and point mutations have been identified in *AVR-Pib* in 60 rice blast isolates from Guangdong, Hunan, Liaoning, Jilin and Heilongjiang of Provinces in China, which resulted in a loss of avirulence [18]. Only TE insertion has been observed in *AVR-Pib* in 248 *M. oryzae* isolates from the Philippines [17].

Further clarification of the diversification and evolution of the *AVR* gene is useful for the prediction of the effectiveness and durability of *R* genes for resistance breeding. Here, we wished to: (i) detect the diversity of nucleotide sequences of *AVR-Pib* alleles of *M. oryzae* under field conditions; (ii) determine the avirulence function of *AVR-Pib* variations to the *Pib* gene; (iii) reveal the molecular diversification principles of *AVR-Pib* alleles in *M. oryzae* in Yunnan Province. Our results provide useful information for rice-blast disease controlling and resistant breeding in China.

2. Results

2.1. Effectiveness of the Pib Gene and Frequency of AVR-Pib Alleles

The efficacy of the *Pib* gene was examined by pathogenicity assays. A total of 223 of the 366 *M. oryzae* isolates tested were avirulent to the *Pib* gene-containing rice monogenic line IRBLb-B (Table 1). The percentage of avirulent isolates to *Pib* was 60.9%, whereas the remaining 143 isolates were virulent to the *Pib* gene (Table 1). The percentage of the avirulent isolate was 100, 75.9, 75.0, 57.6, 53.6, and 48.2% in northwestern, central, northeastern, southeastern, southwestern and western Yunnan Province, respectively. Among 366 isolates, 44.3% of isolates with the *AVR-Pib* allele were amplified by *AVR-Pib*-specific primers (AVR-Pib F1/AVR-Pib R1) (Table 1; Figure S1), and three genotypes (L1 with 1231 bp, L2 with 3100 bp and L3 with both 1231 bp and 3100 bp) of *AVR-Pib* alleles in 162 isolates were amplified (Table 1; Figure S1). The highest percentage of amplification of *AVR-Pib* was 73.3% in the rice blast isolates collected from northwestern Yunnan Province, whereas the lowest percentage was 36.1% from northeastern Yunnan Province (Table 1). The percentage of *AVR-Pib* was 46.3, 36.1, 73.3, 51.5, 67.9 and 39.0% in central, northeastern, northwestern, southeastern, southwestern and western Yunnan Province, respectively. The percentage of *AVR-Pib* was 47.0 and 42.4% in *Xian/Indica* (*XI*) rice- and *Geng/Japonica* (*GJ*) rice production areas in Yunnan. The genotype of L1, L2 and L3 alleles of *AVR-Pib* was detected in 104, 53 and 5 isolates, with percentages of 28.4%, 14.5% and 1.3%, respectively (Table 1). The genotype of L1, L2 and L3 alleles of *AVR-Pib* was detected in the *XI* rice production area, whereas L3 was absent in the *GJ* rice production area (Table 1).

Table 1. Frequency of *AVR-Pib* genotypes and avirulent isolates of *Magnaporthe oryzae* collected from Yunnan, China to IRBLb-B.

Locations	No. of Isolates	PCR Detection [a]				Pathogenicity Assay [b]
		Genotype and No. of Isolates and Frequency (%)				No. of Avirulence Isolates and Frequency (%)
		L1	L2	L3	Total Isolates and Frequency (%)	
Central	54	22 (40.7)	3 (5.6))	0	25 (46.3) ^{B}	41 (75.9) ^{AB}
Northeastern	72	23 (31.9)	3 (4.2)	0	26 (36.1) ^{B}	54 (75.0) ^{B}
Northwestern	15	11 (73.3)	0	0	11 (73.3) ^{A}	15 (100) ^{A}
Southeastern	33	10 (30.3)	6 (18.2)	1 (3.0)	17 (51.5) ^{B}	19 (57.6) ^{C}
Southwestern	28	9 (32.1)	8 (28.6)	2 (7.1)	19 (67.9) ^{A}	15 (53.6) ^{C}
Western	164	29 (17.7)	33 (20.1)	2 (1.2)	64 (39.0) ^{B}	79 (48.2) ^{C}
Total	366	104 (28.4)	53 (14.5)	5 (1.4)	162 (44.3)	223 (60.9)
XI	149	31 (20.8)	34 (22.8)	5 (3.4)	70 (47.0) *	69 (46.3) **
GJ	217	73 (33.6)	19 (8.8)	0	92 (42.4) *	154 (71.0) **
Total	366	104 (28.4)	53 (14.5)	5 (1.4)	162 (44.3)	223 (60.9)

[a] L1 indicates the *AVR-Pib* genotype with the expected size (1231 bp), L2 and L3 indicates the *AVR-Pib* genotype with TE insertion (L2 with 3100 bp and L3 with both 1231 bp and 3100 bp). The frequencies in bracket. The superscript of A and B indicates the significant difference at 0.01 level, and * indicates non significant. [b] Indicates pathogenicity assay on monogenic line IRBLb-B containing *Pib*. *XI* and *GJ* indicates *Xian/Indica* and *Geng/Japonic*, respectively. The frequencies in bracket. The superscript of A, B and C indicates the significant difference at 0.01 level; ** indicates significant difference at 0.01 level.

2.2. Virulence Function of AVR-Pib Variations against the Pib Gene

Twelve *AVR-Pib* haplotypes (H01 to H12) (Table 2), excluding the original *AVR-Pib* allele (GenBank accession number, KM887844), were detected on the nucleotide sequence assemblies of 90 isolates of L1 alleles containing a 719-bp 5′-region, 225-bp coding DNA sequence (CDS) and 302-bp 3′-region of *AVR-Pib* (Table 2; Figure S2). Moreover, insertion of Pot2 (at position −275) and Pot3 (at position −240) was identified based on the DNA sequence assemblies of six and 30 isolates (Table 2; Figure 1), respectively, and the amplicon size difference between L1 and L2 (Figure S1). The 12 novel *AVR-Pib* haplotypes (H01–H12) were identified compared to previous published alleles [3,9]. Alignment of DNA sequence assemblies of the *AVR-Pib* allele from 90 isolates revealed 18 mutation sites, including six mutant sites in the CDS region which were not in the signal–peptide region (Table 2; Figures S2 and S3). Six mutant sites in the CDS region led to changes in amino acids (Table 3). The CDS sequence assemblies of the *AVR-Pib* allele among the 126 isolates (including L1, L2 and L3) were predicted to produce four AVR-Pib proteins (Table 3). Among them, amino acid variations were predicted to occur at six positions (Table 3). Amino-acid variations at F54L in H05 and H06, E46V, F53S and F54V in H07 were found; these isolates of the corresponding haplotypes were avirulent on the monogenic line IRBLb-B (with *Pib*). Meanwhile, the amino acid variations at F47L, I49T and R50G were found in one isolate with H08, which was virulent on the monogenic line IRBLb-B (with *Pib*) (Table 3) and the amino acid variations at F47, I49 and R50 in H01, H02, H03, H04, H05, H06, H07, H09, H10 and H12; these isolates were avirulent on the monogenic line IRBLb-B (with *Pib*), whereas the amino acid variations at 47L, 49T and 50G in H08, as well as the isolate, were virulent on IRBLB-b (Table 3). This finding suggested that the amino acids F47, I49 and R50 were crucial for the avirulence function of *AVR-Pib*.

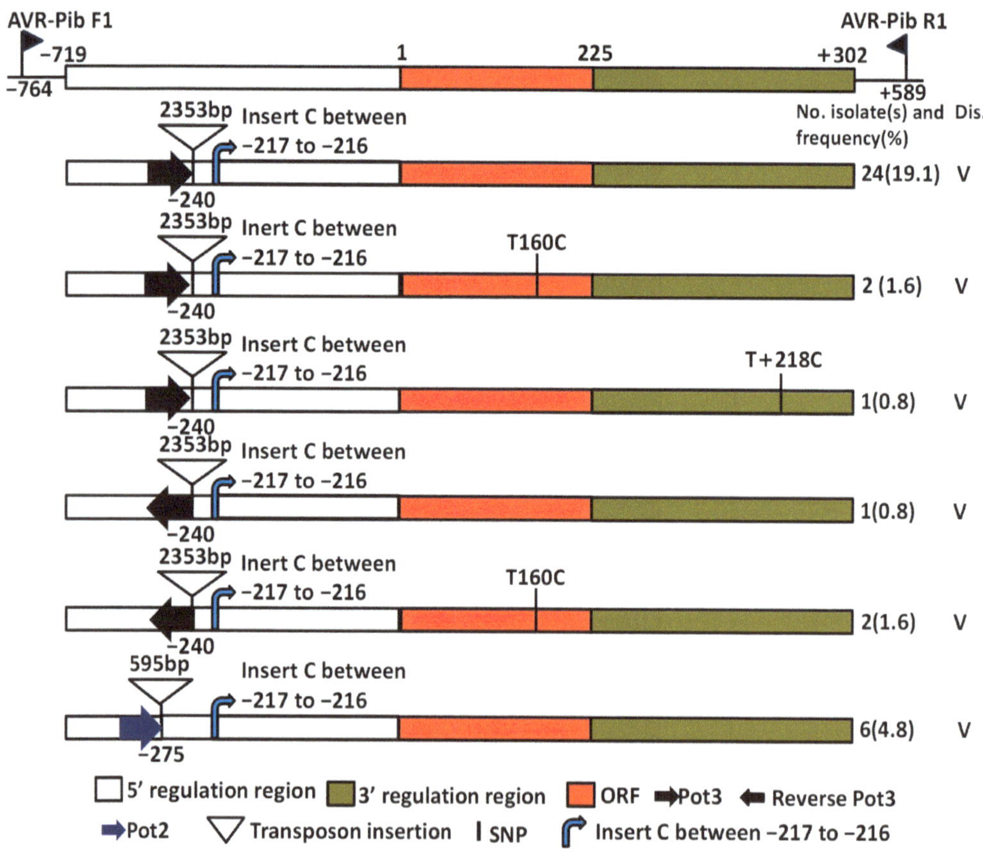

Figure 1. Characterization of allelic variation at *AVR-Pib*. The functional nucleotide polymorphic maps of the six natural alleles in 126 isolates of *M. oryzae* in Yunnan Province. Dis. indicates disease reaction on monogenic line IRBLb-B (containing *Pib*), V indicates the isolates were virulent to IRBLb-B.

Table 2. Haplotypes of *AVR-Pib* loci in *Magnaporthe oryzae* field populations of Yunnan, China.

Haplotype	No. of Isolates	% of Total	Variant Locus [a]																	
			5′ UTR							CDS Regions							3′ UTR			
			−338	Between −325 and −326	Between −239 and −240	Between −216 and −217	−192	−175	Between −210 and −211	−93	137	141	146	148	158	160	+70	+154	+218	+232
KM887844			T	-	-	-	C	C	-	T	A	T	T	C	A	T	C	G	A	C
H01	33	26.2	.	.	.	C	.	.	.	.	.	.	.	.	.	.	.	.	.	.
H02	4	3.2	.	.	.	C	.	.	.	A	.	.	.	.	.	.	.	T	.	.
H03	4	3.2	.	.	.	C	T	.	.	.	.	.	.	.	.	.	.	.	.	.
H04	1	0.8	.	ACTTA	.	C	.	.	.	.	.	.	.	.	.	.	T	.	.	.
H05	8	6.3	C	.	.	C	.	.	.	.	.	.	.	.	.	C	.	.	.	.
H06	1	0.8	C	.	ACGTTA	C	.	.	.	.	.	.	.	.	.	C	.	.	.	.
H07	3	2.4	.	.	.	C	.	T	.	.	T	.	.	.	C	G	.	.	.	A
H08	1	0.8	.	.	.	C	.	.	ACA	.	.	A	C	G	.	.	.	.	.	.
H09	13	10.3	.	ACTTA	.	C	.	.	.	.	.	.	.	.	.	.	.	.	.	.
H10	10	7.9	.	AGTTA	.	C	.	.	.	.	.	.	.	.	.	.	.	.	.	.
H11	2	1.6	.	ATTA	.	C	.	.	.	.	.	.	.	.	.	.	.	.	.	.
H12	10	7.9	.	.	.	C	.	.	ACA	.	.	.	.	.	.	.	.	.	.	.
Pot2	6	4.8		−275 insert Pot2	.	C	.	.	.	.	.	.	.	.	.	.	.	.	.	.
Pot3 rev-A	1	0.8		−240 insert Pot3	.	C	.	.	.	.	.	.	.	.	.	.	.	.	.	.
Pot3 rev-B	2	1.6		−240 insert Pot3	.	C	.	.	.	.	.	.	.	.	.	C	.	.	.	.
Pot3-A	24	19.1		−240 insert Pot3	.	C	.	.	.	.	.	.	.	.	.	.	.	.	.	.
Pot3-B	2	1.6		−240 insert Pot3	.	C	.	.	.	.	.	.	.	.	.	C	.	.	.	.
Pot3-C	1	0.8		−240 insert Pot3	.	C	.	.	.	.	.	.	.	.	.	.	.	.	C	.

[a] Indicates the same with KM887844 (GenBank Accession No.). The KM887844 of *AVR-Pib* was obtained from GenBank. rev: indicates reverse insertion of Pot3 in *AVR-Pib*.

Table 3. Variation of the *AVR-Pib* loci proteins in *M. oryzae* populations of Yunnan, China.

Haplotype	Variant Locus [a]						Disease Reaction [b]
	46	47	49	50	53	54	
KM887844	E	F	I	R	Y	F	
H01	.	.	.	.	.	.	24R + 5M + 4?
H02	.	.	.	.	.	.	3R + 1M
H03	.	.	.	.	.	.	4R
H04	.	.	.	.	.	.	1R
H05	.	.	.	.	.	L	7R + 1?
H06	.	.	.	.	.	L	1R
H07	V	.	.	.	S	V	3R
H08	.	L	T	G	.	.	1S
H09	.	.	.	.	.	.	11R + 2M
H10	.	.	.	.	.	.	9R + 1M
H11	.	.	.	.	.	.	2S
H12	.	.	.	.	.	.	9R + 1M
Pot2	.	.	.	.	.	.	7S
Pot3 rev-A [c]	.	.	.	.	.	.	1S
Pot3 rev-B	.	.	.	.	.	L	2S
Pot3-A	.	.	.	.	.	.	22S + 2M
Pot3-B	.	.	.	.	.	.	2S
Pot3-C	.	.	.	.	.	L	1S

[a] Indicates the same with KM887844. [b] Indicates pathogenicity assay on the monogenic lines IRBLb-B containing the resistant gene of *Pib*. R, M and S indicate disease reaction were resistant, moderate resistant and susceptible, respectively; Ex. 24R indicated 24 isolates were avirulent to IRBLb-B; and ? indicates unknown. [c] rev: indicates reverse insertion of Pot3 in *AVR-Pib*.

The different haplotypes of H01, H02, H03, H04, H09, H10, H11 and H12 had no change on amino acid sequence (Table 3). Three-dimensional protein structures built by homology modeling (SWISS-MODEL; https://swissmodel.expasy.org/, accessed on 18 February 2019) showed the different protein structures of these four (H1, H5, H7 and H8) AVR-Pib variants (Figure S4). Isolates of H01 (amino acids that were the same as that with a GenBank accession number of KM887844), H02, H03, H04, H05, H06, H07, H09, H10 and H12 haplotypes hold *AVR-Pib* because these isolates were avirulent to the *Pib*-containing monogenic line IRBLb-B (Table 3). The isolate of H08 defeated the resistance of *Pib* because this isolate was virulent to the *Pib*-containing monogenic line IRBLb-B (Table 3). Furthermore, Pot2 and Pot3 inserted in the 5′ UTR of the *Pib* gene were identified in six isolates and 30 isolates (Table 3; Figure 1), respectively, and these isolates were virulent to the *Pib*-containing monogenic line IRBLb-B (Table 3). These findings suggested that the insertion of TEs (Pot2 and Pot3) and small segments of the nucleotide in the promoter region, and the nuclear substitution in the ORF region, resulted in a variation of *AVR-Pib* from avirulence to virulence, and that the diverse mutations of the *AVR-Pib* allele of *M. oryzae* were involved.

2.3. Distribution of Haplotypes of AVR-Pib of M. oryzae

Among 12 *AVR-Pib* haplotypes, none were identical to the original *AVR-Pib* (GenBank accession number, KM887844) (Table 2). Eight haplotypes, as well as the Pot2 and Pot3 insertion, were detected in 50 *M. oryzae* isolates from western Yunnan Province (Table 4). Five haplotypes, as well as Pot3 and Pot3 reverse-insertion, were identified in 23 *M. oryzae* isolates from central Yunnan Province. Three haplotypes, as well as Pot2 and Pot3 reverse-insertion, were identified in 15 isolates of *M. oryzae* from northeastern Yunnan Province. Three haplotypes and Pot2 insertion were detected in 19 isolates from northeastern Yunnan Province. Three haplotypes and Pot3 insertion were identified in six *M. oryzae* isolates from southeastern Yunnan Province. Three haplotypes were identified in 13 isolates of *M. oryzae* from northwestern Yunnan Province (Table 4). Eleven and nine haplotypes were detected in *GJ* rice- and *XI* rice production areas, and the diversity index (DI) of haplotypes was

0.84 and 0.79 for these areas, respectively. The DI of *AVR-Pib* was 0.72, 0.71, 0.70, 0.63, 0.59 and 0.54 for southeastern, central, western, southwestern, northeastern and northwestern Yunnan Province, respectively (Table 4).

Table 4. Haplotype distribution of *AVR-Pib* in different Yunnan rice-growing regions.

Haplotype	Regions						Production [c]	
	Central	Northeastern	Northwestern	Southwestern	Southeastern	Western	*XI*	*GJ*
H01	11(47.8) [a]	9(60.0)	2(15.4)	10(52.6)	0	1(2.0)	10(20.8)	23(29.5)
H02	0	0	0	4(21.1)	0	0	4(8.3)	0
H03	4(17.4)	0	0	0	0	0	0	4(5.1)
H04	0	0	0	0	0	1(2.0)	0	1(1.3)
H05	1(4.3)	1(6.7)	3(23.1)	0	2(33.3)	1(2.0)	2(4.2)	6(7.7)
H06	0	0	0	0	1(16.7)	0	1(2.1)	0
H07	0	0	0	0	0	3(6.0)	0	3(3.8)
H08	0	0	0	0	0	1(2.0)	1(2.1)	0
H09	2(8.7)	0	0	0	0	11(22.0)	0	13(16.7)
H10	2(8.7)	0	8(61.5)	0	0	0	0	10(12.8)
H11	0	0	0	0	1(16.7)	1(2.0)	1(2.1)	1(1.3)
H12	0	3(20.0)	0	1(5.3)	0	6(12.0)	6(12.5)	4(5.1)
Pot2	0	1(6.7)	0	4(21.1)	0	1(2.0)	6(12.5)	0
Pot3 rev	2(8.7)	1(6.7)	0	0	0	0	0	3(3.8)
Pot3	1(4.3)	0	0	0	2(33.3)	24(48.0)	17	10(12.8)
Total	23	15	13	19	6	50	48	78
No. of haplotypes	7	5	3	4	4	10	9	11
Index of diversity [b]	0.71	0.59	0.54	0.63	0.72	0.70	0.79	0.84

[a] Number and frequency (in bracket) of isolates of each haplotype. [b] Diversity index was calculated as the frequency of haplotypes types in the *M. oryzae* population following Fontaine's method [28]. Diversity index = $(1 - \sum^{n}_{i=1} p_i^2)$ (where pi is the frequency of the haplotype i in a population). [c] *XI* and *GJ* indicates *Xian*/*Indica* and *Geng*/*Japonic*, respectively.

In brief, the DI of *AVR-Pib* alleles in Yunnan Province was in the order: southeastern > central > western > southwestern > northeastern > northwestern. The DI of *AVR-Pib* alleles in the *GJ*-rice production area was higher than that in the *XI* rice production area. These results indicate that the genetic divergence of *AVR-Pib* of *M. oryzae* in each rice-growing region occurred depending on each field's condition.

Eighteen nucleotide variable sites in *AVR-Pib* alleles were identified (Table 2; Figures S2 and S3). A haplotype network based on sequence variations of 90 isolates of L1 alleles was developed (Figure 2). Four main lineage branches (A to D) of *AVR-Pib* were divided among 90 field isolates (Figure 2), and a different evolution of *AVR-Pib* among them was noted. Isolates of B and D lineage branches of *AVR-Pib* were avirulent to IRBLb-B (with *Pib*) (Figure 2; Table 3). Isolates of H11 of the A-evolved branch and H08 of the C-evolved branch were virulent to the rice-blast-resistant gene *Pib*, respectively (Figure 2; Table 3). These data suggested that the A and C branches of *AVR-Pib* had evolved to virulence from avirulent origins *via* base substitution and insertion, and evaded the recognition of the rice-blast-resistance gene *Pib* in field isolates. The virulence of H08 and H11 was identified in southeastern and western Yunnan Province (Table 4). Moreover, TE insertion in rice samples in all regions except northwestern Yunnan Province (Table 4) suggested that virulent evolution of *AVR-Pib* occurred in most rice production areas of Yunnan Province.

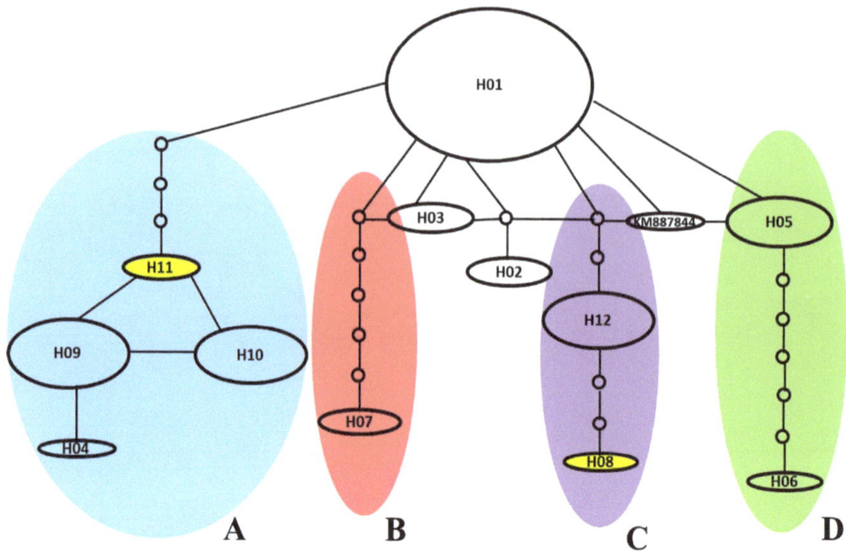

Figure 2. The haplotype network for the 12 *AVR-Pib* alleles. Haplotype network analysis was performed using TCS1.21 (http://darwin.uvigo.es/, accessed on 12 March 2019.). The original *AVR-Pib* allele was designated as the H01 haplotype in the network. Each haplotype was separated by mutational events. The node in the network represents an extinct or a missing haplotype not found among the samples. Each haplotype was separated by mutational events. All haplotypes were displayed as circles. The size of the circles corresponds to the haplotype frequency. The KM887844 (GenBank Accession No.) of *AVR-Pib* was obtained from GenBank. White color indicates avirulent to the *Pib* gene and yellow color indicates virulent to the *Pib* gene. A to D, four major haplotypes of *AVR-Pib* in Yunnan Province of China, are shaded.

2.4. Selection Pressure on AVR-Pib in M. oryzae

The natural-selection pressure on *AVR-Pib* was calculated by Tajima's neutrality test on 126 *AVR-Pib* CDS sequences: the Tajima's *D* value was not significantly different from zero ($D = -1.61687$; $0.10 > P > 0.05$) (Table S1). This result suggested that *AVR-Pib* may suffer neutral selection and evolve neutrally in the population of *M. oryzae*. Furthermore, the results of three positive-selection models kept a higher similarity (Figure S5). The "sliding window" under M8, M8a and M7 models showed values of Ka/Ks (Ka, rate of nonsynonymous substitutions; Ks, rate of synonymous substitutions) across all 74 amino acids (Figure S5). The Ka/Ks value of all sites was >1 under the M8 and M8a model, and the value was 1 under the M7 model for entire residues. These results implied that the sites may have suffered from neutral selection. These findings suggest that the *AVR-Pib* maybe under a neutral evolution.

2.5. Adaption of TE Insertion in AVR-Pib

We wished to confirm the host (rice and non-rice) selection pressure on TE insertion in *AVR-Pib*. A total of 27 isolates from *O. rufipogon* (with *Pib* homologs) [16], *Digitaria sanguinalis*, *Eleusine indica*, *E. coracana* and *Musa nana* Lour, which were stored in our lab, and 5 isolates of the genome sequence from *Lolium perenne* Linn (2 isolates), *Setaria viridis* (Linn.) Beauv. (1 isolate) and *Triticum aestivum* Linn (2 isolates) from Genbank were selected and analyzed (Table S2). The *AVR-Pib* allele was not detected in the isolate from *D. sanguinalis*, *M. nana* Lour, *Lolium perenne* Linn and *Setaria viridis* (Linn.) Beauv. (Table S2). Only the L1 genotypes (with the expected size) of *AVR-Pib* were detected in isolates from *E. indica*, *E. coracana* and *Triticum aestivum* Linn, suggesting that these isolates did not have a TE insertion in the *AVR-Pib* allele. Three genotypes of L1, L2 (with TE insertion) and L3 (with TE insertion) of *AVR-Pib* were detected in 18 isolates from the *Pib* homolog-containing *O. rufipogon*, and the isolate of YN441 (with the H9 haplotype of *AVR-Pib* and identical with the original haplotype of KM887844) was virulent to *O. rufipogon* (Figure 3). These findings showed that the diversification of *AVR-Pib* of *M. oryzae* was dependent upon the *Pib* homolog-containing *O. rufipogon*, and that the variation in TE insertion in *AVR-Pib* could be selected and adapted to rice and other Gramineae species.

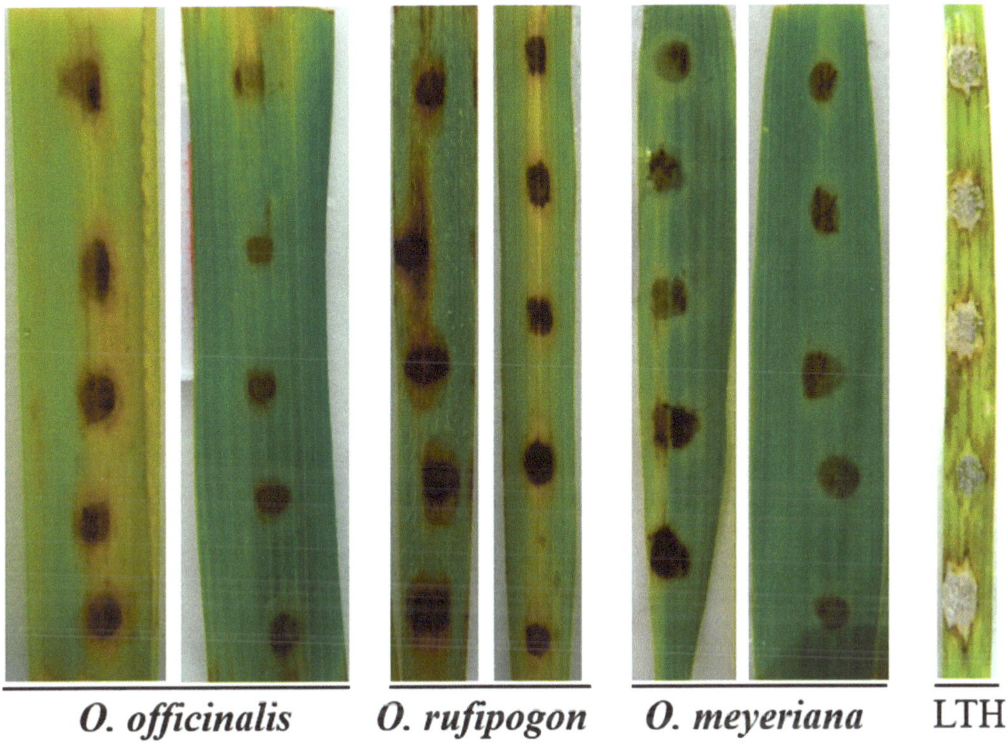

Figure 3. Disease reaction of the identification isolate of YN441 (with the H9 haplotype of *AVR-Pib* which was identical with the original haplotype of KM887844) on *Oryza officinalis*, *O. rufipogon* and *O. meyeriana*. LTH: LijiangxinTuanHeigu.

2.6. Phylogeny of Pib Allele Partial to CDS Regions

Fifty-seven sequences of *Pib* were obtained from GenBank (Table S3). Eleven of them were from five wild-rice species (seven from *O. rufipogon*, one from *O. meyeriana*, one from *O. officinalis*, one from *O. longistaminata* and one from *O. nivara*), and 46 accessions from *O. sativa*, including the original *Pib* (GenBank accession number, AB013448.1) (Table S3). These sequences were aligned. A minimum-evolution phylogenetic tree was constructed based on the nucleotide sequences of exon 1 of *Pib* in 34 accessions and partial regions of exon-3 nucleotide sequences (from 7633 to 8484 of AB013448.1) of *Pib* in 49 accessions, respectively (Figure 4). Exon 1 of *Pib* in wild-rice species (*O. rufipogon*, *O. meyeriana*, *O. officinalis* and *O. longistaminata*) was close to that in *O. sativa* (Figure 4B). The DQ317978.1 group of the wild rice *O. rufipogon* shared >90% identity with the nucleotide sequences of the JN564624.1 group of *Indica*. Two major clades emerged in one part of exon 3 of *Pib* (Figure 4C). One clade contained two wild-rice species (*O. rufipogon* and *O. nivara*) and *O. sativa*. The EF642422.1 group of the wild-rice species *O. nivara* shared >90% identity in nucleotide sequences with the EF642423.1 group of *O. sativa*. The EF642442.1 group of

O. rufipogon shared >75% identity of nucleotide sequences with the EF642423.1 group of *O. sativa*. The other clade contained *O. rufipogon* and *O. sativa*. The EF642440.1 group of the wild-rice species *O. rufipogon* shared >75% identity of nucleotide sequences with the EF642433.1 group of *O. sativa*. The isolate of YN441 (with the H9 haplotype of *AVR-Pib*) was virulent to *O. rufipogon, O. meyeriana*, and *O. officinalis* (Figure 3). These results suggested that different regions of the *Pib* gene may have suffered different selection pressures in the host rather than domestication.

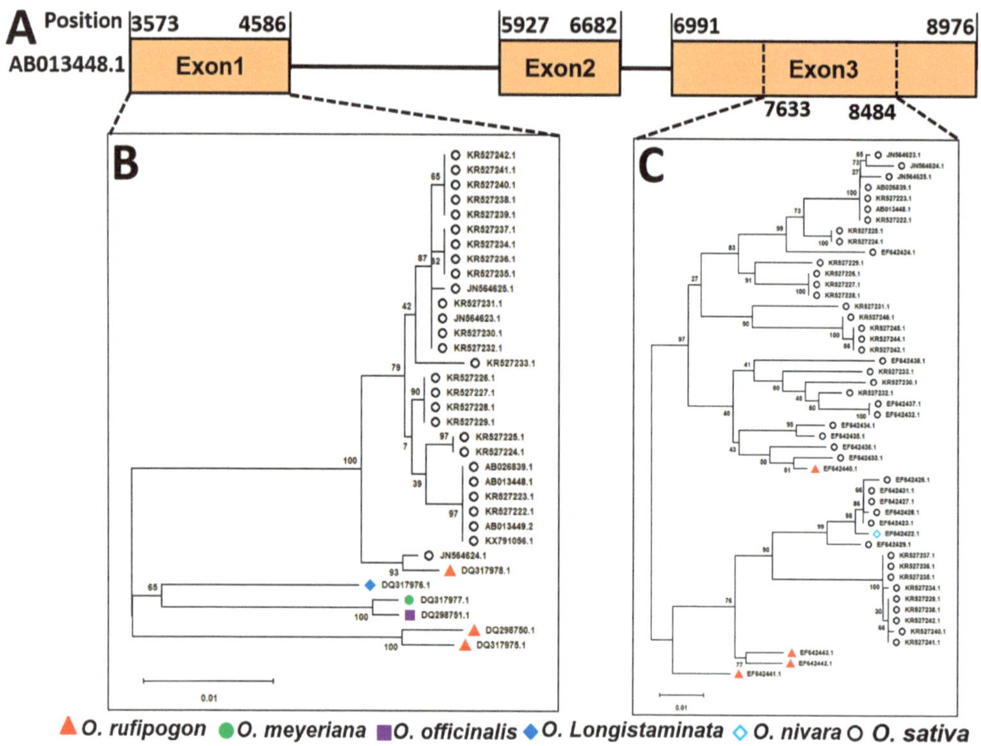

Figure 4. Phylogenetic tree constructed with the nucleotide sequences of *Pib* gene and different partial CDS regions from wild rice and *O. sativa* using minimum evolution method of MEGA X. The numbers associated with individual branches indicate confidence levels based on 1000 bootstrap replicates. (**A**), structure of *Pib* from AB013448.1 (GenBank ID); (**B**), the phylogenetic tree constructed based on the nucleotide sequences of exon1 of *Pib* regions from 34 accessions. (**C**), the phylogenetic tree constructed based on the nucleotide sequences of partial exon3 (from 7633 to 8484 of AB013448.1) of *Pib* from 49 accessions. All accessions of *Pib* were obtained from GenBank.

3. Discussion

We identified 12 new haplotypes, as well as Pot2 and Pot3 insertion in the *AVR-Pib* DNA sequences among rice blast isolates from different rice-growing areas in Yunnan Province. The many virulent isolates to *Pib*-containing rice varieties implied that *Pib* was overcome in these rice-growing regions because of the massive exploitation of *Pib* in China.

Pib alleles have been used widely and have shown strong resistance to disease in China [14]. Complete deletions have occurred in *AVR-Pib* sequences among field isolates of *M. oryzae* from various rice-growing regions of Guangdong, Hunan and Liaoning Provinces [18]. Moreover, TE insertion has occurred in *AVR-Pib* in *M. oryzae* isolates from south and northeast China [18] and the Philippines [17]. These data are consistent with our results. The L1 genotype of *AVR-Pib* identified in rice blast isolates collected from rice fields implied that *Pib* has been effective in preventing rice blast. Li and colleagues showed that rice cultivars with *Pib* were resistant to 74.9% of isolates (282 isolates) from Yunnan Province [8]. The corresponding value was 2.1% in Guangdong Province (146 isolates were tested) [28], and the percentage resistance was <31% in Hunan Province [29]. These results show that *Pib* alleles had poor effects in these rice production areas. The further inspection of variation in *AVR-Pib* DNA sequences in these isolates could reveal the molecular evolutionary patterns of *AVR-Pib* and predict the durability and effectiveness of *Pib* allele-mediated resistance under field conditions in rice production regions.

AVR-Pia, *AVR-Pii* and *AVR-Pita1* located on telomere regions tend to be unstable, and effective mutants in these genes were identified [21,30,31]. The retrotransposon (MINE) insertion in the *ACE1* gene [23] and Pot3 insertion in *AVR-Pita1* [32,33] and *AVR-Pizt* [22] caused new virulent alleles. TE (Pot2 and Pot3) insertion, complete absence, segmental deletion and a point mutation were found in *AVR-Pib* alleles, all of which lead to a gain of virulence [18]. Three expression patterns were identified among different haplotypes of *AVR-Pib* [18]. Recently, insertion of a Pot3 transposon in *AVR-Pib* was shown to mediate the loss of function of *AVR-Pib* in all 248 isolates collected from the Philippines [17]. These findings showed that rice blasts can use transposons to suppress the expression of *AVR* genes to defeat the rice blast resistance gene. The *AVR-Pib* allele was identified in nearly half of rice blast isolates (44.3%) in Yunnan Province (Table 1). This percentage was higher than that in rice blast isolates in Jilin and Heilongjiang Provinces, but lower than those in Guangdong, Hainan and Liaoning Provinces [18]. Meanwhile, 28.6% of isolates contained a TE insertion in *AVR-Pib*. Among them, 21.4% isolates contained a Pot3-element insertion and 2.4% of isolates contained a reverse Pot3-element insertion, and 4.8% isolates contained a Pot2-element insertion in *AVR-Pib* of *M. oryze* from Yunnan Province (Table 4; Figure 1). These insertions resulted in the variation from avirulence to virulence to the corresponding *R* gene. Several nucleotide variations in *AVR-Pib* alleles were identified, which led to variations in amino acids and implied that *AVR-Pib* alleles suffer from strong selection pressure in rice production regions of Yunnan Province. Whereas, mutations that impact gene expression are further acknowledged by transcript level.

We observed no TE insertions in *AVR-Pib* of isolates from *E. indica* or *E. coracana*, and TE insertion in *AVR-Pib* was selected by the host, data that are consistent with the results of Zhang et al. [18]. Pot2 and Pot3 insertions were identified in the *XI*-rice-growing areas, whereas only Pot3 insertion was identified in *GJ*-rice-growing areas of Yunnan Province. TE insertion of *AVR-Pib* was noted in all rice-growing regions except northwestern Yunnan Province, and Pot3 insertion was distributed mainly in western Yunnan Province. These results showed that the virulent *AVR-Pib* alleles were involved in most of the rice-growing regions of Yunnan Province. Hence, the monitoring of these virulent alleles in field populations is important for employing *Pib*-containing rice varieties.

Various mutations were identified in CDS regions of *AVR-Pib*, and 12 *AVR-Pib* haplotypes were found based on 18 variant nucleotides among 90 isolates of L1 alleles collected from Yunnan Province (Table 2). Six new variant amino acids of the *AVR-Pib* loci variants were identified in the 90 *M. oryze* isolates in the present study, and resulted in the identification of four novel haplotypes. A more holonomic network was constructed based on the new variations among different alleles of *AVR-Pib*. The putative and secreted proteins of AVR-Pib in 126 isolates were identified (Table 3), and they were in accordance with the results of Zhang et al. [18]. Nine isolates had variations at the amino acid position F54L; three isolates had variations at the amino acid positions of E46V, Y53S and F54V; one isolate had variations at the amino acid positions of F47L, I49T and R50G (Table 3). These isolates were virulent to the monogenic line IRBLb-B (with *Pib*), suggesting that these amino acids are crucial for avirulent function. Isolates of H11 with insertion of ATTA in the 5′ UTR may change *AVR-Pib* expression and cause a loss of the avirulent function (Tables 2 and 3; Figure S3).

In the course of interactions and co-evolution between pathogens and plants, the *R* genes of plants can discern the cognate *AVR* genes of pathogens and inspire immunity [1].

The genetic variation of the *AVR* genes of the pathogen is dependent upon the *R* genes of the host and changeable environmental conditions. The DI of *AVR-Pib* was higher in *GJ* rice production areas than that of *XI* rice production areas (Table 4). Different variations were observed in *AVR-Pib* between *XI* rice- and *GJ* rice production areas (Table 4). These findings imply that the adaptive mutations of *AVR-Pib* occurred in Yunnan Province under natural conditions, and these results were similar in previous studies [18].

Yunnan Province is abundant in genetic resources of rice. The wild species of *O. officinalis*, *O. meyeriana* and *O. rufipogon* also coexist in this province [34]. More than 5000 rice accessions germplasms have been conserved in Yunnan Province. Among them, 12 out of 227 accessions carried the *Pib* resistance gene screened by the resistance gene identification using different isolates [34]. *Pib* gene homologs were identified in wild rice *O. rufipogon* from Yuanjiang County [16], and four genotypes (L0 to L3) of *AVR-Pib* were detected in *M. oryzae* and *O. rufipogon* in Yuanjiang County. TE insertion of L2 and L3 genotypes of *AVR-Pib* was absent in the isolates from *D. sanguinalis* and *M. nana* Lour. These results suggest that the adaptive variation of *AVR-Pib* is involved during interactions and co-evolution between *AVR-Pib* of *M. oryzae* and *Pib* of *O. rufipogon*. The Tajima's *D* value of −1.61687 (Table S1) indicates that *AVR-Pib* loci may suffer from a purifying selection by the corresponding *R* gene in Yunnan rice production areas.

Massive variations and stepwise mutations in *AVR-Pib* of rice blast isolates were observed in Yunnan Province (Table 2; Figure 2), which suggests that there is an abundant diversity of rice accessions and *M. oryzae* isolates in Yunnan Province. Pot2 insertion in *AVR-Pib* was found in western, southwestern and northeastern Yunnan Province. Pot3-reversed insertion was found in central and northeastern Yunnan Province, which was not observed in the previous studies [18]. Moreover, Pot3 insertion occurred mainly in western Yunnan Province. Pot3 insertion of *AVR-Pib* was found in *GJ* rice and *XI* rice production areas, whereas Pot2- and Pot3-reversed insertion was found only in the *XI* rice production area and *GJ* rice production area, respectively (Table 4). The virulent haplotype of H11 was detected in the *XI* rice production area and *GJ* rice production area, and H08 was detected in the *XI* rice production area. These data showed a high variation of *AVR-Pib* in different rice-growing regions, which may be due to rice variety and the environment.

The stepwise mutations that result in a loss of avirulence function have been identified in *AVRL567* [35] and *AVR-Pik* [36–39]. Based on the *Pib* homologs identified by Yang et al. [16], and our result for *AVR-Pib* in the present study, the potential interactions and co-evolution of *AVR-Pib* alleles in *M. oryzae* and *Pib* alleles of rice were constructed (Figure S6). The *AVR-Pib* homolog L1 (H01) originated from an ancestral *M. oryzae* gene. The *Pib* allele (87-bp deletion in exon 1 of *Pib*) in *O. rufipogon* could not recognize the L1 alleles of *AVR-Pib*. Thus, the other *Pib* allele (gained 87 bp in exon 1 of *Pib*) in cultivated rice evolved to recognize the L1 alleles (H01) of *AVR-Pib*, whereas the altered alleles L2 and L3 evolved to virulence from avirulent origins by TE insertion, base substitution (H08) and segment insertion (H11) to avoid recognition by *Pib* (Table 2; Figure S6). These actions indicated a stepwise evolution of *AVR-Pib* as well as *Pib* interaction and co-evolution. Intriguingly, the *AVR-Pib* alleles H08 and H11 were derived from H01, and could escape recognition by *Pib* (Table 2; Figure S6), but several extinct or missing haplotypes were not identified in the sample (Figure 2). These findings imply that: (i) the *AVR-Pib* loci of *M. oryzae* evolved gradually during the interaction and coevolution between the *Pib* loci of *M. oryzae* in field conditions; (ii) the genome organization of the *AVR-Pib* locus is much more intricate than anticipated.

4. Materials and Methods

4.1. Blast Isolates, Rice Accessions, Culture and Pathogenicity Identification

The rice blast fungus single spores were obtained from infected leaves or panicles incubated on moist filter paper in a Petri dish at room temperature for 24 h according to Jia et al. [38], and the tests made fulfilled the Kock's postulates for all isolates. The seedlings of the rice monogenic line IRBLb-B (which contains *Pib*) and a susceptible cultivar Lijiangxintuanheigu (LTH; does not contain *Pib*) were used for pathogenicity assays (the seeds were acquired originally from Cailin Lei). Four thousand isolates were collected from six rice-growing regions from 1997 to 2012 in Yunnan Province, and a total of 366 isolates are selected from six rice-growing regions as representative isolates. The total of 366 isolates of *M. oryzae* in the present study were the same in the published paper [38]. The storing and culturing methods of the isolates were described in the published paper [38]. Disease reactions were referred to as the method of Jia et al. [39]. In a few words, when rice seedlings were in the 3- to 4-leaf stage, they were inoculated with a spore suspension (1–5 × 10^5 spores/mL containing 0.05% Tween 20). After innoculation, rice seedlings were put into a plastic bag and sealed securely to keep a high relative humidity of 90–100% at 25 °C for 24 h in the dark. Then, the plants were shifted into a greenhouse for 6 days to develop the disease lesion extension fully.

Disease reactions were monitored externally on the second youngest leaf based on the number and degree of lesions using a 0-to-5 disease scale (Figure S7). The disease scale method was described in the published paper [38]. Five seedlings at a time, repeating twice, were arranged in the experiment. In addition, the average value of disease scales was used to discriminate resistance versus susceptibility. The disease reaction of wild rice species *O. rufipogon, O. meyeriana* and *O. officinalis* (which were conserved in Yunnan Academy of Agricultural Sciences) was determined using a detached-leaf method, as described by Jia et al. [39]. One virulent blast isolate was used for inoculation. Disease reactions were evaluated 5–7 days after inoculation.

4.2. DNA Extraction, PCR Amplification and DNA Sequencing

A culturing method of vegetative mycelia of *M. oryzae* isolates was described in the published paper [38]. The genomic DNA of each isolate was extracted from vegetative mycelia by CTAB method [40]. The primers AvrPibF1 (5′-GGACAAGGGAGGCAAATCT AAC-3′) and AvrPibR1 (5′-ATGCCGACAATGCGAGGTAT-3′) were used to amplify the *AVR-Pib* allele, as well as for sequencing according to the method of Zhang et al. [18]. Each PCR reaction was amplified in a total reaction volume of 50 µL containing the following components: 25 µL of 2× Taq PCR MasterMix (Tiangen Biotech, Beijing, China), 1 µL (10 µM) of each primer, 2 µL of template DNA, and 21 µL of ddH_2O (provided in the Tiangen kit). PCR procedure was conducted in a C1000 Touch™ thermal cycler (Bio-Rad Laboratories, Hercules, CA, USA) in the following steps: initial denaturation at 94 °C for 3 min, followed by 29 cycles at 94 °C for 45 s, 55 °C for 45 s and 72 °C for 2.5 min, and a final extension at 72 °C for 5 min. Each reaction was repeated twice. The size of the PCR products was valued by a DNA marker (DL2000, Tiangen Biotech). Amplicons were sequenced twice by Life Technologies Biotechnology (Shanghai, China).

4.3. Data Analyses

DNA sequences of *AVR-Pib* were assembled and aligned using DNASTAR v7.1.0 (www.dnastar.com/, 15 January 2019). DnaSP v5.10.01 [41] was used for the calculation of polymorphic sites (π), the number of DNA haplotypes and the sliding window. TCS1.21 (http://darwin.uvigo.es/, accessed on 1 February 2019) [42] was used for analyses of the haplotype network of *AVR-Pib*. The DI (haplotype diversity index) was counted in *M. oryzae* populations following the method of Fontaine et al. [43]:

$$DI = (1 - \sum^{n}{}_{i=1}{p_i}^2)$$

where p_i is the frequency of haplotype i in a population. Tajima's neutrality test was conducted using MEGA X (www.megasoftware.net, accessed on 12 February 2019) [28]. The Selection Server program (http://selecton.tau.ac.il, accessed on 19 February 2019) was used for analyses of purifying the selection. The purifying of the selected sites of AVR-Pib was identified by used three models: M8 (positive selection enabled, beta + w $\geq$ 1), M7 (beta, null model) and M8a (beta + w = 1, null model). Then, the sliding window of purifying the selected sites of AVR-Pib was drawn under the M8, M7 and M8a models by Excel™ (Microsoft, Redmond, WA, USA). MEGA X [28] was used for the construction of phylogenetic trees by the minimum evolution method [44]. The SWISS-MODEL (http://swissmodel.expasy.org, accessed on 18 February 2019) with ProMod v3.7.0 was used to build the protein homology model. The significant difference of distribution of *AVR-Pib* alleles and avirulence isolates of *M. oryzae* in each region was analyzed by Excel software with CHITEST.

5. Conclusions

We detected twelve novel haplotypes in the field population by using 90 isolates and a transposable element (TE) insertion in 36 of 126 isolates, constructing a complex network of *AVR-Pib* alleles and assessing the efficacy of *Pib* alleles in rice production areas of Yunnan; we also analyzed the adaption of TE insertion of *AVR-Pib* in the isolates from a different host. Our findings support the hypothesis that functional *AVR-Pib* possesses varied sequence structures and can escape surveillance by hosts via multiple variation manners. Haplotype H08 and H11 can overcome all detected *Pib* alleles to date, and Pot insertion can change the avirulent function of *AVR-Pib*. Despite the H08, H11 haplotypes and TEs insertions have low frequencies; the monitoring of these alleles in field populations is critical because of their high risk for *Pib*-holding rice varieties. The TE insertion was not detected in the *AVR-Pib* allele in the isolates from *E. indica, E. coracana,*and *Triticum aestivum* Linn, while three genotypes of *AVR-Pib* were detected in isolates from *O. rufipogon*. The selected and adapted variation of TE insertion in *AVR-Pib* is the occurrence of the long-term co-evolution between *M. oryzae* and hosts (rice and other Gramineae species).

Supplementary Materials: The following supporting information can be downloaded at: https://www.mdpi.com/article/10.3390/ijms242115542/s1. Table S1. Tajima's Neutrality Test of AVR-Pib in *M. oryzae*. Table S2. Genotype of *AVR-Pib* in isolates from different hosts. Table S3. Lists accessions of *Pib* were obtained from GenBank used in this study. Figure S1. Detected alleles of *AVR-Pib* by PCR assays. The amplicons of *AVR-Pib*F1/R1 were used to distinguish four alleles: L1 and L2 (one fragment of different size), and L3 (two fragments of L1 and L2), L0 (no amplification); The isolates 1 to 8 are YJW-1-1d②, YJW-1-1e②, YJW-2-1a, YJW-2-1b, YJW-2-1c, YJW-2-1d, YJW-2-1e and D-1-1g, respectively. M: DNA marker DL2000.L1 indicates the *AVR-Pib* genotype with the expected size (1231 bp), L2 and L3 indicates the *AVR-Pib* genotype with TE insertion (L2 with 3100 bp, L3 with both of 1231 bp and 3100 bp). Figure S2. Diversification of *AVR-Pib* in avirulent isolates. Distribution of variation of the *AVR-Pib* alleles was analyzed using sliding window. X-axis shows the distribution of variation within the full region, including signal peptide and exon of *AVR-Pib*. Lower pane indicates the corresponding schematic presentation of the signal peptide and exon of *AVR-Pib*. Window length: 1; Step size: 1. π value corresponds with the level of variation at each site because it is the sum of pair-wise differences divided by the number of pairs within the population. Figure S3. Part of insertion of sequences of *AVR-Pib*. A, indicates ACTTA, AGTTA, ATTA insert, respectively; B, indicates ACGTTA insert; C, indicates ACA insert; D, indicates C insert. Figure S4. SWISS-MODEL homology modelling, built with PROMOD v. 3.70 and method: X-ray. Amino acid variations at R50G in H1 (same as KM887844, H2, H3, H4, H9, H10 and H11), H5 (same as H6), H7 and H8 protein haplotypes of AVR-Pib, respectively. A: Ramachandran Plots. B: the 50th amino acid of AVR-Pib. Figure S5. Sliding window of positive-selection sites of the AVR-Pib alleles under M8, M8a and M7 models. The Y-axis indicates the ratio of the rate of nonsynonymous substitution (Ka) to the rate of synonymous substitution (Ks) (Ka/Ks); the X-axis indicates the position of the AVR-Pib amino acids in the site. Figure S6. Possible scenario for *M. oryzae AVR-Pib* alleles-rice *Pib* alleles interactions and co-evolution. *AVR-Pib* homolog L1 genotypes (H01) were derived from an ancestral *M. oryzae* gene. L1 genotypes are not recognized by *Pib* in wild rice. In response to this situation, an 87bp insertion in exon1 of *Pib* in cultivated rice evolved and can recognize L1 genotypes (H01) of *AVR-Pib* alleles. Then, another *AVR-Pib* alleles with TE insertion (L2/L3 genotypes), H08 and H11, were derived that cannot be recognized by *Pib* in cultivated rice. Figure S7. Disease reaction of *Magnapothe oryze* isolate on rice leaves. Number of 0 to 5 on the label on top of the Figure indicates 0-to-5 disease scale. A value of 0 to 1 is classified as "resistant", 2 denotes "moderately resistant", and 3 to 5 is classified as "susceptible".

Author Contributions: J.L. was the leading investigator of this research program. J.L. planned and designed the research; J.L., Q.W. and L.L. performed the majority of experiments with the help of C.L.; L.L., Z.S. and C.L. contributed reagents, materials, and analysis tools; J.L. and Q.W. analyzed the data; J.L. and Q.W. wrote the paper with suggestions from L.L., C.L. and Z.S. All authors have read and agreed to the published version of the manuscript.

Funding: This work was supported by the Key Projects of Yunnan Province (2017FA013 and 202301AS070003), the National Natural Science Foundation of China (31860481), the Key Research and Development Program of Yunnan Province (202102AE090003 and 2019IB007), the program of the Innovative Research Team of Yunnan Province (202005AE160003), the Key Laboratory of Green Prevention and Control of Agricultural Transboundary Pests of Yunnan Province (202305AG340007). The funding organizations had no role in the designing, data collection, analysis, and interpretation of data and in writing the manuscript.

Institutional Review Board Statement: Not applicable.

Informed Consent Statement: Not applicable.

Data Availability Statement: All relevant data are presented within the paper and its Supplementary Files. The nucleotide sequences of novel *AVR-Pib* alleles from these isolates have been deposited in GenBank (accession numbers: OR361637 to OR361654).

Acknowledgments: The authors thank Professor Zaiquan Cheng (Institute of Biotechnology and Genetic Resources, Yunnan Academy of Agricultural Sciences) and Professor Yiqing Guo (Yunnan Agricultural University) for useful discussions.

Conflicts of Interest: The authors declare no conflict of interest.

Abbreviations

AVR: Avirulence gene; DI: Diversity index; GJ: Geng/Japonica; Ka: The rate of nonsynonymous substitution; Ks: The rate of synonymous substitution; LTH: Lijiangxintuanheigu; R: Resistance; M: Moderate resistant; S: Susceptible; TE: Transposable element; XI: Xian/Indica.

References

1. Woolhouse, M.; Webster, J.; Domingo, E.; Charlesworth, B.; Levin, B. Biological and biomedical implications of the co-evolution of pathogens and their hosts. *Nat. Genet.* **2002**, *32*, 569–577. [CrossRef] [PubMed]
2. Meng, X.; Xiao, G.; Telebanco-Yanoria, M.J.; Siazon, P.M.; Padilla, J.; Opulencia, R.; Bigirimana, J.; Habarugira, G.; Wu, J.; Li, M.; et al. The broad-spectrum rice blast resistance (*R*) gene *Pita2* encodes a novel R protein unique from *Pita*. *Rice* **2022**, *13*, 19. [CrossRef] [PubMed]
3. Wang, B.; Daniel, J.; Wang, Z. The arms race between *Magnaporthe oryzae* and rice: Diversity and interaction of Avr and R genes. *J. Integr. Agric.* **2017**, *16*, 2746–2760. [CrossRef]
4. Ma, J.; Zhang, G.; Zhang, L.; Deng, L.; Wang, Y.; Wang, Y. Analysis of blast-resistance of rice germplasm and pathogenicity of rice blast fungus *Magnaporthe oryzae* in Heilongjiang Province. *J. Plant Prot.* **2017**, *44*, 209–216. (In Chinese)
5. Du, Y.; Ruan, H.; Shi, N.; Gan, L.; Yang, X.; Chen, F. Pathogenicity analysis of *Magnaporthe grisea* against major *Pi*-genes and main rice varieties in Fujian Province. *J. Plant Prot.* **2016**, *43*, 442–451. (In Chinese)
6. Zhang, S.; Zhong, X.; Qiao, G.; Shen, L.; Zhou, T.; Peng, Y. Difference in virulence of *Magnaporthe oryzae* from Sichuan, Chongqing and Guizhou. *Southwest China J. Agric. Sci.* **2017**, *30*, 359–365. (In Chinese)
7. Yang, J.; Chen, S.; Zeng, L.; Li, Y.; Chen, Z.; Zhu, X. Evaluation on resistance of major rice blast resistance genes to *Magnaporthe grisea* isolates collected from indica rice in Guangdong Province, China. *Chin. J. Rice Sci.* **2008**, *22*, 190–196. (In Chinese)
8. Li, J.; Li, C.; Chen, Y.; Lei, C.; Ling, Z. Evaluation of twenty-two blast resistance genes in Yunnan using monogenetic rice lines. *Acta Phytophylacica Sin.* **2005**, *32*, 113–119. (In Chinese)
9. Xiao, G.; Yang, J.; Zhu, X.; Wu, J.; Zhou, B. Prevalence of ineffective haplotypes at the rice blast resistance (*R*) gene loci in Chinese elite hybrid rice varieties revealed by sequence-based molecular diagnosis. *Rice* **2020**, *13*, 6. [CrossRef]
10. Miyamoto, M.; Ando, I.; Rybka, K.; Kodama, O.; Kawasaki, S. High resolution mapping of the indica-derived rice blast resistance genes. I. *Pi-b*. *Mol. Plant-Microbe Interact.* **1996**, *9*, 6–13. [CrossRef]
11. Monna, L.; Miyao, A.; Zhong, H.S.; Yano, M.; Iwamoto, M.; Umehara, Y.; Umehara, Y.; Kurata, N.; Hayasaka, H.; Sasaki, T. Saturation mapping with subclones of YACs: DNA marker production targeting the rice blast disease resistance gene, *Pi-b*. *Theor. Appl. Genet.* **1997**, *94*, 170–176. [CrossRef]

12. Wang, Z.; Yano, M.; Yamanouchi, U.; Iwamoto, M.; Monna, L.; Hayasaka, H.; Katayose, Y.; Sasaki, T. The *Pib* gene for rice blast resistance belongs to the nucleotide binding and leucine-rich repeat class of plant disease resistance genes. *Plant J.* **1999**, *19*, 55–64. [CrossRef] [PubMed]
13. Qi, C.; Xu, X.; Ma, J.; Wang, S.; Tian, P.; Meng, L. Distribution of blast resistance genes *Pib*, *Pita*, *Pi5*, *Pi25* and *Pi54* in mini-core collection of Chinese rice germplasm. *J. Plant Genet. Resour.* **2019**, *20*, 1240–1246, 1254. (In Chinese)
14. Shi, K.; Lei, C.; Cheng, Z.; Xu, X.; Wang, J.; Wan, J. Distribution of two blast resistance genes *Pita* and *Pib* in major rice cultivars in China. *J. Plant Genet. Resour.* **2009**, *10*, 21–26. (In Chinese)
15. Li, J.; Wang, T.; Xu, M. Identification of *Pi-ta* and *Pi-b* genes for rice blast resistance of rice landraces from Yunnan Province. *Chin. J. Rice Sci.* **2012**, *26*, 593–599. (In Chinese) [CrossRef]
16. Yang, M.; Cheng, Z.; Cheng, S.; Qian, J.; Yin, M.; Wu, C. Cloning and analysis of *Pi-ta* and *Pib* gene homologues from Yuannan Yuanjiang type of common wild rice (*Oryza rufipogon* Griff). *Plant Physiol. Commun.* **2007**, *43*, 483–486. (In Chinese)
17. Olukayode, T.; Quime, B.; Shen, Y.C.; Yanoria, M.J.; Zhang, S.; Yang, J.; Zhu, X.; Shen, W.C.; von Tiedemann, A.; Zhou, B. Dynamic insertion of Pot3 in *AvrPib* prevailing in a field rice blast population in the Philippines led to the high virulence frequency against the resistance gene *Pib* in rice. *Phytopathology* **2019**, *109*, 870–877. [CrossRef]
18. Zhang, S.; Wang, L.; Wu, W.; He, L.; Yang, X.; Pan, Q. Function and evolution of *Magnaporthe oryzae* avirulence gene *AvrPib* responding to the rice blast resistance gene *Pib*. *Sci. Rep.* **2015**, *5*, 11642. [CrossRef]
19. Ray, S.; Singh, P.K.; Gupta, D.K.; Mahato, A.K.; Sarkar, C.; Rathour, R.; Singh, N.K.; Sharma, T.R. Analysis of *Magnaporthe oryzae* genome reveals a fungal effector, which is able to induce resistance response in transgenic rice line containing resistance gene, *Pi54*. *Front. Plant Sci.* **2016**, *7*, 1140. [CrossRef]
20. Wu, J.; Kou, Y.; Bao, J.; Li, Y.; Tang, M.; Zhu, X.; Ponaya, A.; Xiao, G.; Li, J.; Li, C.; et al. Comparative genomics identifies the *Magnaporthe oryzae* avirulence effector *AvrPi9* that triggers *Pi9*-mediated blast resistance in rice. *New Phytol.* **2015**, *206*, 1463–1475. [CrossRef]
21. Yoshida, K.; Saitoh, H.; Fujisawa, S.; Kanzaki, H.; Matsumura, H.; Yoshida, K.; Tosa, Y.; Chuma, I.; Takano, Y.; Win, J.; et al. Association genetics reveals three novel avirulence genes from the rice blast fungal pathogen *Magnaporthe oryzae*. *Plant Cell.* **2009**, *21*, 1573–1591. [CrossRef] [PubMed]
22. Li, W.; Wang, B.; Wu, J.; Lu, G.; Hu, Y.; Zhang, X.; Zhang, Z.; Zhao, Q.; Feng, Q.; Zhang, H.; et al. The *Magnaporthe oryzae* avirulence gene *AVR-Pizt* encodes a predicted secreted protein that triggers the immunity in rice mediated by the blast resistance gene *Piz-t*. *Mol. Plant-Microbe Interact.* **2009**, *22*, 411–420. [CrossRef] [PubMed]
23. Fudal, I.; Bohnert, H.U.; Tharreau, D.; Lebrun, M.H. Transposition of MINE, a composite retrotransposon, in the avirulence gene *ACE1* of the rice blast fungus *Magnaporthe grisea*. *Fungal Genet. Biol.* **2005**, *42*, 761–772. [CrossRef]
24. Farman, M.L.; Leong, S.A. Chromosome walking to the *AVR1-CO39* avirulence gene of *Magnaporthe grisea*: Discrepancy between the physical and genetic maps. *Genetics* **1998**, *150*, 1049–1058. [CrossRef]
25. Orbach, M.J.; Farrall, L.; Sweigard, J.A.; Chumley, F.G.; Valent, B. A telomeric avirulence gene determines efficacy for the rice blast resistance gene *Pi-ta*. *Plant Cell* **2000**, *12*, 2019–2032. [CrossRef] [PubMed]
26. Kang, S.; Sweigard, J.A.; Valent, B. The *PWL* host specificity gene family in the blast fungus *Magnaporthe grisea*. *Mol. Plant-Microbe Interact.* **1995**, *8*, 939–948. [CrossRef]
27. Sweigard, J.A.; Carroll, A.M.; Kang, S.; Farrall, L.; Chumley, F.G.; Valent, B. Identification, cloning, and characterization of *PWL2*, a gene for host species specificity in the rice blast fungus. *Plant Cell* **1995**, *7*, 1221–1233.
28. Kumar, S.; Stecher, G.; Li, M.; Knyaz, C.; Tamura, K. MEGA X: Molecular Evolutionary Genetics Analysis across computing platforms. *Mol. Biol. Evol.* **2018**, *35*, 1547–1549. [CrossRef]
29. Xing, J.; Jia, Y.; Peng, Z.; Shi, Y.; He, Q.; Shu, F.; Zhang, W.; Zhang, Z.; Deng, H. Characterization of molecular identity and pathogenicity of rice blast fungus in Hunan Province of China. *Plant Dis.* **2017**, *101*, 557–561. [CrossRef]
30. Chuma, I.; Isobe, C.; Hotta, Y.; Ibaragi, L.; Futamata, N.; Kusaba, M.; Yoshida, K.; Terauchi, R.; Fujita, Y.; Nakayashiki, H.; et al. Multiple translocation of the *AVR-Pita* effector gene among chromosomes of the rice blast fungus *Magnaporthe oryzae* and related species. *PLoS Pathog.* **2011**, *7*, e1002147. [CrossRef]
31. Dai, Y.; Jia, Y.; Correll, J.; Wang, X.; Wang, Y. Diversification evolution of the avirulence gene *AVR-Pita1* in field isolates of *Magnaporthe oryzae*. *Fungal Genet. Biol.* **2010**, *47*, 973–980. [CrossRef]
32. Kang, S.; Lebrun, M.H.; Farrall, L.; Valent, B. Gain of virulence caused by insertion of a Pot3 transposon in a *Magnaporthe grisea* avirulence gene. *Mol. Plant-Microbe Interact.* **2001**, *14*, 671–674. [CrossRef]
33. Zhou, E.; Jia, Y.; Singh, P.; Correll, J.; Lee, F.N. Instability of the *Magnaporthe oryzae* avirulence gene *AVR-Pita* alters virulence. *Fungal Genet. Biol.* **2007**, *44*, 1024–1034. [CrossRef]
34. Li, J. Breeding of Yunnan rice; identification of rice blast resistance genes in *japonica* rice in Yunnan. In *Yunnan Rice*; Jiang, Z., Ed.; Yunnan Science and Technology Press: Kunming, China, 1995; pp. 178–189. (In Chinese)
35. Wang, C.A.; Guncar, G.; Forwood, J.K.; Teh, T.; Catanzariti, A.; Lawrence, G.J.; Loughlin, F.E.; Mackay, J.P.; Schirra, H.J.; Anderson, P.A.; et al. Crystal structures of flax rust avirulence proteins AvrL567-A and -D reveal details of the structural basis for flax disease resistance specificity. *Plant Cell* **2007**, *19*, 2898–2912. [CrossRef]
36. Kanzaki, H.; Yoshida, K.; Saitoh, H.; Fujisaki, K.; Hirabuchi, A.; Alaux, L.; Fournier, E.; Tharreau, D.; Terauchi, R. Arms race co-evolution of *Magnaporthe oryzae AVR-Pik* and rice *Pik* genes driven by their phyical interactions. *Plant J.* **2012**, *72*, 894–907. [CrossRef]

37. Wu, W.; Wang, L.; Zhang, S.; Li, Z.; Zhang, Y.; Lin, F.; Pan, Q. Stepwise arms race between *AvrPik* and *Pik* alleles in the rice blast pathosystem. *Mol. Plant-Microbe Interact.* **2014**, *27*, 759–769. [CrossRef]
38. Li, J.; Wang, Q.; Li, C.; Bi, Y.; Fu, X.; Wang, R. Novel haplotypes and networks of *AVR-Pik* alleles in *Magnaporthe oryzae*. *BMC Plant Biol.* **2019**, *19*, 204. [CrossRef]
39. Jia, Y.; Valent, B.; Lee, F.N. Determination of host responses to *Magnaporthe grisea* on detached rice leaves using a spot inoculation method. *Plant Dis.* **2003**, *87*, 129–133. [CrossRef]
40. Tai, T.; Tanksley, S.D. A rapid and inexpensive method for isolation of total DNA from dehydrated plant tissue. *Plant Mol. Biol. Rep.* **1990**, *8*, 297–303. [CrossRef]
41. Rozas, J.; Sánchez-Del, B.J.; Messeguer, X.; Rozas, R. DnaSP, DNA polymorphism analyses by the coalescent and other methods. *Bioinformatics* **2003**, *19*, 2496–2497. [CrossRef]
42. Clement, M.; Posada, D.; Crandall, K. TCS: A computer program to estimate gene genealogies. *Mol. Ecol.* **2000**, *9*, 1657–1659. [CrossRef]
43. Fontaine, C.; Lovett, P.N.; Sanou, H.; Maley, J.; Bouvet, J.-M. Genetic diversity of the shea tree (*Vitellaria paradoxa* C.F. Gaertn), detected by RAPD and chloroplast microsatellite markers. *Heredity* **2004**, *93*, 639–648. [CrossRef]
44. Rzhetsky, A.; Nei, M.A. Simple method for estimating and testing minimum evolution trees. *Mol. Biol. Evol.* **1992**, *9*, 945–967.

International Journal of *Molecular Sciences*

MDPI

Article

Knocking Out *OsAAP11* to Improve Rice Grain Quality Using CRISPR/Cas9 System

Yihao Yang †, Yi Zhang †, Zixing Sun, Ziyan Shen, Youguang Li, Yifan Guo, Yuntong Feng, Shengyuan Sun, Min Guo, Zhi Hu and Changjie Yan *

Jiangsu Key Laboratory of Crop Genetics and Physiology/Key Laboratory of Plant Functional Genomics of the Ministry of Education/Jiangsu Key Laboratory of Crop Genomics and Molecular Breeding, Yangzhou University, Yangzhou 225009, China; yihao.yang@yzu.edu.cn (Y.Y.); 007041@yzu.edu.cn (S.S.)

* Correspondence: cjyan@yzu.edu.cn

† These authors contributed equally to this work.

Abstract: The demand for rice grain quality, particularly in terms of eating and cooking quality, is increasingly concerning at present. However, the limited availability of rice-quality-related gene resources and time-consuming and inefficient traditional breeding methods have severely hindered the pace of rice grain quality improvement. Exploring novel methods for improving rice grain quality and creating new germplasms is an urgent problem that needs to be addressed. In this study, an amino-acid-transporter-encoding gene *OsAAP11* (*Os11g0195600*) mainly expressed in endosperm was selected as the target for gene editing using the CRISPR/Cas9 system in three *japonica* genetic backgrounds (Wuyungeng30, Nangeng9108, and Yanggeng158, hereafter referred to as WYG30, NG9108, and YG158). We successfully obtained homozygous *osaap11* mutants without transgenic insertion. Subsequently, we conducted comprehensive investigations on the agronomic traits, rice grain quality traits, and transcriptomic analysis of these mutants. The results demonstrate that loss of *OsAAP11* function led to a reduced amino acid content and total protein content in grains without affecting the agronomic traits of the plants; meanwhile, it significantly increased the peak viscosity, holding viscosity, and final viscosity values during the cooking process, thereby enhancing the eating and cooking quality. This study not only provides valuable genetic resources and fundamental materials for improving rice grain quality but also provides novel technical support for the rapid enhancement of rice grain quality.

Keywords: CRISPR/Cas9; *OsAAP11*; grain quality; rice

Citation: Yang, Y.; Zhang, Y.; Sun, Z.; Shen, Z.; Li, Y.; Guo, Y.; Feng, Y.; Sun, S.; Guo, M.; Hu, Z.; et al. Knocking Out *OsAAP11* to Improve Rice Grain Quality Using CRISPR/Cas9 System. *Int. J. Mol. Sci.* **2023**, *24*, 14360. https://doi.org/10.3390/ijms241814360

Academic Editor: Li Zhu

Received: 28 August 2023
Revised: 16 September 2023
Accepted: 18 September 2023
Published: 21 September 2023

1. Introduction

The improvement in people's living standards has led to an increasing consumer focus on rice grain quality, making it a crucial breeding objective. The genetic mechanism underlying rice grain quality is highly intricate, resulting in sluggish progress in its improvement [1,2]. Therefore, it is extremely important and urgent to identify the key genes governing rice grain quality and devise a novel approach to enhance rice grain quality quickly.

Nitrogen is an essential element for the growth and development of rice, which determines the yield and quality of rice [3,4]. In soil, nitrogen exists in two forms: organic nitrogen (mainly ammonium and nitrates) and inorganic nitrogen (mainly amino acids, peptides, and proteins) [5–7]. Amino acids are an important organic nitrogen source and the main form of nitrogen transport in rice. The main sources of amino acids in plants include the direct uptake of proton co-transporters driven by H^+-ATPase in roots from the soil, the conversion of nitrate or ammonium salts in roots, and synthesis occurring in leaves following the transport of xylem nitrate [8,9]. Subsequently, amino acids are loaded into the phloem and transported to the reservoir tissue through sieve tubes. During reproductive growth, organic nitrogen is exported through the seed coat to the interior of the seed, where it is stored as a storage protein in the endosperm or embryo/cotyledon [8,9]. Therefore, it is highly significant to investigate genes associated with amino acid biochemical pathways for enhancing rice yield and grain quality.

The transport of amino acids from the rhizosphere to the seed is facilitated by a series of membrane-localized amino acid transporters (AATs) [10]. In general, plant AATs can be divided into two major families: the amino acid/auxin permease (AAAP) superfamily and the amino acid polyamine and choline transport (APC) superfamily. Within the AAAP superfamily, there are eight subfamilies, including amino acid permeases (AAPs), lysine–histidine-like transporters (LHTs), proline transporters (ProTs), γ-aminobutyric acid transporters (GATs), auxin transporters (AUXs), aromatic and neutral amino-acid-like transporters (ANTs), amino acid transporters-like (ATLs) and amino acid vacuolar transporters (AVTs). The APC superfamily consists of subfamilies such as cationic amino acid transporters (CATs), bidirectional acid transporters (BATs), and L-type amino acid transporters (LATs) [11,12]. Currently, the accumulated knowledge about rice AATs is limited to the understanding of the amino acid permeases (AAPs) and lysine-histidine-like transporters (LHT) subfamilies, which are involved in amino acid absorption and transport from source to sink and ultimately affect rice yield and rice quality [13].

OsAAP1 is expressed in the root tip, lateral root, leaf sheath, stem, axillary bud, young panicle, and leaf vascular bundle. It exhibits a wide range of amino acid transport capabilities and positively regulates tillering, panicle number, yield, and nitrogen use efficiency in rice [14,15]. *OsAAP3* is expressed in roots, stems, leaves, leaf sheaths, and panicles with a broad spectrum of amino acid transport abilities and is particularly notable for its basic amino acid transport ability. The knockout mutant prevents the excessive accumulation of amino acids at tiller nodes while significantly increasing the number of tillers and effective panicles per plant without affecting the nutritional quality of rice [14,16]. *OsAAP4* is primarily expressed in roots, leaves, leaf sheaths, stems, and ears, predominantly facilitating neutral amino acid transfer, which positively regulates tillering number, yield per plant, and nitrogen utilization in rice [17]. *OsAAP5* is mainly expressed in the root system including the tiller base as well as leaf sheaths and young panicles; it primarily transports basic and neutral amino acids. Knocking out this gene can effectively enhance biomass production along with increasing the number of tillers and yield per plant in rice [18]. The expression of *OsAAP6* is predominantly observed in the roots, glume vascular bundle, capital, internode, node, and endosperm. Overexpression of *OsAAP6* enhances the roots' capacity to uptake amino acids, facilitating their transport through vascular bundle tissues to the endosperm. This ultimately improves the protein biosynthesis ability in the endosperm and indirectly influences key gene expressions related to starch biosynthesis [19–21]. *OsAAP7* and *OsAAP16* have a broad-spectrum ability to transport amino acids [14]. The expression of *OsAAP10* was relatively high in the stem and endosperm, whereas a significant reduction in grain protein content was observed upon *OsAAP10* knockout. Additionally, alterations were also detected in both the quantity and composition of starch [20]. Additionally, *OsLHT1* is expressed in various plant tissues including roots, stems, flag leaves, flag leaf sheaths, and young panicles. *OsLHT1* functions as a highly specific amino acid transporter with high affinity and demonstrates effective transport of aspartate, asparagine, and glutamate. Mutations in *OsLHT1* result in a significant reduction (40–66%) in root amino acid uptake along with impaired transportation of synthesized amino acids to the above-ground parts of rice. These alterations significantly impact growth and development processes while also leading to decreased biomass production and grain yield [21–24].

Although extensive research has been conducted on rice AATs genes, particularly the AAPs family genes, the majority of studies have focused on their impact on yield-related traits such as tillering number, grain number per panicle, and overall plant yield. Unfortunately, there has been limited attention given to their effect on quality-related traits. It is worth noting that numerous studies have indicated a potential negative correlation between rice yield and quality [1,2]. Therefore, it is crucial for us to address the urgent issue of enhancing rice quality without compromising agronomic traits. In this study, we selected *OsAAP11* as our target gene due to its relatively high expression in rice endosperm. By utilizing CRISPR/Cas9 technology, we performed gene knockout experiments in three different *japonica* rice cultivars to investigate their effects specifically on quality-related traits. This study aims to establish a foundation for comprehending the biological functions of amino acid transporter genes while offering theoretical support for improving rice quality.

2. Results

2.1. RT-qPCR Analysis of OsAAP11

In order to explore the expression characteristics of *OsAAP11* in various tissues of rice, particularly in grains at different stages of maturity, we employed the RT-qPCR technique to assess the expression levels of *OsAAP11* in roots, stems, leaves, leaf sheaths, and grains at distinct developmental stages across three diverse *japonica* rice cultivars (WYG30, NG9108, and YG158). Overall, *OsAAP11* exhibited similar expression patterns across different genetic backgrounds. It was relatively lowly expressed in rice roots, stems, leaves, and sheaths but highly expressed in grains at 5 days after flowering (DAF). Furthermore, gene expression decreased as the grains developed (Figure 1). These findings suggest that *OsAAP11* plays a pivotal role during early endosperm development.

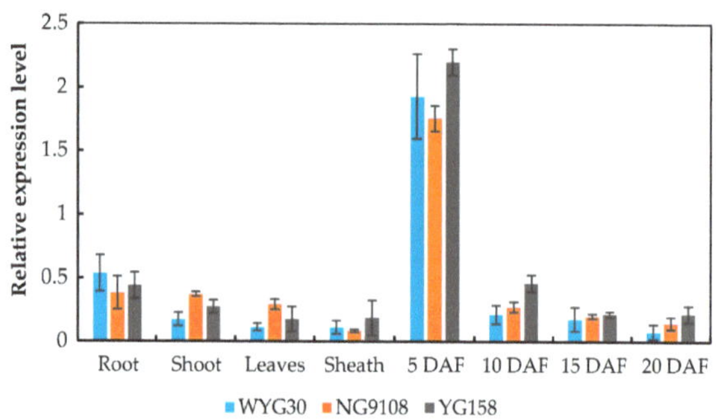

Figure 1. RT-qPCR analysis of *OsAAP11* expression.

2.2. Generation of osaap11 Mutants

In order to investigate the effect of *OsAAP11* (*Os11g0195600*) on rice grain quality, we employed the CRISPR/Cas9 system for targeted knockout, with the specific target site designed within the first exon of *OsAAP11* (44 bp downstream from ATG) (Figure 2A). Genome-wide off-target analysis was further conducted using CRISPR-GE (http://skl.scau.edu.cn/ (accessed on 22 January 2022)) and revealed a certain risk of off-target effects in the 5 'UTR regions of *Os07g0185300* and *Os01g0276700* (Figure 2B).

In the T_0 generation, a total of 18, 20, and 20 independent transgenic plants were successfully obtained from three distinct genetic cultivars (WYG30, NG9108, and YG158). Among them, 15, 18, and 16 transgenic plants exhibited hygromycin resistance with positive rates of approximately 83%, 90%, and 80%. Subsequently, amplification sequencing was performed on regions with high off-target scores, but no gene editing events were detected (Table S1). The T_0 *hyg*-positive individual plants obtained were subsequently transferred to T_1 lines and subjected to screening using *cas9* primers and *hygromycin* primers in order to identify transgene-free plants. Target site sequence amplification and sequencing were subsequently performed on the screened transgene-free plants. Finally, two types of homozygous mutations (−53 bp and −59 bp) were identified in the WYG30 background, while homozygous mutations (−4 bp and −1 bp) were found in both the NG9108 and YG158 backgrounds (Figure 2C–E). The aforementioned nucleotide mutations have the potential to cause frameshift mutations in amino acids.

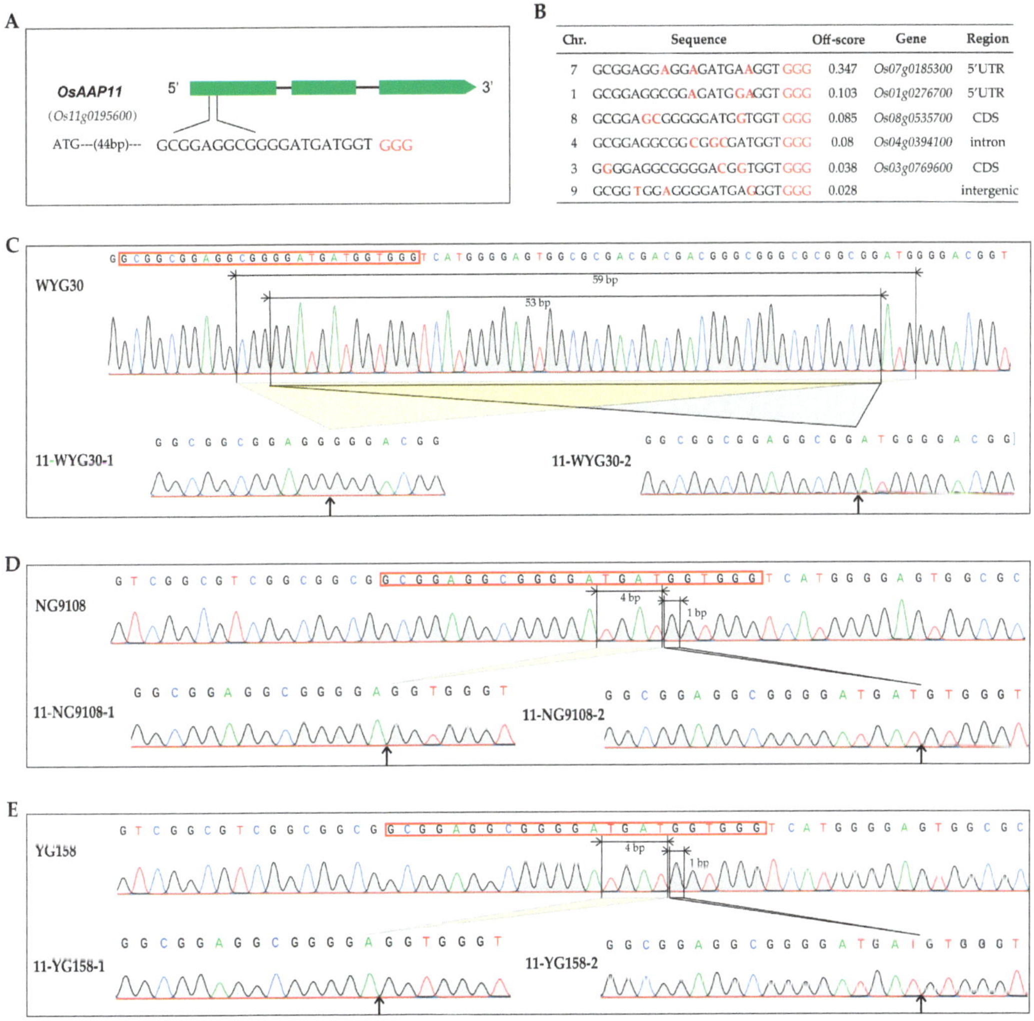

Chr.	Sequence	Off-score	Gene	Region
7	GCGGAGGAGGAGATGAAGGT GGG	0.347	*Os07g0185300*	5′UTR
1	GCGGAGGCGGAGATGGAGGT GGG	0.103	*Os01g0276700*	5′UTR
8	GCGGAGCGGGGGATGGTGGT GGG	0.085	*Os08g0535700*	CDS
4	GCGGAGGCGGCGGCGATGGT GGG	0.08	*Os04g0394100*	intron
3	GGGGAGGCGGGGACGGTGGT GGG	0.038	*Os03g0769600*	CDS
9	GCGGTGGAGGGGATGAGGGT GGG	0.028		intergenic

Figure 2. Generation of *osaap11* mutants using CRISPR/Cas9 system. (**A**) Schematic map of the genomic region of *OsAAP11*. The protospacer-adjacent motif (PAM) is indicated by red letters. (**B**) Off-target sequence analysis. The different nucleotides are marked in bold. (**C**) Mutation detection in T_1. Sanger chromatogram of intact WYG30 and mutated *OsAAP11* target motif. 11-WYG30-1 carried 59 bp deletion and 11-WYG30-2 carried 53 bp deletion. (**D**) Mutation detection in T_1. Sanger chromatogram of intact NG9108 and mutated *OsAAP11* target motif. 11-NG9108-1 carried 4 bp deletion and 11-NG9108-2 carried 1 bp deletion. (**E**) Mutation detection in T_1. Sanger chromatogram of intact YG158 and mutated *OsAAP11* target motif. 11- YG158-1 carried 4 bp deletion and 11-YG158-2 carried 1 bp deletion. The sequence marked by a red frame indicates the target motif. Arrows indicate fragments deleted from the WT sequence and the corresponding ligation points in the mutant alleles.

2.3. *Agronomic Traits in osaap11 Mutants*

In order to assess the impact of gene functional loss on agronomic traits in plants, we first confirmed that potential off-targets were not edited in T_1 plants and then conducted a systematic examination of the agronomic characteristics of mutants across different genetic backgrounds. The findings revealed a significant increase in grain length for mutant 11-

WYG30-2. Conversely, mutant 11-NG9108-1 exhibited notable reductions in both plant height and grain length. Additionally, mutant 11-NG9108-2 displayed a substantial decrease in hundred-grain weight. Furthermore, mutant 11-YG158-1 demonstrated a significant decrease in plant height, while mutant 11-YG158-2 exhibited marked increases in both grain length and width (Figure 3, Table 1). Despite observing certain alterations within individual agronomic traits among these mutants, no discernible pattern was evident, suggesting that these phenomena are likely attributed to tissue culture effects.

Figure 3. Whole plant of the wild-type and homozygous mutants. Scale = 10 cm.

Table 1. Analysis of agronomic characteristics of the wild-type and homozygous mutant lines.

	Plant Height (cm)	Tiller Number	Grain Number/Panicle	Fertility (%)	Grain Length (cm)	Grain Width (cm)	Grain Weight/100 (g)
WYG30	100.9 ± 1.3a	10.0 ± 0.9a	192.2 ± 16.0a	93.1 ± 1.4a	7.21 ± 0.05bc	3.65 ± 0.04a	2.77 ± 0.14a
11-WYG30-1	100.9 ± 2.7a	10.0 ± 1.0a	194.3 ± 14.0a	89.2 ± 0.9a	7.12 ± 0.05c	3.70 ± 0.02a	2.66 ± 0.10a
11-WYG30-2	100.2 ± 1.9a	8.3 ± 1.0a	187.0 ± 20.5a	88.9 ± 2.4a	7.26 ± 0.10a	3.68 ± 0.05a	2.70 ± 0.12a
NG9108	94.1 ± 2.1a	10.5 ± 1.4a	176.8 ± 10.7a	90.5 ± 1.7a	7.11 ± 0.10a	3.85 ± 0.03a	2.78 ± 0.07a
11-NG9108-1	85.0 ± 3.3b	9.7 ± 0.6a	183.3 ± 5.9a	88.2 ± 2.5a	7.08 ± 0.03b	3.72 ± 0.07a	2.71 ± 0.10ab
11-NG9108-2	93.1 ± 1.8a	10.7 ± 2.1a	174.3 ± 11.1a	88.7 ± 1.9a	7.11 ± 0.10a	3.79 ± 0.06a	2.67 ± 0.05b
YG158	76.9 ± 3.5a	8.3 ± 0.6a	130.2 ± 6.9a	91.7 ± 2.0a	7.06 ± 0.29b	3.64 ± 0.04b	2.78 ± 0.12ab
11-YG158-1	70.9 ± 2.2b	8.7 ± 1.5a	121.3 ± 9.0a	89.3 ± 1.4a	7.09 ± 0.06b	3.64 ± 0.05b	2.71 ± 0.06b
11-YG158-2	76.6 ± 2.1a	8.7 ± 2.1a	134.3 ± 11.6a	89.7 ± 4.8a	7.31 ± 0.12a	3.72 ± 0.03a	2.82 ± 0.06a

Different lowercase letters represent significant differences at $p < 0.05$ determined using one-way ANOVA with Duncan's multiple range test.

2.4. Grain Protein Content in osaap11 Mutants

To investigate the impact of *OsAAP11* knockout on rice grain protein content, the grain powders of *osaap11* mutants with different genetic backgrounds were analyzed for total protein and storage protein contents using the Kjeldahl method. As depicted in Figure 4, compared with the wild type, *osaap11* mutants in the WYG30 background exhibited significant reductions of 17.9% and 19.3% in total protein content, while those in the NG9108 background showed decreases of 3.3% and 4.3%. In the YG158 background, *osaap11* mutants displayed declines of 13.9% and 14.6%. The results demonstrate that *OsAAP11* knockout can significantly reduce the total grain protein content to a similar extent within the same genetic background, albeit with slight variations observed across different genetic backgrounds. Notably, WYG30 exhibited the most pronounced reduction, approaching 20%, while NG9108 displayed the smallest decrease, both falling below 5%.

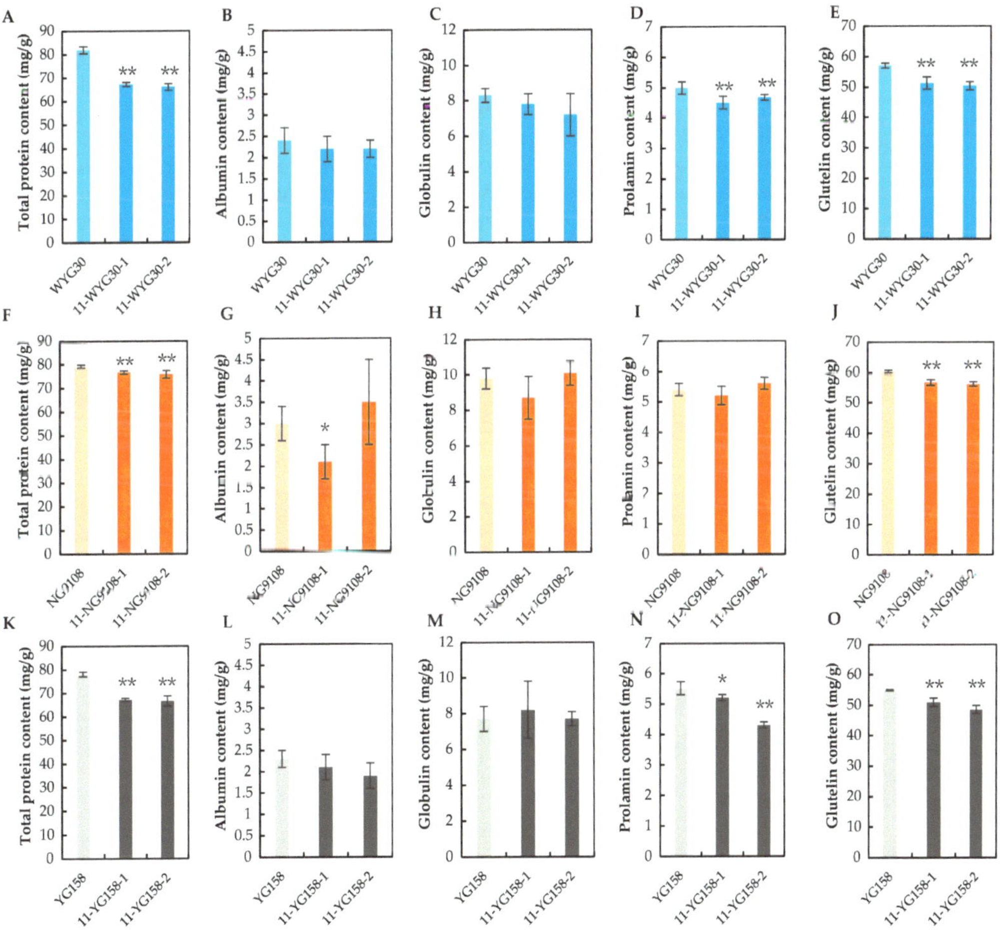

Figure 4. Grain protein contents in wild-type and *osaap11* mutants. (**A**,**F**,**K**) The total protein content in the wild type and the mutants. (**B**,**G**,**L**) The albumin content in the wild type and the mutants. (**C**,**H**,**M**) The globulin content in the wild type and the mutants. (**D**,**I**,**N**) The prolamin content in the wild type and the mutants. (**E**,**J**,**O**) The glutelin content in the wild type and the mutants. *p*-value < 0.05 * and < 0.01 **, calculated using independent-samples *t*-test.

Further analysis of the contents of four kinds of storage protein revealed that most mutants exhibited lower levels of albumin, globulin, and prolamin compared with the wild type, without achieving statistical significance. Conversely, the glutelin content was substantially reduced in all mutants relative to the wild type to a similar degree as that observed for total protein (Figure 4). The above results indicate that the loss of *OsAAP11* function led to a partial reduction in the rice storage protein, and the glutelin content was the primary factor.

2.5. Amino Acid Content in osaap11 Mutants

It has been shown that members of the AAPs family are responsible for the transmembrane transport of specific amino acids in plants [13]. To investigate the effect of *OsAAP11* knockout on the amino acid contents in rice grains, the amino acids were extracted and further determined. As depicted in Figure 5A,C,E, the content of each amino acid decreased to varying degrees in the mutants in the three genetic backgrounds, particularly the contents of aspartate, glutamate, valine, and arginine with high contents. The total amino acid content significantly declined across different backgrounds, which was similar to the trend observed for rice protein content. The decrease rate was smallest (4.4% and 7.6%) for mutants in the NG9108 background, while it exceeded 10% for the WYG30 and YG158 backgrounds, respectively (Figure 5B,D,F). Furthermore, *OsAAP11* knockout led to a reduction ranging from 0.5 to 1.8 mg/g for basic amino acids; from 0.8 to 4.2 mg/g for acidic amino acids; and from 0.4 to 7.8 mg/g for neutral amino acids (Figure 5B,D,F). In conclusion, the loss of *OsAAP11* function mostly affects neutral amino acids and, in general, the contents of amino acids in rice grains.

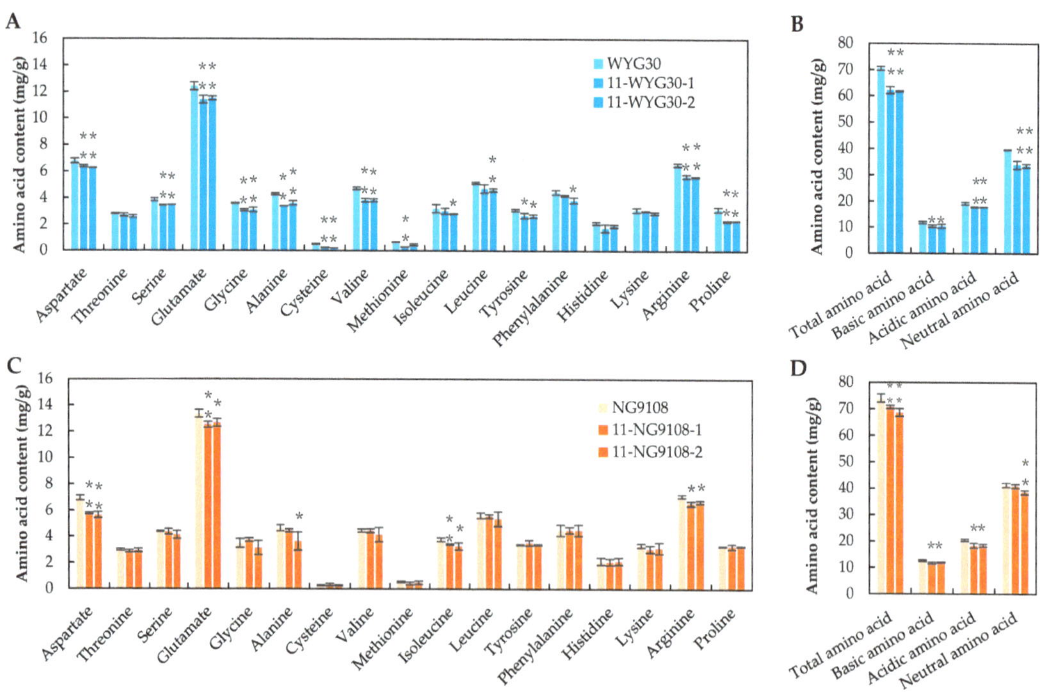

Figure 5. *Cont.*

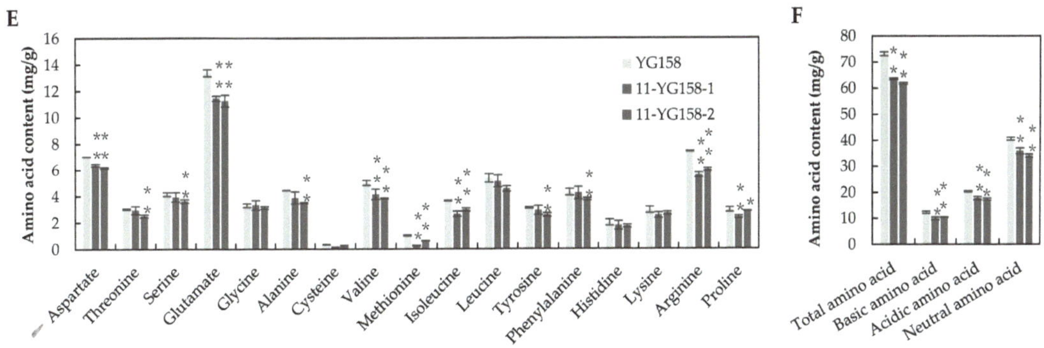

Figure 5. Comparison of amino acid contents between the wild-type and homozygous mutant lines. (**A**,**C**,**E**) The amino acid contents in the wild types and the mutants. (**B**,**D**,**F**) The total, basic, acidic, and neutral amino acid contents in the wild types and the mutants. p-value < 0.05 * and < 0.01 **, calculated using independent-samples t-test.

2.6. Starch Content and Physicochemical Properties of osaap11 Mutants

In order to investigate the potential impact of *OsAAP11* on the physicochemical properties of rice starch, we initially assessed the total starch and amylose contents in each mutant. As depicted in Figure 6A,B, the total starch contents in 11-WYG30-2, 11-YG158-1, and 11-YG158-2 exhibited significant increases compared with their corresponding wild types, with increments of 1.7%, 3.3%, and 3.7%, respectively. Furthermore, mutants 11-WYG30-1 and 11-WYG30-2 displayed notable decreases in amylose contents of 2.1% and 1.8%, respectively. The most probable cause of this phenomenon is the rice endosperm's role in maintaining the carbon and nitrogen balance. These findings suggest that the absence of *OsAAP11* not only affected the amino acid and protein contents in rice grains but also played a crucial indirect regulatory role in determining starch quantity and composition.

Starch viscosity analysis serves as an effective tool for evaluating rice eating and cooking quality by simulating the rice cooking process to measure starch gelatinization characteristics [25]. The following figure illustrates that, in comparison with the corresponding wild types, nearly all mutants exhibited higher peak viscosity, holding viscosity, and final viscosity while demonstrating reduced setback values, indicating that loss of *OsAAP11* function is highly likely to influence the palatability of rice (Figure 6C).

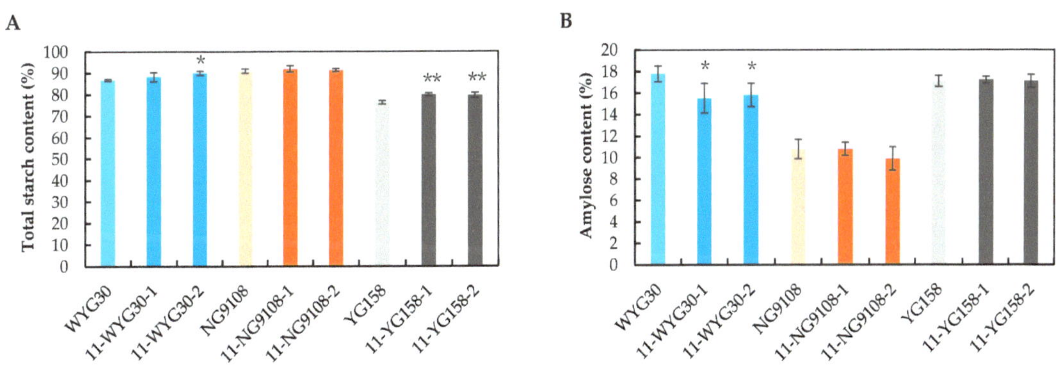

Figure 6. *Cont.*

C

Genotype	Peak viscosity (cP)	Holding viscosity (cP)	Breakdown value (cP)	Final viscosity (cP)	Setback viscosity (cP)	Pasting temperature(℃)
WYG30	3386±98 c	2170±57 c	1216±38 b	3496±98 c	109±24 a	76.8±0.1 a
11-WYG30-1	3543±22 b	2230±110 b	1362±98 a	3609±17 b	66±16 b	77.0±0.2 a
11-WYG30-2	3931±154 a	2681±37 a	1250±92 ab	3973±72 a	42±20 b	76.8±0.1 a
NG9108	2792±74 c	1855±119 c	938±45 b	2757±121 b	-35±49 a	76.2±0.8 a
11-NG9108-1	3919±93 b	2455±54 b	1464±50 a	3515±40 a	-404±93 b	75.7±0.4 a
11-NG9108-2	4082±89 a	2678±98 a	1405±117 a	3541±57 a	-541±32 c	75.8±0.4 a
YG158	3253±61 b	2657±144 b	596±83 a	3827±81 b	575±43 a	77.7±0.4 a
11-YG158-1	3664±77 a	3022±83 a	642±116 a	4130±111 a	466±150 a	76.6±0.8 a
11-YG158-2	3526±81 a	2962±76 a	564±36 a	4046±95 a	519±37 a	78.1±1.2 a

Figure 6. Starch contents and physicochemical properties of osaap11 mutants. (**A**) Comparison of total starch contents between the wild-type and homozygous mutant lines (100% = 1 g rice flour). (**B**) Comparison of the amylose contents between the wild-type and homozygous mutant lines (100% = 1 g rice flour). (**C**) RVA (Rapid Visco Analyser) profile of the wild-type and homozygous mutant lines. p-value < 0.05 * and < 0.01 **, calculated using independent-samples t-test. Different lowercase letters represent significant differences at $p < 0.05$ determined using one-way ANOVA with Duncan's multiple range test.

2.7. Taste Scores of osaap11 Mutants

In order to investigate the impact of *OsAAP11* gene mutation on the eating and cooking quality of cooked rice, the taste scores of all mutant types were subsequently assessed [26]. The findings reveal a significant enhancement in eating and cooking quality for all mutants across the three genetic backgrounds, with taste score increases ranging from 3.9% to 17.2% and an average increment of 9.8%. Notably, YG158 exhibited the most pronounced effect, resulting in a more than 10% improvement in the taste score for cooked rice, while WYG30 demonstrated a moderate improvement effect, and NG9108 showed the least enhancement (Figure 7).

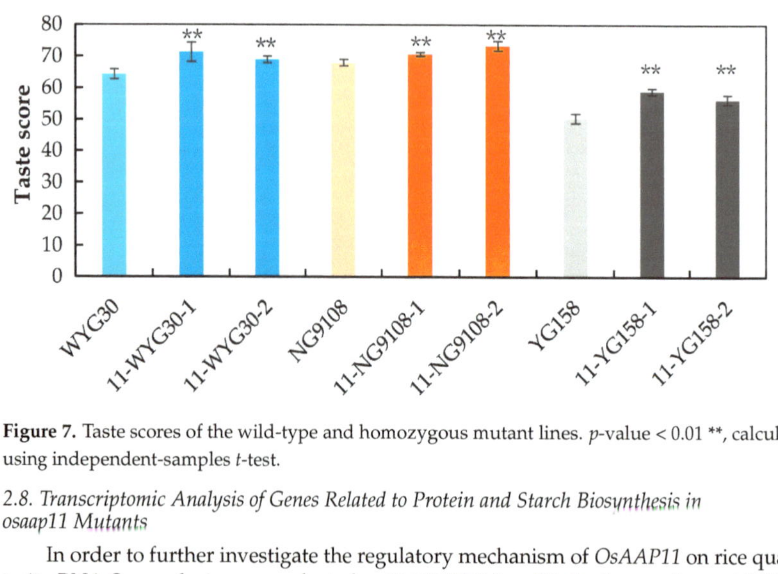

Figure 7. Taste scores of the wild-type and homozygous mutant lines. p-value < 0.01 **, calculated using independent-samples t-test.

2.8. Transcriptomic Analysis of Genes Related to Protein and Starch Biosynthesis in osaap11 Mutants

In order to further investigate the regulatory mechanism of *OsAAP11* on rice quality traits, RNA-Seq analysis was conducted on 11-WYG30-1 and its background parent WYG30.

A total of 1596 differentially expressed genes were identified based on two criteria: a fold change greater than |log2(Fold change)| > 1 and a significance level of p-value < 0.05. Among these genes, 1108 were significantly upregulated, while 547 were significantly downregulated. Gene Ontology (GO) functional enrichment analysis revealed that these differentially expressed genes were primarily associated with nutrient reservoir activity, cytoplasmic ribosomes, aleurone grain, transmembrane transport activity, intrinsic components of the plasma membrane, and substrate-specific transmembrane transporter activity (Figure 8A). Kyoto Encyclopedia of Genes and Genomes (KEGG) pathway enrichment analysis indicated that the differentially expressed genes mainly participated in ribosome, photosynthesis-antenna protein, flavonoids, and flavonols biosynthesis; alanine, aspartate, and glutamate metabolism; taurine and hypotaurine metabolism; nitrogen metabolism; starch and sucrose metabolism; betaine biosynthesis; glycosaminoglycan degradation; and so on (Figure 8B).

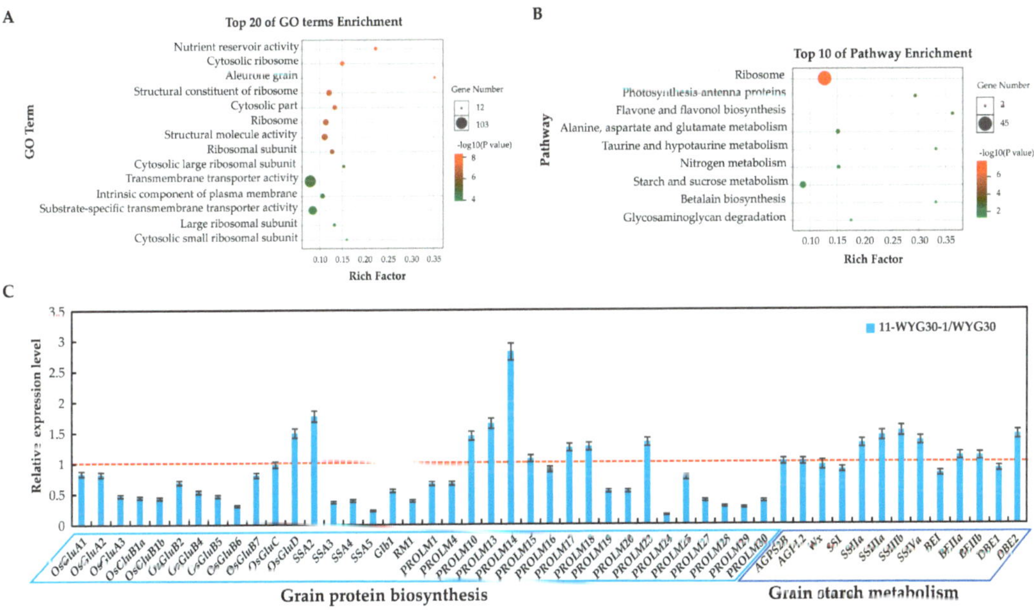

Figure 8. Transcriptomic analysis of genes related to protein and starch biosynthesis in *osaap11* mutants. (**A**,**B**) GO functional and KEGG pathway enrichment analyses of the DEGs. (**C**) The relative expression analysis for the genes related to protein and starch biosynthesis in *osaap11* mutants.

To investigate whether alterations in storage protein and starch accumulation were reflected by changes in mRNA levels, we further examined the expressions of 49 key genes associated with storage substances in endosperm at 5 DAF. This encompassed 36 genes related to grain protein synthesis and 13 genes related to grain starch metabolism. The findings revealed that most genes involved in storage protein synthesis were downregulated, particularly *OsGluA1-OsGluB5*, *RM1*, and *Glb1*, which exhibited high endogenous expression levels. The genes associated with starch metabolism remained largely unaltered, while a few genes involved in starch synthesis exhibited slight upregulation (Figure 8C).

3. Discussion

Numerous studies have extensively investigated the impact of grain protein content on the eating and cooking quality of rice. Xie et al. conducted a study where they introduced DTT into the cooking water to enzymatically hydrolyze proteins [27]. This resulted in reduced rice hardness and a significant increase in viscosity in non-glutinous rice varieties. These findings suggest that rice protein not only affects texture but also plays a crucial role in determining rice viscosity. In another study, Yang et al. analyzed the levels of total protein and storage proteins in nearly 100 *japonica* rice cultivars. They then conducted correlation analyses with the taste scores of the cooked rice. The results revealed a strong negative correlation between total protein content, glutelin content, and taste score, with correlation coefficients of −0.89 and −0.57, respectively [28]. Furthermore, Park et al. utilized a collection of RIL populations to identify a stable rice grain protein content QTL*qPro9* on chromosome 9. Interestingly, this QTL was found to be closely associated with a QTL*qTV9* related to taste score, with an additive effect that was in the opposite direction [29]. The aforementioned studies directly or indirectly indicate that high levels of grain protein negatively impact the taste characteristics of rice overall.

Under normal circumstances, the grain protein content in high-quality rice is often lower than 7% [28]. However, in actual production practice, the grain protein content in ordinary rice cultivars is generally high due to the high nitrogen application level and variety characteristics. This high grain protein content seriously affects the eating and cooking quality of rice. Previous genetic studies have primarily concentrated on identifying QTLs associated with rice protein content. Among these QTLs, only *qPC1* and *qGPC-10* have been successfully cloned [19,30]. Therefore, the scarcity of gene resources and methods available for improving rice grain protein-related traits severely hampers the further enhancement of rice grain quality.

Plant AAPs regulate the process of organic nitrogen release from the parental grain coat to the grain endoplasm, thereby influencing the storage protein content in either the endosperm or embryo/cotyledon. In Arabidopsis thaliana, *AtAAP1* is involved in transporting amino acids to the embryo, and the amino acid content in the endosperm of the mutant is significantly decreased, and the protein level is also significantly decreased [31]. *AtAAP8* can transport amino acids for early embryonic growth, and its mutant production is significantly reduced [32]. The T-DNA insertion line of *AtAAP2* also results in a significant reduction in protein content within dried seeds [33]. In peas, *VfAAP1* overexpression lines could significantly increase seed protein content under different nitrogen levels [34]. The *OsAAP6* gene in rice plays a positive role in regulating rice protein content. Overexpression of *OsAAP6* enhances the root's ability to absorb certain amino acids, promotes the expression of key genes related to storage proteins, and ultimately results in an expanded proteome and increased rice protein content [19]. Therefore, regulating the expression of rice *AAPs* presents a viable approach to improve rice protein content and improve rice grain quality.

In this study, we observed a high expression of *OsAAP11* in the endosperm of rice 5 days after flowering, indicating its crucial role in early rice endosperm development. Subsequently, CRISPR/Cas9 gene-editing technology was employed to knock out *OsAAP11* in three different genetic backgrounds. The results revealed significant reductions in amino acid and total protein contents across almost all mutants, accompanied by increased peak viscosity, holding viscosity, final viscosity, and improved rice taste scores. Remarkably, these improvements were achieved without substantial changes in agronomic traits. This study not only provides valuable genetic resources for enhancing rice quality but also presents a viable approach for rapid improvement.

4. Materials and Methods

4.1. Experimental Materials and Plant Method

The experimental materials used in this study were the Wuyungeng 30, Nangeng 9108, and Yanggeng 158 rice varieties suitable for cultivation in Jiangsu Province. These materials were cultivated in Yangzhou, Jiangsu Province, China, and Lingshui, Hainan Province, China. Each genotype was represented by a single plant with two rows of 10 plants per row. The spacing between plants was maintained at 20 cm × 20 cm. Standardized water and fertilizer management practices were followed.

The PC1300-Cas9 knockout vector and SK-gRNA intermediate vector were generously provided by Dr. Kejian Wang, a renowned researcher at the China Rice Research Institute.

4.2. Target Site Design and Mutation Detection

The coding sequence of *OsAAP11* (*Os11g0195600*) was obtained from the RGAP website (http://rice.plantbiology.msu.edu/ (accessed on 22 January 2022)). In the first exon, a targeted gene sequence was designed. The target sequence for *OsAAP11* was GGAGGCGGGGATGATGGTGGG, which is located 44 bp downstream of the transcription start codon ATG. The CRISPR—GE (http://skl.scau.edu.cn/offtarget/ (accessed on 22 January 2022)) website was utilized for the analysis of off-target effects for the target sequence. The vector construction method and process were carried out with reference to the literature [20]. The NCBI (http://www.ncbi.nlm.nih.gov (accessed on 27 January 2022)) database was utilized to identify the appropriate design primer *OsAAP11*-jc-F and *OsAAP11*-jc-R sequences (*OsAAP11*-jc-F: CAACCCTACACGCCTCTCTC; *OsAAP11*-jc-R: AGGTTGGCGTACTGGATGAC). Subsequently, the CTAB (cetyltrimethylammonium bromide) method was employed for DNA extraction from transgenic plants, followed by PCR amplification sequencing to determine the types of mutations.

4.3. Transgenic Component Detection

The transgenic T_0 plants in our study were obtained using the agrobacterium-mediated method at Biorun Bio-Company, Wuhan, China (https://plant.biorun.com/ (accessed on 10 May 2022)). In the T_1 generation population, DNA was sampled and extracted from a single plant, and exogenous transgenic components were detected using specific primers targeting the *cas9* coding gene (Cas9-F and Cas9-R) as well as the *hygromycin* gene (Hyg-F and Hyg-R) (Table S1). The absence of transgenic components in plants was confirmed by the absence of corresponding bands.

4.4. Analysis of Protein Contents and Amino Acid Contents in Rice Grains

The total nitrogen content in rice flour and the protein content of each component were analyzed with the kjeltec automatic nitrogen analyzer (FOSS, Hilleroad, Denmark) and the protein content was calculated using 6.25 as the nitrogen conversion coefficient.

The steps for extracting 4 kinds of storage protein were as follows: An amount of 1 g of rice flour was weighed. Albumin was extracted using 1 mL of ddH_2O. Globulin was extracted using 1 mL of a solution containing 50 mM KH_2PO_4, 0.5 mM NaCl, and 1 mM EDTA-2Na (pH = 6.8). Alcohol-soluble protein was extracted using 1 ml of 75% ethanol. Glutenin was extracted from the EDTA-2Na solution using a mixture of 0.1 mM NaOH and 1 mM concentration. Each component protein was shaken at a temperature of 4 °C for a duration of six hours, followed by centrifugation at a speed of 10,000 rpm for fifteen minutes, which was repeated three times.

An amount of 10 mg of rice flour was weighed into a 2 mL spiral cap tube and 300 nM of L-(+)-Norleucine was added for subsequent correction purposes. An amount of 500 µL of 6 N HCl was hydrolyzed in an oven at 110 °C for 24 h, followed by centrifugation at 13,000 rpm for 10 min and complete drying in a water bath at 57 °C. The residues were dissolved in an amino acid diluent and centrifuged at 10,000 rpm for another 10 min. The supernatant was filtered using a nylon membrane syringe with a pore size of 0.45 µm and transferred to an automatic sample vial for analysis using an automatic amino acid analyzer (Biochrom30+, Cambridge, England).

4.5. Analysis of Amylose and Total Starch Contents in Rice Grains

The amylose content determination was carried out as follows: Amounts of 0.1 g of the sample to be tested and a standard sample were weighed, and then 1 mL of ethanol solution was added into a 10 mL centrifuge tube. Next, 9 mL of a 1 mol/L NaOH solution was added into the same centrifuge tube and shaken vigorously. The tube was wrapped with a sealing film and heated in boiling water for 10 min. It was allowed to cool naturally, and then 1 mL of the mixture was diluted into another 10 mL centrifuge tube. An amount of 1 mL of acetic acid solution and 2 mL of iodine reagent were added to this new tube. It was filled up with water until it reached a total volume of 10 mL, shaken well, and let sit for another 10 min. Finally, the spectrophotometer (BioTek, Winooski, Vermont, United States) was zeroed using sodium hydroxide solution before measuring the absorbance at a wavelength of 720 nm.

The determination of total starch content involved the preparation of an anthrone–sulfuric acid solution by dissolving 0.4 g of anthrone in 100 mL of H_2SO_4. Amounts of 0.1 g of the sample to be tested and a standard sample were weighed and placed in separate 50 mL centrifuge tubes. Then, 30 mL of 52% perchloric acid was added to each tube for complete shaking and dissolution. The residue at the pipe mouth was washed with water, followed by adjusting the volume to 50 mL and diluting it by a factor of 25. Additionally, the standard starch solution was further diluted into concentrations of 12.5%, 25%, 50%, 75%, and 100%. Each sample (1 mL) was transferred into a separate 10 mL centrifuge tube, followed by adding 6 mL of the anthrone–sulfuric acid reagent for thorough mixing. After cooling, the tubes were placed in boiling water for 5 min before being taken out to cool down to room temperature. Finally, the absorbance at 640 nm was measured using a spectrophotometer after zero adjustment with distilled water.

4.6. Analysis of Starch Viscosity Characteristics

In accordance with the AACC (American Association of Cereal Chemists) procedures, the RVA TecMaster (Perkinelmer, Sydney, Australia) and its accompanying TCW (3.17.3.509) (Thermocline for Windows) software was utilized for the analysis. An amount of 3 grams of rice flour with a water content of 13% was weighed, followed by the addition of 25 g of water. The mixture was thoroughly stirred in an aluminum can before being subjected to analysis using the machine. The procedure consisted of the following cycles: 1 min at 50 °C, 2.5 min at 95 °C, 1.4 min at 50 °C, and a total duration of 3.8 min, including both heating and cooling processes. Throughout this period, the RVA (Rapid Visco Analyser) blades rotated at an initial speed of 960 rpm for the first ten seconds and subsequently maintained a constant speed of 160 rpm for the remaining time interval. The RVA spectral characteristic values were expressed through parameters such as peak viscosity, holding viscosity, final viscosity, breakdown value (peak viscosity minus holding viscosity), setback value (final viscosity minus peak viscosity), as well as other relevant characteristics.

4.7. Analysis of Rice Taste Scores

An amount of 30 g of milled rice was cleaned with tap water, water was added according to a 1:1.3 ratio, and it was sealed with filter paper and rubber band, soaked for 30 min, and steamed in a pot for 30 min, the power was turned off, and it was stewed for 10 min. When finished, the rice was turned with a spoon to prevent it from sticking to the sides of the pot. Then, the air in the cooling box was blown for 10 min, the lid of the matching can was closed, and then it was left to stand for 90 min, the machine was preheated in advance for 30 min, the software was opened, and the Japanese *Japonica* rice mode was entered for calibration. The cooled rice was weighed to 8.00 g in a metal ring, with positive and negative pressure for 10 s, and the STA1A eating instrument (Sasaki Corporation, Aomori, Japan) was used to determine the taste score of the rice [26]. The working principle of this instrument is converting various physicochemical parameters of rice into "taste" scores based on correlations between the near-infrared reflectance (NIR) measurements of key constituents (e.g., amylose, protein, moisture, and fat acidity) and preference sensory scores.

4.8. Investigation of Main Agronomic Characteristics of Rice Plants

The main panicles of 5 mutants and 5 wild-type rice plants were selected at the mature stage for the investigation of height, tiller, number of grains per panicle, grain length, grain width, 100-grain weight, and fertility. Grain length and width were analyzed using the seed-measuring instrument (Model SC-G, Wanshen, Hangzhou, China).

4.9. Transcriptome Analysis

Total RNA was isolated from the endosperm at 15 DAF of WYG30 and 11-WYG30-1, using TRIzol reagent (Invitrogen, Shanghai, China) and purified using an RNeasy Plant Mini Kit (Qiagen, Venlo, Netherlands). Each sample contained over 25 grains, with 3 biological replicates, and was used for paired-end library construction. Sequencing was performed on an Illumina platform of the Gene Denovo company, China (https://www.genedenovo.com/ (accessed on 28 May 2023)). HTSeq v0.9.1 was used to count the read numbers mapped to each gene. The DESeq R package (1.18.0) was used for analyzing the differentially expressed genes. The Go functional enrichment and KEGG pathway enrichment analyses were performed using the OmicShare tools, a free online platform for data analysis (https://www.omicshare.com/tools (accessed on 28 May 2023)).

4.10. RNA Extraction, cDNA Preparation, and qRT-PCR

Total RNA was extracted from endosperm at 15 DAF using an RNA extraction kit (Tiangen, Beijing, China). First-strand cDNA was synthesized using a reverse transcription kit (Vazyme, Nanjing, China). Quantitative reverse transcriptase (qRT-PCR) was performed with a CFX96 Real-Time PCR System (Bio-Rad, Hercules, CA, USA) using an SYBR qPCR Master Mix (Vazyme, Nanjing, China). All assays were performed with at least three biological replicates. The rice actin gene served as the internal control to normalize gene expression. All primers are listed in Table S1.

4.11. Data Analysis

The experiment was conducted with three biological replicates for each sample. Microsoft Excel 2016 was utilized for data collection, while the statistical analysis software SPSS 15.0 was employed for variance analysis. All experimental data are presented as means ± SD.

Supplementary Materials: The following supporting information can be downloaded at https://www.mdpi.com/article/10.3390/ijms241814360/s1.

Author Contributions: Y.Y., Y.Z. and C.Y. designed this research and analyzed the data; Z.S. (Zixing Sun), Z.S. (Ziyan Shen), Y.L., Y.F. and Y.G. performed this research; Y.Y. and C.Y. wrote this manuscript; S.S., Z.H., and M.G. contributed to the writing. All authors have read and agreed to the published version of the manuscript.

Funding: This research was funded by the National Natural Science Foundation of China (32301827), the programs of the Jiangsu Province Government (BK20210798, BE2022365-2, BE2022335, JBGS [2021]001, and BE2021334-1), the China Postdoctoral Science Foundation (2023M733005), and the Priority Academic Program Development of Jiangsu Higher Education Institutions (PAP).

Institutional Review Board Statement: Not applicable.

Informed Consent Statement: Not applicable.

Data Availability Statement: Not applicable.

Conflicts of Interest: The authors declare no conflict of interest.

References

1. Ren, D.; Ding, C.; Qian, Q. Molecular bases of rice grain size and quality for optimized productivity. *Sci. Bull.* **2023**, *68*, 314–350. [CrossRef] [PubMed]
2. Gong, D.; Zhang, X.; He, F.; Chen, Y.; Li, R.; Yao, J.; Zhang, M.; Zheng, W.; Yu, G. Genetic Improvements in Rice Grain Quality: A Review of Elite Genes and Their Applications in Molecular Breeding. *Agronomy* **2023**, *13*, 1375. [CrossRef]
3. Fang, L.; Ma, L.; Zhao, S.; Cao, R.; Jiao, G.; Hu, P.; Wei, X. Alanine aminotransferase (*OsAlaAT1*) modulates nitrogen utilization, grain yield, and quality in rice. *J. Genet. Genom.* **2022**, *49*, 510–513. [CrossRef] [PubMed]

4. Hu, B.; Wang, W.; Chen, J.; Liu, Y.; Chu, C. Genetic improvement toward nitrogen-use efficiency in rice: Lessons and perspectives. *Mol. Plant* **2023**, *16*, 11. [CrossRef]
5. Salmenkallio, M.; Sopanen, T. Amino Acid and Peptide uptake in the scutella of germinating grains of barley, wheat, rice, and maize. *Plant Physiol.* **1989**, *89*, 1285–1291. [CrossRef]
6. Paungfoo-Lonhienne, C.; Lonhienne, T.; Rentsch, D.; Robinson, N.; Christie, M.; Webb, R.; Gmage, H.; Carroll, B.; Schenk, P.; Schmidt, S. Plants can use protein as a nitrogen source without assistance from other organisms. *Proc. Natl. Acad. Sci. USA* **2008**, *105*, 4524–4529. [CrossRef]
7. Xu, G.; Fan, X.; Miller, A. Plant nitrogen assimilation and use efficiency. *Annu. Rev. Plant Biol.* **2012**, *63*, 153–182. [CrossRef]
8. Bush, D. Proton-coupled sugar and amino acid transporters in plants. *Annu. Rev. Plant Physiol. Plant Mol. Biol.* **1993**, *44*, 513–542. [CrossRef]
9. Tegeder, M.; Rentsch, D. Uptake and partitioning of amino acids and peptides. *Mol. Plant* **2010**, *3*, 997–1011. [CrossRef]
10. Mechthild, T. Transporters involved in source to sink partitioning of amino acids and ureides: Opportunities for crop improvement. *J. Exp. Bot.* **2014**, *65*, 1865–1878.
11. Zhao, H.; Ma, H.; Yu, L.; Wang, X.; Zhao, J. Genome-wide survey and expression analysis of amino acid transporter gene family in rice (*Oryza sativa* L.). *PLoS ONE* **2012**, *7*, e49210. [CrossRef] [PubMed]
12. Lalonde, S.; Wipf, D.; Frommer, W. Transport mechanisms for organic forms of carbon and nitrogen between source and sink. *Annu. Rev. Plant Biol.* **2004**, *55*, 341–372. [CrossRef] [PubMed]
13. Guo, N.; Zhang, S.; Gu, M.; Xu, G. Function, transport, and regulation of amino acids: What is missing in rice? *Crop J.* **2021**, *3*, 530–542. [CrossRef]
14. Taylor, M.; Reinders, A.; Ward, J. Transport function of rice amino acid permeases (AAPs). *Plant Cell Physiol.* **2015**, *56*, 1355–1363. [CrossRef] [PubMed]
15. Ji, Y.; Huang, W.; Wu, B.; Fang, Z.; Wang, X. The amino acid transporter AAP1 mediates growth and grain yield by regulating neutral amino acid uptake and reallocation in *Oryza sativa*. *J. Exp. Bot.* **2020**, *71*, 4763–4777. [CrossRef] [PubMed]
16. Lu, K.; Wu, B.; Wang, J.; Zhu, W.; Nie, H.; Qian, J.; Huang, W.; Fang, Z. Blocking amino acid transporter OsAAP3 improves grain yield by promoting outgrowth buds and increasing tiller number in rice. *Plant Bio. J.* **2018**, *16*, 1710–1722. [CrossRef]
17. Fang, Z.; Wu, B.; Ji, Y. The Amino Acid Transporter OsAAP4 Contributes to Rice Tillering and Grain Yield by Regulating Neutral Amino Acid Allocation through Two Splicing Variants. *Rice* **2021**, *14*, 2. [CrossRef]
18. Wang, J.; Wu, B.; Lu, K.; Wei, Q.; Qian, J.; Chen, Y.; Fang, Z. The Amino Acid Permease 5 (OsAAP5) Regulates Tiller Number and Grain Yield in Rice. *Plant Physiol.* **2019**, *180*, 1031–1045. [CrossRef]
19. Peng, B.; Kong, H.; Li, Y.; Wang, L.; Zhong, M.; Sun, L.; Gao, G.; Zhang, Q.; Luo, L.; Wang, G.; et al. OsAAP6 functions as an important regulator of grain protein content and nutritional quality in rice. *Nat. Commun.* **2014**, *5*, 4847. [CrossRef]
20. Wang, S.; Yang, Y.; Guo, M.; Zhong, C.; Yan, C.; Sun, S. Targeted mutagenesis of amino acid transporter genes for rice quality improvement using the CRISPR/Cas9 system. *Crop J.* **2020**, *8*, 457–464. [CrossRef]
21. Liu, D.; Gong, W.; Bai, Y.; Luo, J.; Zhu, Y. OsHT, a rice gene encoding for a plasma-membrane localized histidine transporter. *J. Integr. Plant Biol.* **2005**, *47*, 92–99. [CrossRef]
22. Wang, X.; Yang, G.; Shi, M.; Hao, D.; Wei, Q.; Wang, Z.; Fu, S.; Su, Y.; Xia, J. Disruption of an amino acid transporter LHT1 leads to growth inhibition and low yields in rice. *BMC Plant Biol.* **2019**, *19*, 268. [CrossRef] [PubMed]
23. Guo, N.; Hu, J.; Yan, M.; Luo, L.; Tegeder, M.; Xu, G. *Oryza sativa* Lysine-Histidine-type Transporter 1 functions in root uptake and root-to-shoot allocation of amino acids in rice. *Plant J.* **2020**, *103*, 395–411. [CrossRef]
24. Guo, N.; Gu, M.; Hu, J.; Qu, H.; Xu, G. Rice OsLHT1 Functions in Leaf-to-Panicle Nitrogen Allocation for Grain Yield and Quality. *Front. Plant Sci.* **2020**, *11*, 1150. [CrossRef] [PubMed]
25. Tong, C.; Chen, Y.; Tang, F.; Xu, F.; Huang, Y.; Chen, H.; Bao, J. Genetic diversity of amylose content and RVA pasting parameters in 20 rice accessions grown in Hainan, China. *Food Chem.* **2014**, *161*, 239–245. [CrossRef] [PubMed]
26. Champagne, E.; Richard, O.; Bett, K.; Grimm, C.; Vinyard, B.; Webb, B.; Mcclung, A.; Barton, F.; Lyon, B.; Moldenhauer, L.; et al. Quality evaluation of U.S. medium-grain rice using a Japanese taste analyzer. *Cereal Chem.* **1996**, *73*, 290–294.
27. Xie, L.; Chen, N.; Duan, B.; Zhu, Z.; Liao, X. Impact of proteins on pasting and cooking properties of waxy and non-waxy rice. *J. Cereal Sci.* **2008**, *47*, 372–379. [CrossRef]
28. Yang, Y.; Shen, Z.; Li, Y.; Xu, C.; Xia, H.; Zhuang, H.; Sun, S.; Guo, M.; Yan, C. Rapid improvement of rice eating and cooking quality through gene editing toward glutelin as target. *J. Integr. Plant Biol.* **2022**, *64*, 1860–1865. [CrossRef]
29. Park, S.; Park, H.; Baek, M.; Jeong, J.; Cho, Y.; Lee, G.; Lee, C.; Suh, J.; Kim, C.; Kim, S. Improving the glossiness of cooked rice, an important component of visual rice grain quality. *Rice* **2019**, *12*, 87. [CrossRef]
30. Yang, Y.; Guo, M.; Sun, S.; Zou, Y.; Yin, S.; Liu, Y.; Tang, S.; Gu, M.; Yang, Z.; Yan, C. Natural variation of *OsGluA2* is involved in grain protein content regulation in rice. *Nat. Commun.* **2019**, *10*, 1949. [CrossRef]
31. Perchlik, M.; Tegeder, M. Improving Plant Nitrogen Use Efficiency through Alteration of Amino Acid Transport Processes. *Plant Physiol.* **2017**, *175*, 235–247. [CrossRef] [PubMed]
32. Santiago, J.; Tegeder, M. Connecting Source with Sink: The Role of Arabidopsis AAP8 in Phloem Loading of Amino Acids. *Plant Physiol.* **2016**, *171*, 508–521. [CrossRef] [PubMed]

33. Zhang, L.; Tan, Q.; Lee, R.; Trethewy, A.; Lee, Y.; Tegeder, M. Altered xylem-phloem transfer of amino acids affects metabolism and leads to increased seed yield and oil content in Arabidopsis. *Plant Cell* **2010**, *22*, 3603–3620. [CrossRef] [PubMed]
34. Rolletschek, H.; Hosein, F.; Miranda, M.; Heim, U.; Gotz, K.; Schlereth, A.; Borisjuk, L.; Saalbach, I.; Wobus, U.; Weber, H. Ectopic Expression of an Amino Acid Transporter (VfAAP1) in Seeds of Vicia narbonensis and Pea Increases Storage Proteins. *Plant Physiol.* **2005**, *137*, 1236–1249. [CrossRef] [PubMed]

International Journal of
Molecular Sciences

MDPI

Article

Genome-Wide Association Study Identifies a Plant-Height—Associated Gene *OsPG3* in a Population of Commercial Rice Varieties

Shasha Peng [1,2,†], Yanchen Liu [2,†], Yuchen Xu [1,2], Jianhua Zhao [3], Peng Gao [3], Qi Liu [2], Shuangyong Yan [4], Yinghui Xiao [1,*], Shi-Min Zuo [3] and Houxiang Kang [2,*]

[1] College of Agronomy, Hunan Agricultural University, Changsha 410128, China; pengshasha0914@gmail.com (S.P.); ippxuyuchen@163.com (Y.X.)
[2] State Key Laboratory for Biology of Plant Diseases and Insect Pests, Institute of Plant Protection, Chinese Academy of Agricultural Sciences, Beijing 100193, China; liuyc@big.ac.cn (Y.L.); qiliu6786@gmail.com (Q.L.)
[3] Jiangsu Key Laboratory of Crop Genomics and Molecular Breeding/Zhongshan Biological Breeding Laboratory/Key Laboratory of Plant Functional Genomics of the Ministry of Education, Agricultural College of Yangzhou University, Yangzhou 225009, China; dx120180075@stu.yzu.edu.cn (J.Z.); yzugaopeng@163.com (P.G.); smzuo@yzu.edu.cn (S.-M.Z.)
[4] Tianjin Key Laboratory of Crop Genetic Breeding, Tianjin Crop Research Institute, Tianjin Academy of Agriculture Sciences, Tianjin 300112, China; bioponser@gmail.com
* Correspondence: xiaoyh@hunau.edu.cn (Y.X.); kanghouxiang@caas.cn (H.K.)
† These authors contributed equally to this work.

Abstract: Plant height is one of the most crucial components of plant structure. However, due to its complexity, the genetic architecture of rice plant height has not been fully elucidated. In this study, we performed a genome-wide association study (GWAS) to determine rice plant height using 178 commercial rice varieties and identified 37 loci associated with rice plant height (LAPH). Among these loci, in LAPH2, we identified a polygalacturonase gene, *OsPG3*, which was genetically and functionally associated with rice plant height. The rice plant exhibits a super dwarf phenotype when the knockout of the *OsPG3* gene occurs via CRISPR-Cas9 gene-editing technology. RNA-Seq analysis indicated that *OsPG3* modulates the expression of genes involved in phytohormone metabolism and cell-wall-biosynthesis pathways. Our findings suggest that *OsPG3* plays a vital role in controlling rice plant height by regulating cell wall biosynthesis. Given that rice architecture is one of the most critical phenotypes in rice breeding, *OsPG3* has potential in rice's molecular design breeding toward an ideal plant height.

Keywords: rice plant height; genome-wide association study; polygalacturonase; CRISPR/Cas9 gene editing; cell wall

Citation: Peng, S.; Liu, Y.; Xu, Y.; Zhao, J.; Gao, P.; Liu, Q.; Yan, S.; Xiao, Y.; Zuo, S.-M.; Kang, H. Genome-Wide Association Study Identifies a Plant-Height—Associated Gene *OsPG3* in a Population of Commercial Rice Varieties. *Int. J. Mol. Sci.* **2023**, *24*, 11454. https://doi.org/10.3390/ijms241411454

Academic Editor: Jong Seong Jeon

Received: 15 June 2023
Revised: 5 July 2023
Accepted: 11 July 2023
Published: 14 July 2023

1. Introduction

Rice (*Oryza sativa* L.) is a crucial crop that serves as a staple food for more than half of the world's population. The increasing global population, which is expected to reach 8.5 billion by 2030 and 9.7 billion by 2050, poses a significant challenge to food production [1]. Additionally, agricultural land degradation, farmland reduction, and environmental contamination further exacerbate the issue of food production [2,3]. Therefore, improving rice yields has been a major focus of agricultural research.

Plant height (PH) is one of the most important agronomic traits of rice. The introduction of semi-dwarf varieties of rice and wheat during the 'Green Revolution' in the 1960s was a significant breakthrough for crop yields [4,5]. Semi-dwarfism, characterized by shorter PH and stronger stems, has been a valuable trait in rice breeding because it enhances yield and reduces lodging [6,7]. Thus, understanding the genetic mechanisms underlying rice PH is crucial for crop improvement.

Previous research has demonstrated that PH is primarily regulated by numerous hormones [4], including gibberellin (GA) [5–9], brassinosteroid (BR) [10–12], indole-3-acetic acid (Aux/IAA) [13–15], and strigolactones (SLs) [16,17]. GA plays a vital role in promoting stem and internode elongation. For instance, the deficiency of the GA receptor GIBBERELLIN INSENSITIVE DWARF1(GID1) resulted in a dwarf phenotype in rice [18]. BR, a class of poly-hydroxysteroid plant hormones, significantly impacts the growth and development of rice plants [19,20]. *OsBRI1*, which is homologous to the Arabidopsis *BRI1* gene, encodes a putative BR receptor, and the loss-of-function of OsBRI1 prevents internode elongation [21]. Aux/IAA, which predominates among plant auxins, is crucial for various plant growth and development activities, including cell division, differentiation, elongation, floral and vascular development, tropism, and embryogenesis. The overexpression of *OsIAA1* resulted in decreased auxin sensitivity and plant height [22].

Besides plant-hormones-related pathways, several other rice PH-associated genes were identified. These genes are involved in various processes, including the development of cell walls, cytosolic glutamine synthesis, RNA editing, cell division, and fatty acid metabolism [23–27]. Plant cell walls comprise polysaccharides, including cellulose and pectin, which are important for mechanical strength, flexibility, and wall permeability [28]. The cell-wall-associated receptor-like kinase OsWAK10 and its variants regulate cell wall signaling and control the magnitude of secondary wall cellulose synthesis, thereby controlling rice PH [29].

The genome-wide association study (GWAS) is an efficient method for identifying loci associated with specific traits [30]. Over the past two decades, GWAS has been widely used to identify quantitative trait loci (QTLs) and genes associated with rice's numerous agronomic traits [31,32]. In this study, we conducted a GWAS to investigate the genetic basis of rice PH and identified dozens of PH-associated loci. Further haplotype analysis allowed us to focus on a candidate gene, *OsPG3*, which encodes polygalacturonase (PG), a pectin-degrading enzyme. The gene knockout mutant *ospg3* exhibits a decreased rice PH phenotype. We performed phylogenetic and structural analysis to predict its potential functions and detected its expression profile using multiple tissue samples throughout rice's development stages. In addition, RNA-Seq analysis indicated that *OsPG3* modulates the expression of genes involved in phytohormones metabolism and cell wall biosynthesis pathways. In short, our results suggest that *OsPG3* is associated with rice PH, and the suitable *OsPG3* haplotypes, through further screening, will help the development of rice varieties with ideal rice PH.

2. Results

2.1. Variation in Plant Height in 178 Homozygous Commercial Rice Varieties

To explore the genetic variation of rice PH, we collected a set of 178 varieties comprised of 109 *indica* and 69 *japonica* rice subpopulations (Table S1). Then, we planted these accessions in the field in Yangzhou (Southeast China) and measured rice PH at maturate stage (Table S1). The rice PH exhibited a normal distribution ranging from 69.67 cm to 190.67 cm, with an average of 111.74 ± 19.45 cm (Figure 1A, Table S2). Consistent with previous research suggesting that *indica* rice generally has a higher average PH than *japonica* rice [33], we also observed a significantly higher average PH in *indica* rice compared to *japonica* rice, and the highest cultivar was from the *indica* rice subpopulation (Figure 1B). These results indicate that similar to the rice diversity germplasm resources [32], there is also diversity in the rice PH phenotype in the selected commercial rice variety population.

2.2. Identification of the Locus Associated with Plant Height (LAPH) and PH Candidate Genes through Genome-Wide Association Study

To identify the loci associated with plant height (LAPHs) in rice, we performed GWAS of the PH phenotypes and the genotypes of 42,469 single-nucleotide polymorphisms (SNPs) with a minor allele frequency (MAF) at least 0.05% in 178 commercial rice varieties. We used the mixture linear model (MLM) for the analysis and identified 314 SNPs significantly associated with PH (Figure 1C and Table S3). Based on the average size of the rice linkage disequilibrium (LD) decay [34], these 314 SNPs could be further divided into 37 non-redundant LAPH regions and distributed on rice chromosomes 1, 2, 3, 4, 5, 8, 9, 10, and 12, respectively. Among these LAPH regions, ten co-localized with previously reported quantitative trait loci (QTLs) or genes linked to rice PH phenotype (Figure 1C and Table S4). For instance, LAPH9 co-localized with the GA biosynthesis gene, *SD1* [7], and LAPH29 co-localized with the semi-dominant QTL, *Ideal Plant Architecture 1* (*IPA1*) [35].

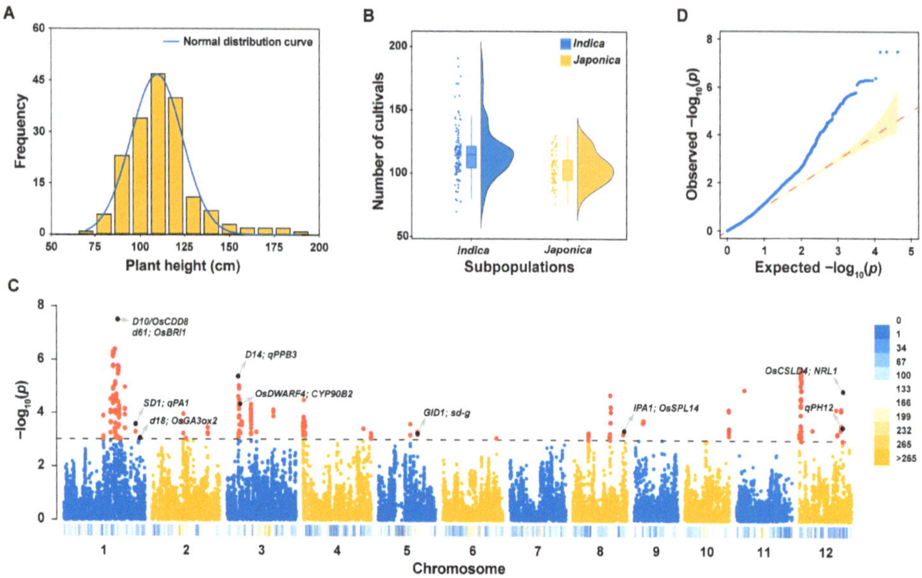

Figure 1. PH phenotypic distribution and genome-wide association scan in 178 commercial rice varieties. (**A**) Histogram illustrating PH diversity in the rice population. (**B**) PH differences between the two rice subpopulations, with each dot representing a rice variety. Boxplot lines indicate the average PH value for the two subpopulations. (**C**) Manhattan plot of the genome-wide *p*-values from rice PH GWAS. The x-axis shows the SNPs along each chromosome, while the y-axis represents -log10 (*p*-value) for the association. Red dots indicate SNPs with *p*-values $< 1 \times 10^{-3}$. Grey arrows indicate the PH-associated loci co-located with known QTLs or genes. (**D**) Quantile–quantile plot of rice PH GWAS.

It is noteworthy that 27.03% (10 out of 37) of the LAPH regions identified in this study were on chromosome 1 (as shown in Figures 1C and 2A and Table S4). Among those loci, we selected one previously unreported locus, LAPH2, for further investigation. The candidate genomic region spanned from 25,224 to 25,704 kb based on the LD block analysis around the top SNP (Figure 2A), which yielded 56 annotated genes (Table S5). Out of these genes, the number of genes annotated as retrotransposon proteins, expressed proteins, and hypothetical proteins were 6, 16, and 3, respectively. According to the type of protein encoded in this region, we focus on a gene, LOC_Os01g45060 (named *OsPG3*), that encodes a polygalacturonase (PG) for further study. The PG gene family is a large gene family in plants; it has been reported to be responsible for various cell-separation processes [36,37]. We then cloned and sequenced *OsPG3* in seven accessions with high PH and seven accessions with low PH (Figure 2B and Table S6). A total of 25 polymorphisms for *OsPG3* were identified across the 14 accessions (Figure 2B). Among them, 19 polymorphisms are distributed in the non-coding region, and 6 SNPs are located in the exon region of *OsPG3*. Among the six SNPs, three resulted in missense mutations.

Specifically, two missense mutations occurred in the first exon, and one occurred in the second exon (Figure 2B).

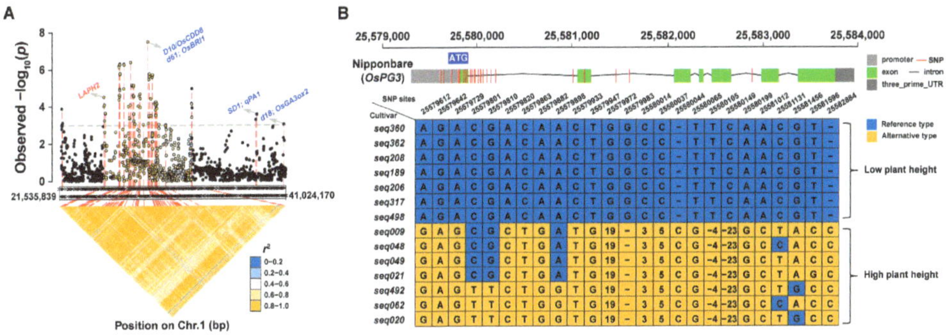

Figure 2. *OsPG3* is genetically linked to rice PH. (**A**) Genome-wide association signals and LD heatmap of *OsPG3* region on rice chromosome 1. The red-labeled loci are *OsPG3* locus associated with PH. Blue-labeled loci represent the three known QTLs or genes related to rice PH. (**B**) *OsPG3* sequence variation in different rice cultivars. The top part displays the *OsPG3* gene structure, while the bottom panel represents the detailed sequence variations in fourteen cultivars.

2.3. Phylogenetic Analysis of PG Genes

Sequence alignment and conserved domains analysis were performed using the Basic Local Alignment Search Tool (BLAST) (https://blast.ncbi.nlm.nih.gov/Blast.cgi/, accessed on 23 March 2023) and Simple Modular Architecture Research Tool (SMART) [38] (https://smart.embl.de/, accessed on 24 March 2023). The analysis revealed that OsPG3 belongs to the pectin lyase-like superfamily and contains five or more parallel beta-helix repeat (PbH1) motifs that are conserved in this protein family (Figure S1). To investigate the relationships and gene structure of the PG genes, we utilized 44 PG genes [37,39] from rice and 49 PG genes [40] from *Arabidopsis* to construct a phylogenetic tree, which allowed us to classify these PG genes into 5 clades (Figure 3A). These five clades are consistent with the classification of *Arabidopsis* PGs, which were also divided into five clades (A to E) [37]. In the phylogenetic tree, *OsPG3* is in Clade A, where it is closely related to seven other rice PG genes (Figure 3A,B). It is noteworthy that one of these genes, LOC_Os01g19170 (hereafter named *OsPG1* or *PSL1*), has been shown to play a role in modifying rice cell wall structure. Loss-of-function mutation of this gene results in leaf rolling and leaf tip necrosis, as well as reduced plant height compared to wild-type plants [41,42]. These results suggest that *OsPG3* may also play a role in regulating the plant cell wall structure and influencing rice PH. To further analyze the conserved motifs of the rice PG genes, 44 rice PG protein sequences were aligned using the online tool MEME [43], and a total of 10 conserved motifs (Motif 1-10) were identified. Among these motifs, motifs 1, 2, 3, 4, 8, 9, and 10 represented the highly conserved domain PLN02218 and PL-6 of PG genes involved in clades A-D (as shown in Figure 3B). Motifs 1, 5, and 8 were present in all 44 PG protein sequences except for LOC_Os06g01760 (Figure 3B). Additionally, motif 7 was found exclusively in members of clade E (Figure 3B). Notably, *OsPG3* is classified in Clade A and contains all the PG protein motifs, except for motifs 7 and 8.

2.4. OsPG3 Is Essential for Maintaining Rice Plant Height

To investigate the role of *OsPG3* in rice growth and development, we analyzed its spatial and temporal expression pattern. The total RNA was extracted from the roots, stems, and leaves of ZH11 at the seedling stage, tillering stage, and mature stage, respectively. RT-qPCR showed that, in three different growth stages, *OsPG3* transcripts were abundant in the stems rather than in the roots and leaves (Figure 4A). Gene-expression patterns are usually associated with their function [44], and the transcription levels of *OsPG3* in the stem were exceptionally higher than those in the root and leaf, indicating that *OsPG3* might play an important role in stem development.

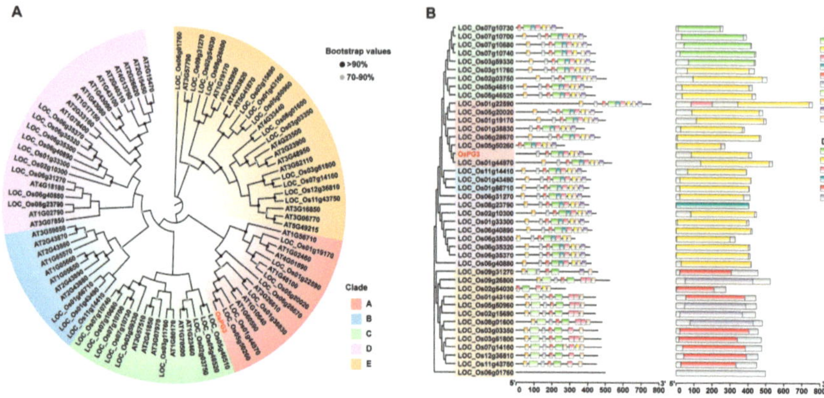

Figure 3. Phylogenetic analysis of PG Genes. (**A**) Phylogenetic analysis of PG gene families between rice and *Arabidopsis*. AT represents *Arabidopsis thaliana*; Os represents *Oryza sativa*. (**B**) MEME motif analysis identified domains of PG genes in rice.

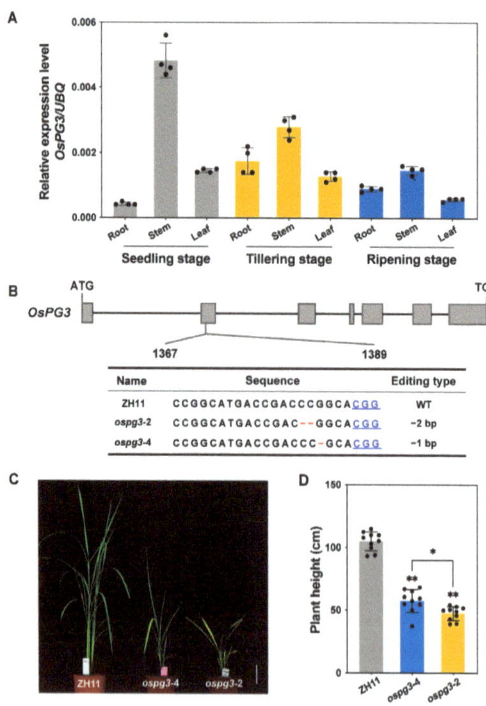

Figure 4. *OsPG3* gene knockout results in a dwarf phenotype in rice. (**A**) *OsPG3* transcription levels in various rice tissues. 'Seedling', 'tillering', and 'ripening' indicate different growth stages. (**B**) Editing types of *ospg3* mutant. The red-labeled '−−' and '−' indicate two base and one base deletions, respectively. (**C**) Phenotype comparison between wild-type ZH11 and two *ospg3* mutants. Scale bar = 10 cm. (**D**) Histogram of PH differences between wild-type ZH11 and two *ospg3* mutants. '**' represents very significant differences ($p < 0.01$), and '*' represents significant differences ($p < 0.05$).

Based on the results of the GWAS and expression analysis, it was hypothesized that *OsPG3* is involved in rice plant growth. To further validate the function of *OsPG3*, we generated a CRISPR/Cas9 knockout mutant *ospg3* using the *japonica* rice cultivar Zhonghua 11 (ZH11) through *Agrobacterium*-mediated transformation. Two independent homozygous lines, *ospg3*-2 and *ospg3*-4, were used for further investigation (as shown in Figure 4B). The gene-knockout mutant exhibited a plant height that was over 40% shorter than the corresponding wild-type plants at the reproductive stage (Figure 4C,D). In addition, we observed a significant decrease in plant height and shoot length in the *ospg3* mutant compared to the wild-type rice at the seedling stage, as shown in Figure S2. These results indicate that the *OsPG3* gene plays a crucial role in rice normal growth and development, as its knockout resulted in a super-dwarf phenotype.

2.5. OsPG3 Modulates the Expression of Genes Involved in Phytohormone Metabolism and Cell-Wall-Biosynthesis Pathways

In order to investigate the role of *OsPG3* in regulating rice PH, we performed a global transcriptional comparison between the knockout mutant *ospg3* and wild-type ZH11 plants using RNA-Seq analysis. A volcano plot was generated to visualize the differentially expressed genes (DEGs) (Figure 5A). There were 440 DEGs between ZH11 and the *ospg3* knockout lines, of which 266 DEGs were up-regulated and 174 DEGs were down-regulated. The GO enrichment analysis indicated that most of the differentially expressed genes (DEGs) were annotated to biological process terms (Figures 5B and S3), with the most common term being the metabolic process, indicating that *OsPG3* has a wide-ranging effect on metabolic activities. GO analysis also revealed that the DEGs were involved in cell wall biogenesis, cellulose biosynthetic process, plant-type cell wall biogenesis, and plant-type cell wall organization or biogenesis (Figures 5B and S3), indicating that *OsPG3* affects cell wall biosynthesis. In addition, the RNA-Seq data showed that the expression levels of genes related to development were apparently decreased (Figure S4). To validate our findings from RNA-Seq analysis, we conducted qRT-PCR analysis on the expression levels of the secondary cell wall and Auxin/IAA-related genes using the wild-type ZH11 and knockout mutant *ospg3*-4 and *ospg3*-2 plants; detailed results are presented in Figure 5C. Notably, the expressions of three cellulose synthase genes, CESA4 (LOC_Os01g54620), CESA7 (LOC_Os10g32980), and CESA9 (LOC_Os09g25490), which are part of a cellulose-synthesizing complex involved in the synthesis of the secondary cell wall [45], were down-regulated in the *ospg3* mutant (Figures 5C and S4). Moreover, the expression of *Brittle Culm1* (LOC_ Os03g30250), which encodes a COBRA-like protein precursor that regulates the biosynthesis of secondary cell walls [46], was also down-regulated in the *ospg3* mutant (Figures 5C and S4). We further analyzed the DEGs using the KEGG database to identify enriched pathways (Figure 5D). We found that the main enriched pathways were the biosynthesis of secondary metabolites, metabolic pathways, and phenylpropanoid biosynthesis, indicating that *OsPG3* might regulate rice plant architecture through these metabolic processes (Figure 5E).

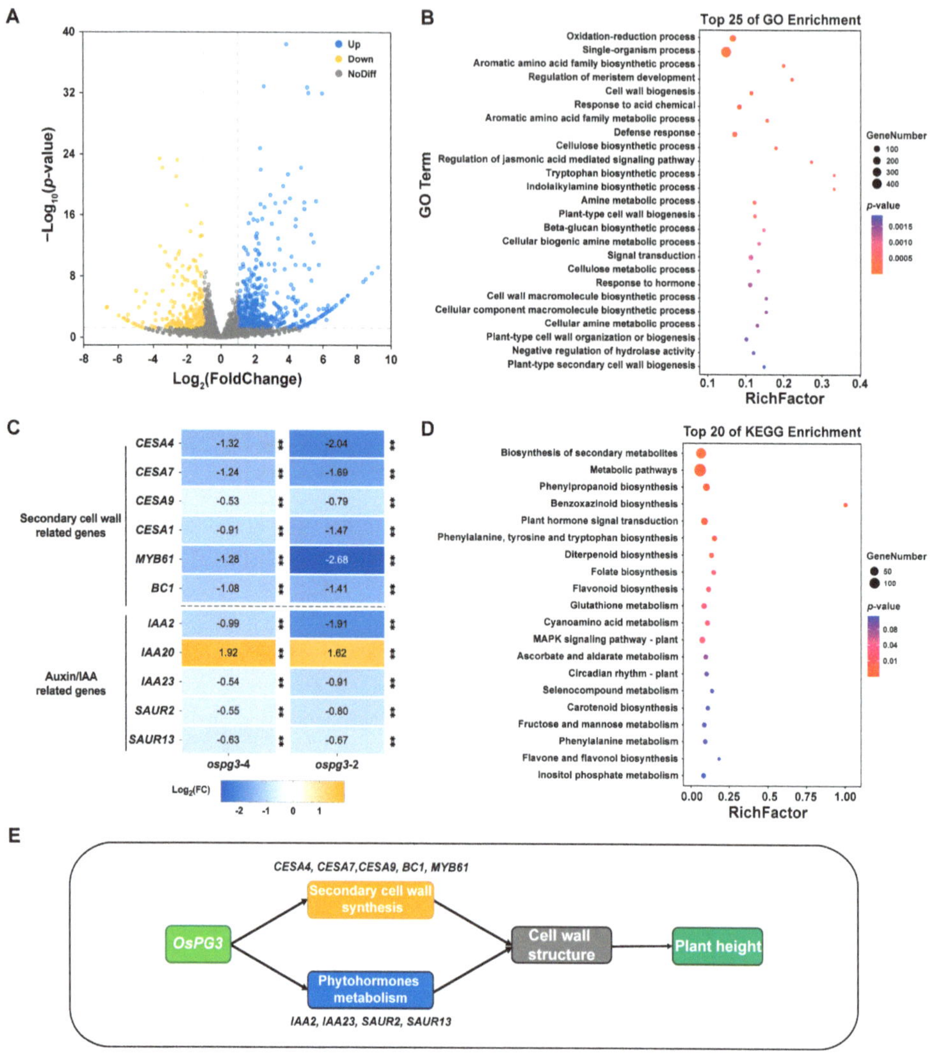

Figure 5. Differentially expressed genes in rice stem identified by RNA-Seq comparing wild-type ZH11 and *ospg3* mutant. (**A**) Volcano plot of differentially expressed genes. (**B**) Top 25 Gene Ontology (GO) enrichment categories. (**C**) Heat map showing the expression profiles of typical cell wall biosynthesis and regulatory genes and Auxin/IAA-related genes in *ospg3* knockout mutant. Yellow represents up-regulation, and blue represents down-regulation (** $p < 0.01$, *t*-test). (**D**) KEGG pathway enrichment of differentially expressed genes (DEGs). (**E**) The proposed model for *OsPG3* mediating plant height in rice.

3. Discussion

Plant height is not only a crucial factor of plant architecture but also an important agronomic trait that is directly linked to yield potential [47]. Over the past few decades, several genes involved in regulating PH have been identified, many of which are involved in plant hormone regulation [48,49]. Notably, some additional dwarfing genes were found to participate in other pathways, such as cell wall development, cytosolic glutamine synthesis, and cell division [24,26,50]. In this study, we observed significant variations in PH among a set of 178 commercial rice varieties at the maturate stage (Figure 1A). We found that *indica* rice has a higher average PH than *japonica* rice across rice subpopulations (Figure 1B). The large PH variations in this rice population provided an ideal resource for dissecting the genetic basis of rice PH.

GWAS is an efficient method for identifying genomic regions associated with a given agronomic trait, including rice PH [51–53]. In this study, we performed a GWAS using the PH phenotypes of 178 commercial rice varieties, and the genotype data consisted of 42,469 SNPs (Figure 1C,D). We identified 37 LAPHs (Figure 1C, Table S4), of which 10 co-localized with the previously reported QTLs related to rice PH [7,9,11,13,18,21,23,35,54,55]. Additionally, our GWAS analysis revealed that most of the significant SNPs associated with plant height (PH) are located on chromosomes 1, 3, and 12, while there are few significant SNPs on chromosomes 6, 7, and 11. These chromosomes contain genes previously reported to be related to PH, such as *OsMPH1* on chromosome 6, which, when overexpressed, increases PH by elongating internode cell length [56] and loss-of-function mutations in the rice homeobox gene *OSH15* on chromosome 7, which affect the internode architecture and result in dwarf plants [57]. However, we utilized a set of 178 commercial rice varieties in this study; the diversity of the commercial rice varieties may not be as comprehensive as that of natural populations. Finally, we identified a gene, *OsPG3* (LOC_Os01g45060), in LAPH2 on chromosome 1, with nucleotide polymorphisms linked to the rice PH phenotype (Figure 2A,B). *OsPG3* is annotated as a polygalacturonase (PG), a pectin-degrading enzyme involved in the cell wall modification process.

The cell wall not only provides mechanical strength to plant tissues but also regulates plant growth and development [58]. The function and regulation of plant cell wall hydrolytic enzymes have been well-studied [28]. These enzymes can alter the cell wall extensibility and cellular adhesion, leading to cell wall loosening, cell elongation, root tip sloughing, and fruit softening [39,58]. Plant PGs, among these enzymes, belong to the large Glycoside Hydrolase Family 28 (GH28), a member of the Glycoside Hydrolase (GH) superfamily in organisms [37,59]. PGs have been identified in various plants, including *Arabidopsis*, apples, and peaches [60–62]. In *Arabidopsis*, PGs encoded by *ADPG1* and *ADPG2* are essential for cell-separation and -expansion events during reproductive development [63]. Overexpression of *PGX1*, encoding a PG involved in cell expansion, led to enhanced hypocotyl elongation in etiolated *Arabidopsis* seedlings [64]. In rice, *PSL1* (*OsPG1*) has been shown to play a role in modifying the cell wall's structure. Loss of function mutation of this gene results in leaf rolling and leaf tip necrosis, as well as reduced plant height compared to wild-type plants [41,42]. In our study, through phylogenetic and gene structure analyses, we identified *OsPG3* as a typical PG gene and classified it in Clade A of the PG gene family (Figures 3A,B and S1).

The spatial–temporal expression profile showed that *OsPG3* was primarily expressed in the stems (Figure 4A). Loss-of-function of *OsPG3* resulted in a super dwarf phenotype (Figure 4C,D), indicating that *OsPG3* is essential for maintaining the height of rice plants. RNA-Seq analysis supported the *OsPG3* functions in the regulation of phytohormones metabolism and cell wall biosynthesis pathways (Figure 5). Thus, identifying and utilizing suitable natural alleles or creating appropriate alleles through gene editing are potential approaches for further utilizing *OsPG3* in rice molecular design breeding toward ideal plant height.

4. Materials and Methods

4.1. Plant Materials and Growth Conditions

A total of 178 homozygous commercial rice varieties were used for the plant height analysis, including 109 *indica* and 69 *japonica* varieties (as listed in Table S1). These varieties were cultivated using standard local practices during the natural growing season (April to September) of 2019 in Yangzhou, located in the Jiangsu province of China (32.39° N 119.42° E). Plant heights were measured after grain maturity; the data can be found in Tables S1 and S2.

4.2. Genome-Wide Association Study

GWAS analysis was performed using the mixture linear (MLM) method in TASSEL software (V5.2.87) [65]. A total of 42,469 SNPs with a minor allele frequency (MAF) > 0.05 were used for GWAS. The SNP data used in this study were personally obtained from a submitted paper titled 'A natural variation of sheath blight-resistance receptor-like kinase 1 (SBRR1) improves sheath blight resistance in rice.' It will be publicly available soon. The kinship matrix (K value) and the population structure Q value were used as random effects. The effective numbers of independent SNPs and significance thresholds were calculated following a previously described method [66]; the *p*-value threshold used in this study is 1E-03. Manhattan and Q-Q plots were generated using the CMplot *R* package [67]. Based on the average size of the LD decay blocks in rice [34], a 200 kb interval with at least two significantly associated SNPs was used for candidate gene analysis.

4.3. DNA Extraction and Gene Sequencing

Fourteen rice varieties, including seven with low plant height and seven with high plant height, were chosen for candidate gene sequencing. Rice seedling DNA was extracted using the CTAB method. The rice genome of the MSU7.0 version (http://rice.plantbiology.msu.edu/, accessed on 17 March 2022) was used as a reference for primer design (primers in Table S7). PCR and Sanger sequencing were then performed to obtain the sequences of candidate genes in all 14 rice varieties. The sequences were assembled and aligned using Snapgene6.0.2 and MEGA(V7.0) software [68].

4.4. Phylogenetic and Motif Analysis

Protein domain annotation was based on UniProt (https://www.uniprot.org, OsPG3 protein accession number: Q5VNN6_ORYSJ). A total of 44 rice PG protein sequences and 49 *Arabidopsis* PG protein sequences were obtained from blastp and tblastn searches against the National Center for Biotechnology Information (NCBI, https://www.ncbi.nlm.nih.gov/, accessed on 20 April 2023) database, using the glycosyl hydrolase family 28 (GH28, Pfam accession: PF00295) as the query. The phylogenetic tree was constructed using the neighbor-joining method with 1000 bootstrap pseudoreplicates in MEGA7. The resulting circle tree was visualized using the online tool tvBOT [69]. Conserved motifs in rice PGs protein sequences were identified using the MEME (v5.52) [43] (https://meme-suite.org/meme/tools/meme, accessed on 4 April 2023), with the following parameters: Classic mode, Zero or One Occurance Per Sequence (zoops); the number of motifs expected to be found was set to 25; and the motif width was set from 6 to 100 in 'Advanced options.' The conserved domains analysis was performed using the NLM's Conserved Domains Database (CDD) (https://www.ncbi.nlm.nih.gov/Structure/bwrpsb/bwrpsb.cgi/, accessed on 20 April 2023). The MEME and conserved domain results were visualized using the TBtools software (V1.131) [70].

4.5. Vector Construction and Rice Transformation

The *ospg3* mutant was generated using CRISPR/Cas9 gene-editing technology in the ZH11 background to obtain a loss-of-function mutant. The second exon of *OsPG3* was targeted using sgRNA (5′-CCGGCATGACCGACCCGGCA-3′) designed using the online toolkit for CRISPR-based genome editing (CRISPR-GE) [71] (http://skl.scau.edu.cn/betarget/, accessed on 20 January 2019). The *Agrobacterium*-mediated transformation of rice callus (ZH11) and regeneration of rice plants for *OsPG3* gene editing was conducted following previously published protocols [72]. Homozygous *ospg3* mutants were screened via PCR and Sanger sequencing of the sgRNA target site (Tsingke Biotech, Beijing, China). Primer sequences used for plasmid construction and mutant identification are listed in Table S7.

4.6. Gene Expression Analysis

Spatial and temporal expression profiles of *OsPG3* were studied by collecting stem, leaf, and root tissues from ZH11 at the seedling, tillering, and ripening stages, respectively. RNA extraction and cDNA synthesis were performed using UNIQ-10 Column Trizol Total RNA Isolation Kit from Sangon Biotech (code: B511321, Shanghai, China) and HiScript II 1st Strand cDNA Synthesis Kit from Vazyme (code: R212-01, Nanjing, China), following the manufacturer's instructions. For qRT-PCR, ChamQ SYBR qPCR Master Mix from Vazyme (code: Q311-02, Nanjing, China) was used with gene-specific primers (Table S7) and the following PCR cycling conditions: 95 °C for 30 s for pre-incubation, followed by 40 cycles of 95 °C for 5 s and 60 °C for 30 s for 2-step amplification. The rice *ubiquitin* (*UBQ*, LOC_Os03g13170) was used as an internal control for normalization. Transcript levels relative to *UBQ* were calculated using $2^{-\Delta\Delta Ct}$ methods, three biological replicates, and three technical replicates per sample.

4.7. RNA-Seq and Sequence Analysis

Rice plants were cultivated in pots within a greenhouse environment, and developing stem samples were collected from two-month-old wild-type (ZH11) and *ospg3* knockout mutant plants. Three biological replicates were collected for each sample at each time point. RNA extraction, library construction, and sequencing were carried out following the methods described in a previous study [73]. The clean RNA sequences were aligned to the rice cDNA file (MSUv7.0) using BWA software (V0.7.17), and the aligned data were viewed and sorted using Samtools. The number of reads aligned to each rice gene was calculated using the 'grep' and 'awk' commands in the Linux system. The *R* package DESeq2 was used to identify the differentially expressed genes (DEGs) [74]. The DEGs were determined using a $|\log_2(\text{FoldChange, FC})| \geq 1$ and a *p*-value < 0.05 as the criteria. The DEGs were classified via Gene Ontogeny (GO) annotation (http://www.geneontology.org/, accessed on 12 March 2023). The GO analyses were performed using AgriGO V2.0. The protein pathways were annotated using the Kyoto Encyclopedia of Genes and Genomes (KEGG) database. The online service tool KAAS was used to annotate the KEGG database description (http://www.genome.jp/tools/kaas/, accessed on 13 March 2023), and the resulting annotations were mapped onto the KEGG pathway database using KEGG MAPPER.

4.8. Data Analysis and Figures

The *p*-values were computed using Student's t-test, ANOVA, two-tailed Mann–Whitney U-test, and an unpaired two-sample *t*-test, Pearson correlation (r^2) was computed with GraphPad Prism (version 9.1.1), and $p < 0.05$ was considered a significant difference. The histograms were visualized in GraphPad Prism. The CloudRain plot and RegionalPlot were plotted using *R* package ggplot2 [75].

Supplementary Materials: The following supporting information can be downloaded at: https://www.mdpi.com/article/10.3390/ijms241411454/s1.

Author Contributions: H.K. and S.-M.Z. designed the research. S.P., Y.L., J.Z., P.G. and Y.X. (Yuchen Xu) conducted the research. Y.X. (Yuchen Xu), Q.L., S.Y. and Y.X. (Yinghui Xiao) contributed to the experiments. S.P., H.K. and Y.L. performed GWAS analysis. J.Z., P.G. and S.P. contributed to the field management. S.P., Y.L. and Y.X. (Yuchen Xu) analyzed data. H.K. developed the database and bioinformatics tools. S.P. and H.K. analyzed the data and wrote the manuscript. All authors have read and agreed to the published version of the manuscript.

Funding: This work was financially supported by the National Key R&D Program of China (Grant 2021YFC2600400), the National Natural Science Foundation of China (Grant 32261143468), and the International Science and Technology Innovation Program of the Chinese Academy of Agricultural Sciences (CAASTIP) (CAAS-ZDRW202108). And the Project of Zhongshan Biological Breeding Laboratory.

Institutional Review Board Statement: Not applicable.

Informed Consent Statement: Not applicable.

Data Availability Statement: The data presented in this study are available on request from the corresponding authors.

Acknowledgments: We thank Guoliang Wang (The Ohio State University) for helpful suggestions related to this study.

Conflicts of Interest: The authors declare no conflict of interest.

References

1. Gerland, P.; Hertog, S.; Wheldon, M.; Kantorova, V.; Gu, D.; Gonnella, G.; Williams, I.; Zeifman, L.; Bay, G.; Castanheira, H.; et al. *World Population Prospects 2022: Summary of Results*; Department of Economic and Social Affairs, United Nations: New York, NY, USA, 2022.
2. McClung, C.R. Plant science. Making hunger yield. *Science* **2014**, *344*, 699–700. [CrossRef]
3. Velten, S.; Leventon, J.; Jager, N.; Newig, J. What Is Sustainable Agriculture? A Systematic Review. *Sustainability* **2015**, *7*, 7833–7865. [CrossRef]
4. Ferrero-Serrano, A.; Cantos, C.; Assmann, S.M. The Role of Dwarfing Traits in Historical and Modern Agriculture with a Focus on Rice. *Cold Spring Harb. Perspect. Biol.* **2019**, *11*, a034645. [CrossRef]
5. Sakamoto, T.; Matsuoka, M. Generating high-yielding varieties by genetic manipulation of plant architecture. *Curr. Opin. Biotechnol.* **2004**, *15*, 144–147. [CrossRef]
6. Hedden, P. The genes of the Green Revolution. *Trends Genet.* **2003**, *19*, 5–9. [CrossRef] [PubMed]
7. Sasaki, A.; Ashikari, M.; Ueguchi-Tanaka, M.; Itoh, H.; Nishimura, A.; Swapan, D.; Ishiyama, K.; Saito, T.; Kobayashi, M.; Khush, G.S.; et al. Green revolution: A mutant gibberellin-synthesis gene in rice. *Nature* **2002**, *416*, 701–702. [CrossRef]
8. Liu, C.; Zheng, S.; Gui, J.; Fu, C.; Yu, H.; Song, D.; Shen, J.; Qin, P.; Liu, X.; Han, B.; et al. Shortened Basal Internodes Encodes a Gibberellin 2-Oxidase and Contributes to Lodging Resistance in Rice. *Mol. Plant* **2018**, *11*, 288–299. [CrossRef] [PubMed]
9. Itoh, H.; Ueguchi-Tanaka, M.; Sentoku, N.; Kitano, H.; Matsuoka, M.; Kobayashi, M. Cloning and functional analysis of two gibberellin 3 beta-hydroxylase genes that are differently expressed during the growth of rice. *Proc. Natl. Acad. Sci. USA* **2001**, *98*, 8909–8914. [CrossRef] [PubMed]
10. Mori, M.; Nomura, T.; Ooka, H.; Ishizaka, M.; Yokota, T.; Sugimoto, K.; Okabe, K.; Kajiwara, H.; Satoh, K.; Yamamoto, K.; et al. Isolation and characterization of a rice dwarf mutant with a defect in brassinosteroid biosynthesis. *Plant Physiol.* **2002**, *130*, 1152–1161. [CrossRef]
11. Sakamoto, T.; Morinaka, Y.; Ohnishi, T.; Sunohara, H.; Fujioka, S.; Ueguchi-Tanaka, M.; Mizutani, M.; Sakata, K.; Takatsuto, S.; Yoshida, S.; et al. Erect leaves caused by brassinosteroid deficiency increase biomass production and grain yield in rice. *Nat. Biotechnol.* **2006**, *24*, 105–109. [CrossRef]
12. Qiao, S.; Sun, S.; Wang, L.; Wu, Z.; Li, C.; Li, X.; Wang, T.; Leng, L.; Tian, W.; Lu, T.; et al. The RLA1/SMOS1 Transcription Factor Functions with OsBZR1 to Regulate Brassinosteroid Signaling and Rice Architecture. *Plant Cell* **2017**, *29*, 292–309. [CrossRef] [PubMed]
13. Zhang, S.; Li, G.; Fang, J.; Chen, W.; Jiang, H.; Zou, J.; Liu, X.; Zhao, X.; Li, X.; Chu, C.; et al. The interactions among DWARF10, auxin and cytokinin underlie lateral bud outgrowth in rice. *J. Integr. Plant Biol.* **2010**, *52*, 626–638. [CrossRef] [PubMed]
14. Zhang, S.; Wang, S.; Xu, Y.; Yu, C.; Shen, C.; Qian, Q.; Geisler, M.; Jiang, A.; Qi, Y. The auxin response factor, OsARF19, controls rice leaf angles through positively regulating OsGH3-5 and OsBRI1. *Plant Cell Environ.* **2015**, *38*, 638–654. [CrossRef] [PubMed]
15. Guo, F.; Huang, Y.; Qi, P.; Lian, G.; Hu, X.; Han, N.; Wang, J.; Zhu, M.; Qian, Q.; Bian, H. Functional analysis of auxin receptor OsTIR1/OsAFB family members in rice grain yield, tillering, plant height, root system, germination, and auxinic herbicide resistance. *New Phytol.* **2021**, *229*, 2676–2692. [CrossRef]
16. Jiang, L.; Liu, X.; Xiong, G.; Liu, H.; Chen, F.; Wang, L.; Meng, X.; Liu, G.; Yu, H.; Yuan, Y.; et al. DWARF 53 acts as a repressor of strigolactone signalling in rice. *Nature* **2013**, *504*, 401–405. [CrossRef]
17. Zhou, F.; Lin, Q.; Zhu, L.; Ren, Y.; Zhou, K.; Shabek, N.; Wu, F.; Mao, H.; Dong, W.; Gan, L.; et al. D14-SCF(D3)-dependent degradation of D53 regulates strigolactone signalling. *Nature* **2013**, *504*, 406–410. [CrossRef]
18. Ueguchi-Tanaka, M.; Ashikari, M.; Nakajima, M.; Itoh, H.; Katoh, E.; Kobayashi, M.; Chow, T.Y.; Hsing, Y.I.; Kitano, H.; Yamaguchi, I.; et al. GIBBERELLIN INSENSITIVE DWARF1 encodes a soluble receptor for gibberellin. *Nature* **2005**, *437*, 693–698. [CrossRef]
19. Clouse, S.D. Brassinosteroid signal transduction: From receptor kinase activation to transcriptional networks regulating plant development. *Plant Cell* **2011**, *23*, 1219–1230. [CrossRef]
20. Wang, Z.Y.; Bai, M.Y.; Oh, E.; Zhu, J.Y. Brassinosteroid signaling network and regulation of photomorphogenesis. *Annu. Rev. Genet.* **2012**, *46*, 701–724. [CrossRef]
21. Yamamuro, C.; Ihara, Y.; Wu, X.; Noguchi, T.; Fujioka, S.; Takatsuto, S.; Ashikari, M.; Kitano, H.; Matsuoka, M. Loss of function of a rice brassinosteroid insensitive1 homolog prevents internode elongation and bending of the lamina joint. *Plant Cell* **2000**, *12*, 1591–1606. [CrossRef]
22. Song, Y.; You, J.; Xiong, L. Characterization of OsIAA1 gene, a member of rice Aux/IAA family involved in auxin and brassinosteroid hormone responses and plant morphogenesis. *Plant Mol. Biol.* **2009**, *70*, 297–309. [CrossRef]

23. Luan, W.; Liu, Y.; Zhang, F.; Song, Y.; Wang, Z.; Peng, Y.; Sun, Z. OsCD1 encodes a putative member of the cellulose synthase-like D sub-family and is essential for rice plant architecture and growth. *Plant Biotechnol. J.* **2011**, *9*, 513–524. [CrossRef]
24. Zhou, H.L.; He, S.J.; Cao, Y.R.; Chen, T.; Du, B.X.; Chu, C.C.; Zhang, J.S.; Chen, S.Y. OsGLU1, a putative membrane-bound endo-1,4-beta-D-glucanase from rice, affects plant internode elongation. *Plant Mol. Biol.* **2006**, *60*, 137–151. [CrossRef]
25. Kim, S.R.; Yang, J.I.; Moon, S.; Ryu, C.H.; An, K.; Kim, K.M.; Yim, J.; An, G. Rice OGR1 encodes a pentatricopeptide repeat-DYW protein and is essential for RNA editing in mitochondria. *Plant J.* **2009**, *59*, 738–749. [CrossRef] [PubMed]
26. Asano, K.; Miyao, A.; Hirochika, H.; Kitano, H.; Matsuoka, M.; Ashikari, M. SSD1, which encodes a plant-specific novel protein, controls plant elongation by regulating cell division in rice. *Proc. Jpn. Acad. Ser. B Phys. Biol. Sci.* **2010**, *86*, 265–273. [CrossRef]
27. Liu, H.L.; Yin, Z.J.; Xiao, L.; Xu, Y.N.; Qu, Q. Identification and evaluation of omega-3 fatty acid desaturase genes for hyperfortifying alpha-linolenic acid in transgenic rice seed. *J. Exp. Bot.* **2012**, *63*, 3279–3287. [CrossRef] [PubMed]
28. Hofte, H.; Voxeur, A. Plant cell walls. *Curr. Biol.* **2017**, *27*, R865–R870. [CrossRef] [PubMed]
29. Yue, Z.L.; Liu, N.; Deng, Z.P.; Zhang, Y.; Wu, Z.M.; Zhao, J.L.; Sun, Y.; Wang, Z.Y.; Zhang, S.W. The receptor kinase OsWAK11 monitors cell wall pectin changes to fine-tune brassinosteroid signaling and regulate cell elongation in rice. *Curr. Biol.* **2022**, *32*, 2454–2466.e7. [CrossRef]
30. Uffelmann, E.; Huang, Q.Q.; Munung, N.S.; de Vries, J.; Okada, Y.; Martin, A.R.; Martin, H.C.; Lappalainen, T.; Posthuma, D. Genome-wide association studies. *Nat. Rev. Methods Primers* **2021**, *1*, 59. [CrossRef]
31. Huang, X.; Wei, X.; Sang, T.; Zhao, Q.; Feng, Q.; Zhao, Y.; Li, C.; Zhu, C.; Lu, T.; Zhang, Z.; et al. Genome-wide association studies of 14 agronomic traits in rice landraces. *Nat. Genet.* **2010**, *42*, 961–967. [CrossRef]
32. Zhao, K.; Tung, C.W.; Eizenga, G.C.; Wright, M.H.; Ali, M.L.; Price, A.H.; Norton, G.J.; Islam, M.R.; Reynolds, A.; Mezey, J.; et al. Genome-wide association mapping reveals a rich genetic architecture of complex traits in *Oryza sativa*. *Nat. Commun.* **2011**, *2*, 467. [CrossRef] [PubMed]
33. Zhan, Z.; Ting, M.; Xin, L.; Yizhou, Z.; Mingfei, Y. Research on Difference and Classification of Indica-japonica Subspecies at Asian Cultivated Rice. *North Rice* **2013**, *43*, 66–69. [CrossRef]
34. Mather, K.A.; Caicedo, A.L.; Polato, N.R.; Olsen, K.M.; McCouch, S.; Purugganan, M.D. The extent of linkage disequilibrium in rice (*Oryza sativa* L.). *Genetics* **2007**, *177*, 2223–2232. [CrossRef] [PubMed]
35. Jiao, Y.; Wang, Y.; Xue, D.; Wang, J.; Yan, M.; Liu, G.; Dong, G.; Zeng, D.; Lu, Z.; Zhu, X.; et al. Regulation of OsSPL14 by OsmiR156 defines ideal plant architecture in rice. *Nat. Genet.* **2010**, *42*, 541–544. [CrossRef] [PubMed]
36. Hadfield, K.A.; Bennett, A.B. Polygalacturonases: Many genes in search of a function. *Plant Physiol.* **1998**, *117*, 337–343. [CrossRef]
37. Kim, J.; Shiu, S.H.; Thoma, S.; Li, W.H.; Patterson, S.E. Patterns of expansion and expression divergence in the plant polygalacturonase gene family. *Genome Biol.* **2006**, *7*, R87. [CrossRef]
38. Letunic, I.; Khedkar, S.; Bork, P. SMART: Recent updates, new developments and status in 2020. *Nucleic Acids Res.* **2021**, *49*, D458–D460. [CrossRef]
39. Yokoyama, R.; Nishitani, K. Genomic basis for cell-wall diversity in plants. A comparative approach to gene families in rice and Arabidopsis. *Plant Cell Physiol.* **2004**, *45*, 1111–1121. [CrossRef]
40. Park, K.C.; Kwon, S.J.; Kim, P.H.; Bureau, T.; Kim, N.S. Gene structure dynamics and divergence of the polygalacturonase gene family of plants and fungus. *Genome* **2008**, *51*, 30–40. [CrossRef]
41. Cao, Y.; Zhang, Y.; Chen, Y.; Yu, N.; Liaqat, S.; Wu, W.; Chen, D.; Cheng, S.; Wei, X.; Cao, L.; et al. OsPG1 Encodes a Polygalacturonase that Determines Cell Wall Architecture and Affects Resistance to Bacterial Blight Pathogen in Rice. *Rice* **2021**, *14*, 36. [CrossRef]
42. Zhang, G.; Hou, X.; Wang, L.; Xu, J.; Chen, J.; Fu, X.; Shen, N.; Nian, J.; Jiang, Z.; Hu, J.; et al. PHOTO SENSITIVE LEAF ROLLING 1 encodes a polygalacturonase that modifies cell wall structure and drought tolerance in rice. *New Phytol.* **2021**, *229*, 890–901. [CrossRef]
43. Bailey, T.L.; Johnson, J.; Grant, C.E.; Noble, W.S. The MEME Suite. *Nucleic Acids Res.* **2015**, *43*, W39–W49. [CrossRef] [PubMed]
44. Gibney, E.R.; Nolan, C.M. Epigenetics and gene expression. *Heredity* **2010**, *105*, 4–13. [CrossRef]
45. Tanaka, K.; Murata, K.; Yamazaki, M.; Onosato, K.; Miyao, A.; Hirochika, H. Three distinct rice cellulose synthase catalytic subunit genes required for cellulose synthesis in the secondary wall. *Plant Physiol.* **2003**, *133*, 73–83. [CrossRef] [PubMed]
46. Li, Y.; Qian, Q.; Zhou, Y.; Yan, M.; Sun, L.; Zhang, M.; Fu, Z.; Wang, Y.; Han, B.; Pang, X.; et al. BRITTLE CULM1, which encodes a COBRA-like protein, affects the mechanical properties of rice plants. *Plant Cell* **2003**, *15*, 2020–2031. [CrossRef] [PubMed]
47. Yang, X.C.; Hwa, C.M. Genetic modification of plant architecture and variety improvement in rice. *Heredity* **2008**, *101*, 396–404. [CrossRef]
48. Sakamoto, T.; Matsuoka, M. Identifying and exploiting grain yield genes in rice. *Curr. Opin. Plant Biol.* **2008**, *11*, 209–214. [CrossRef]
49. Tanaka, W.; Yamauchi, T.; Tsuda, K. Genetic basis controlling rice plant architecture and its modification for breeding. *Breed. Sci.* **2023**, *73*, 3–45. [CrossRef]
50. Cai, W.; Hong, J.; Liu, Z.; Wang, W.; Zhang, J.; An, G.; Liang, W.; Persson, S.; Zhang, D. A receptor-like kinase controls the amplitude of secondary cell wall synthesis in rice. *Curr. Biol.* **2023**, *33*, 498–506.e6. [CrossRef]
51. Ma, X.; Feng, F.; Wei, H.; Mei, H.; Xu, K.; Chen, S.; Li, T.; Liang, X.; Liu, H.; Luo, L. Genome-Wide Association Study for Plant Height and Grain Yield in Rice under Contrasting Moisture Regimes. *Front. Plant Sci.* **2016**, *7*, 1801. [CrossRef]

52. Khahani, B.; Tavakol, E.; Shariati, V.; Fornara, F. Genome wide screening and comparative genome analysis for Meta-QTLs, ortho-MQTLs and candidate genes controlling yield and yield-related traits in rice. *BMC Genom.* **2020**, *21*, 294. [CrossRef] [PubMed]
53. Hong, J.; Su, S.; Wang, L.; Bai, S.; Xu, J.; Li, Z.; Betts, N.; Liang, W.; Wang, W.; Shi, J.; et al. Combined genome-wide association study and epistasis analysis reveal multifaceted genetic architectures of plant height in Asian cultivated rice. *Plant Cell Environ.* **2023**, *46*, 1295–1311. [CrossRef]
54. Liu, W.; Wu, C.; Fu, Y.; Hu, G.; Si, H.; Zhu, L.; Luan, W.; He, Z.; Sun, Z. Identification and characterization of HTD2: A novel gene negatively regulating tiller bud outgrowth in rice. *Planta* **2009**, *230*, 649–658. [CrossRef] [PubMed]
55. He, N.; Zhan, G.; Huang, F.; Abou-Elwafa, S.F.; Yang, D. Fine Mapping and Cloning of a Major QTL qph12, Which Simultaneously Affects the Plant Height, Panicle Length, Spikelet Number and Yield in Rice (*Oryza sativa* L.). *Front. Plant Sci.* **2022**, *13*, 878558. [CrossRef] [PubMed]
56. Zhang, Y.; Yu, C.; Lin, J.; Liu, J.; Liu, B.; Wang, J.; Huang, A.; Li, H.; Zhao, T. OsMPH1 regulates plant height and improves grain yield in rice. *PLoS ONE* **2017**, *12*, e0180825. [CrossRef]
57. Sato, Y.; Sentoku, N.; Miura, Y.; Hirochika, H.; Kitano, H.; Matsuoka, M. Loss-of-function mutations in the rice homeobox gene OSH15 affect the architecture of internodes resulting in dwarf plants. *EMBO J.* **1999**, *18*, 992–1002. [CrossRef]
58. Cosgrove, D.J. Expansive growth of plant cell walls. *Plant Physiol. Biochem.* **2000**, *38*, 109–124. [CrossRef]
59. Markovic, O.; Janecek, S. Pectin degrading glycoside hydrolases of family 28: Sequence-structural features, specificities and evolution. *Protein Eng.* **2001**, *14*, 615–631. [CrossRef]
60. Gonzalez-Carranza, Z.H.; Elliott, K.A.; Roberts, J.A. Expression of polygalacturonases and evidence to support their role during cell separation processes in Arabidopsis thaliana. *J. Exp. Bot.* **2007**, *58*, 3719–3730. [CrossRef]
61. Chen, H.; Shao, H.; Fan, S.; Ma, J.; Zhang, D.; Han, M. Identification and Phylogenetic Analysis of the POLYGALACTURONASE Gene Family in Apple. *Hortic. Plant J.* **2016**, *2*, 241–252. [CrossRef]
62. Qian, M.; Zhang, Y.; Yan, X.; Han, M.; Li, J.; Li, F.; Li, F.; Zhang, D.; Zhao, C. Identification and Expression Analysis of Polygalacturonase Family Members during Peach Fruit Softening. *Int. J. Mol. Sci.* **2016**, *17*, 1933. [CrossRef]
63. Ogawa, M.; Kay, P.; Wilson, S.; Swain, S.M. ARABIDOPSIS DEHISCENCE ZONE POLYGALACTURONASE1 (ADPG1), ADPG2, and QUARTET2 are Polygalacturonases required for cell separation during reproductive development in Arabidopsis. *Plant Cell* **2009**, *21*, 216–233. [CrossRef] [PubMed]
64. Xiao, C.; Somerville, C.; Anderson, C.T. POLYGALACTURONASE INVOLVED IN EXPANSION1 functions in cell elongation and flower development in Arabidopsis. *Plant Cell* **2014**, *26*, 1018–1035. [CrossRef] [PubMed]
65. Bradbury, P.J.; Zhang, Z.; Kroon, D.E.; Casstevens, T.M.; Ramdoss, Y.; Buckler, E.S. TASSEL: Software for association mapping of complex traits in diverse samples. *Bioinformatics* **2007**, *23*, 2633–2635. [CrossRef]
66. Li, M.X.; Yeung, J.M.; Cherny, S.S.; Sham, P.C. Evaluating the effective numbers of independent tests and significant p-value thresholds in commercial genotyping arrays and public imputation reference datasets. *Hum. Genet.* **2012**, *131*, 747–756. [CrossRef]
67. Yin, L.; Zhang, H.; Tang, Z.; Xu, J.; Yin, D.; Zhang, Z.; Yuan, X.; Zhu, M.; Zhao, S.; Li, X.; et al. rMVP: A Memory-efficient, Visualization-enhanced, and Parallel-accelerated tool for Genome-Wide Association Study. *Genom. Proteom. Bioinform.* **2021**, *19*, 619–628. [CrossRef] [PubMed]
68. Kumar, S.; Stecher, G.; Tamura, K. MEGA7: Molecular Evolutionary Genetics Analysis Version 7.0 for Bigger Datasets. *Mol. Biol. Evol.* **2016**, *33*, 1870–1874. [CrossRef]
69. Xie, J.; Chen, Y.; Cai, G.; Cai, R.; Hu, Z.; Wang, H. Tree Visualization By One Table (tvBOT): A web application for visualizing, modifying and annotating phylogenetic trees. *Nucleic Acids Res.* **2023**, *51*, W587–W592. [CrossRef]
70. Chen, C.; Chen, H.; Zhang, Y.; Thomas, H.R.; Frank, M.H.; He, Y.; Xia, R. TBtools: An Integrative Toolkit Developed for Interactive Analyses of Big Biological Data. *Mol. Plant* **2020**, *13*, 1194–1202. [CrossRef]
71. Xie, X.; Ma, X.; Zhu, Q.; Zeng, D.; Li, G.; Liu, Y.G. CRISPR-GE: A Convenient Software Toolkit for CRISPR-Based Genome Editing. *Mol. Plant* **2017**, *10*, 1246–1249. [CrossRef]
72. Hood, E.E.; Fraley, R.T.; Chilton, M.D. Virulence of Agrobacterium tumefaciens Strain A281 on Legumes. *Plant Physiol.* **1987**, *83*, 529–534. [CrossRef]
73. Zhang, M.; Liang, X.; Wang, L.; Cao, Y.; Song, W.; Shi, J.; Lai, J.; Jiang, C. A HAK family Na(+) transporter confers natural variation of salt tolerance in maize. *Nat. Plants* **2019**, *5*, 1297–1308. [CrossRef] [PubMed]
74. Love, M.I.; Huber, W.; Anders, S. Moderated estimation of fold change and dispersion for RNA-seq data with DESeq2. *Genome Biol.* **2014**, *15*, 550. [CrossRef] [PubMed]
75. Villanueva, R.A.; Chen, Z. ggplot2: Elegant Graphics for Data Analysis (2nd ed.). *Meas. Interdiscip. Res. Perspect.* **2019**, *17*, 160–167. [CrossRef]

MDPI AG
Grosspeteranlage 5
4052 Basel
Switzerland
Tel.: +41 61 683 77 34

International Journal of Molecular Sciences Editorial Office
E-mail: ijms@mdpi.com
www.mdpi.com/journal/ijms

www.ingramcontent.com/pod-product-compliance
Lightning Source LLC
Chambersburg PA
CBHW041353210926
51850CB00013B/3071